AF897962

Biomass and Solar-Powered Sustainable Digital Cities

Scrivener Publishing
100 Cummings Center, Suite 541J
Beverly, MA 01915-6106

Publishers at Scrivener
Martin Scrivener (martin@scrivenerpublishing.com)
Phillip Carmical (pcarmical@scrivenerpublishing.com)

Biomass and Solar-Powered Sustainable Digital Cities

Edited by

O.V. Gnana Swathika

K. Karthikeyan

Milind Shrinivas Dangate

and

Nicoletta Ravasio

Scrivener Publishing

WILEY

This edition first published 2024 by John Wiley & Sons, Inc., 111 River Street, Hoboken, NJ 07030, USA and Scrivener Publishing LLC, 100 Cummings Center, Suite 541J, Beverly, MA 01915, USA
© 2024 Scrivener Publishing LLC
For more information about Scrivener publications please visit www.scrivenerpublishing.com.

All rights reserved. No part of this publication may be reproduced, stored in a retrieval system, or transmitted, in any form or by any means, electronic, mechanical, photocopying, recording, or otherwise, except as permitted by law. Advice on how to obtain permission to reuse material from this title is available at http://www.wiley.com/go/permissions.

Wiley Global Headquarters
111 River Street, Hoboken, NJ 07030, USA

For details of our global editorial offices, customer services, and more information about Wiley products visit us at www.wiley.com.

Limit of Liability/Disclaimer of Warranty
While the publisher and authors have used their best efforts in preparing this work, they make no representations or warranties with respect to the accuracy or completeness of the contents of this work and specifically disclaim all warranties, including without limitation any implied warranties of merchantability or fitness for a particular purpose. No warranty may be created or extended by sales representatives, written sales materials, or promotional statements for this work. The fact that an organization, website, or product is referred to in this work as a citation and/or potential source of further information does not mean that the publisher and authors endorse the information or services the organization, website, or product may provide or recommendations it may make. This work is sold with the understanding that the publisher is not engaged in rendering professional services. The advice and strategies contained herein may not be suitable for your situation. You should consult with a specialist where appropriate. Neither the publisher nor authors shall be liable for any loss of profit or any other commercial damages, including but not limited to special, incidental, consequential, or other damages. Further, readers should be aware that websites listed in this work may have changed or disappeared between when this work was written and when it is read.

Library of Congress Cataloging-in-Publication Data

ISBN 978-1-394-24934-3

Front cover images supplied by Wikimedia Commons and Adobe Firefly
Cover design by Russell Richardson

Set in size of 11pt and Minion Pro by Manila Typesetting Company, Makati, Philippines

Printed in the USA

10 9 8 7 6 5 4 3 2 1

Contents

Preface **xxi**

1 Additives in Bio-Oil Components for Increased Stability, Quality, and Legal Status **1**
Abhinav Koushik, Jayendra Kasture and Milind Shrinivas Dangate
1.1 Introduction 2
1.2 Background Problems 2
1.3 Pyrolysis Bio-Oil 3
1.3.1 Fast Pyrolysis 3
1.3.2 Chemical Improvement of Bio-Oil From Pyrolysis 4
1.4 Using the Characterization and Pyrolysis Conditions of the Biomass Feedstock, One May the Qualities of the Bio-Oil Produced by Fast Pyrolysis 4
1.4.1 CMD Stands for Computer-Aided Molecular Design 4
1.5 Multi-Stage Computational Framework for Improving Pyrolysis Bio-Oil Through Solvent Design 5
1.5.1 Solvent Design Taking Into Account Economics 5
1.5.2 Problem of Multi-Objective Optimization 6
1.6 Rough Set Theory 6
1.7 Verification 7
1.8 Enhancing the Fuel Qualities of Pyrolysis Bio-Oil Through a Fusion of Computational and Experimental Methods Involves a Systematic Approach 7
1.8.1 Feedstock Selection and Pyrolysis Condition Optimization 7
1.8.2 Characterization of Bio-Oil 8
1.8.3 Blending and Additive Selection 8
1.9 Conclusion 8
References 10

2 Potential Source of Energy: Microalgae and Legal Commercialization **13**
Abhinav Koushik, Jayendra Kasture and Milind Shrinivas Dangate
2.1 Introduction 14
2.2 Algae Cultivation 15
2.3 Open-Air Systems 15
2.3.1 Shutdown Systems 16
2.4 Chemical Composition 16
2.4.1 Chlorophyll 17
2.5 Microalgae 17
2.6 HTL Bio-Crude Upgrading 17
2.6.1 Analytical Pyrolysis (Py-GC-MS) 18
2.7 One- and Two-Dimensional NMR Spectroscopy 18
2.8 Two-Dimensional NMR 19
2.9 Conclusions 19
References 20

3 Renewable Energy Integration for Rural Electrification Legally **21**
Abhinav Koushik, Nasrin I. Shaikh and Milind Shrinivas Dangate
3.1 Introduction 22
3.1.1 Grid and Environment for Communities 22
3.1.2 Configuration of a Hybrid Renewable Energy System 22
3.1.3 Configuration of an AC-Coupled HRES 23
3.1.4 Coupled Hybrid HRES Configuration 23
3.1.5 Components of Hybrid Renewable Energy Systems 23
3.2 Solar Energy Technology 23
3.3 Technique for Solar Radiation Modeling 24
3.4 Iterative Procedures 24
3.5 Methodology for Modeling Load Profiles 24
3.6 Data Collection for Surveys 24
3.7 System of Energy Modeling 25
3.7.1 Modeling of Converter Systems 25
3.7.2 Worldwide Solar Radiation Estimation 25
3.8 Technical Design Approach 26
3.8.1 Technical Design Approach for Distribution of Power on Regional Grids 26
3.8.2 Analysis and Design of a Hybrid Energy System 26
3.8.3 System Powered by Biomass 26

3.8.4 Techno-Economic Examination of Hybrid Energy Systems 27
3.9 Conclusion 27
References 28

4 Fuels from Waste Biomass for the Transport Industry 31
Sakthi Ashok, Abhinav Koushik, Milind Shrinivas Dangate and Nasrin I. Shaikh
4.1 Introduction 31
4.1.1 Background Problems 32
4.1.2 Effects of Climate Change 32
4.2 The Issue with Fossil Fuels 33
4.3 Existing Biofuel and Artificial Fuel Technologies 33
4.4 Pyrolysis Method 34
4.4.1 Technique for Order of Preference by Similarity Method 35
4.4.2 Analytic Hierarchy Process Approach 35
4.5 Esterification 36
4.6 Ketonic Decarboxylation 36
4.7 Hydro-Processing 38
4.7.1 Hydrotreating in Single Stage 38
4.7.2 Hydrotreating in Two Stages 38
4.8 MCDM Categories 39
4.9 Conclusion 39
References 40

5 The Potential Use of Biodiesel in Diesel Engine and Legal Environmental Laws 43
Abhinav Koushik, Jayendra Kasture and Milind Shrinivas Dangate
5.1 Introduction 44
5.1.1 Background Problem 45
5.1.2 Coal 45
5.1.3 The Decline in Coal Consumption 45
5.1.4 Vegetable Oil Derivatives as Fuels 45
5.2 Properties of Vegetable Oils 46
5.2.1 Properties of Vegetable Oil Esters 48
5.3 Charcoal Chemistry 49
5.3.1 Charcoal 49
5.3.2 Charcoal as Fuel 50
5.3.3 Industrial Uses 50

5.4 The Pyrolysis Phase 50
5.5 Cooling Phase 50
5.6 Products Obtained from the Carbonization Process 51
5.7 SS Charcoal's Qualities 51
5.8 Volatile and Fixed Carbon Content 51
5.8.1 Density of Charcoal 52
5.8.2 Gross Calorific Value of Charcoal 52
5.8.3 Pyrolysis Products 52
5.8.4 Production of Charcoal from a Specific Wood Species 53
5.9 Conclusion 54
References 54

6 Bioinspired Machine Learning: A New Era in Sustainable Energy Systems 57
Akshaya Kumar Mandal, Pankaj Kumar Deva Sarma, Satchidananda Dehuri and Nayanjyoti Mazumdar
Abbreviations 58
6.1 Introduction 59
6.2 Historical Backgrounds 61
6.3 Applications of Bio-ML 62
6.3.1 Wind Renewable Energy Systems 63
6.3.2 Solar Renewable Energy Systems 65
6.3.3 Hydro Renewable Energy Systems 66
6.3.4 Tidal Renewable Energy Systems 66
6.3.5 Geothermal Renewable Energy Systems 67
6.3.6 Biomass Renewable Energy Systems 68
6.3.7 Green-Hydrogen Renewable Energy Systems 69
6.4 Advantages of Using Renewable Energy 69
6.5 Bioinspired-Based Smart Energy Management System 70
6.6 Opportunities and Challenges of Future Research 72
6.6.1 Hybridizing Bioinspired Approaches 72
6.6.2 Energy Management Systems (EMSs) 73
6.6.3 Internet of Energy (IoE) 73
6.7 Conclusion 74
Acknowledgments 75
Credit Authorship Contribution Statement 75
Funding 75
Competing Interest 75
References 76

7 **The Impact of the Internet of Things in the Smart City From the Point of View of Energy Consumption Optimization** **81**
Sanjeevikumar Padmanaban, Mostafa Azimi Nasab, Mohsen Hatami, Omid Halimi Milani, Mohammad Ali Dashtaki, Morteza Azimi Nasab and Mohammad Zand
7.1 Introduction 82
7.1.1 Infrastructure of Information Technology for Energy Data Collection 82
7.1.2 Meter Reading Information Collection Service 85
7.1.3 Smart Water Sensor 85
7.1.4 Temperature and Humidity Sensor 86
7.1.5 Daylight Sensor 86
7.1.6 Mobile Device Sensors 87
7.2 Network Technologies and Communications 87
7.2.1 Low-Power Wi-Fi (Halow) Protocol 88
7.2.2 Zigbee Protocol 88
7.2.3 RPL Protocol 90
7.2.4 802.15.4 IEEE Protocol 90
7.2.5 6LoWPAN Protocol 91
7.2.6 MQTT Protocol 91
7.2.7 Management and Storage Systems for Data 93
7.2.8 Cloud Computing 95
7.3 Data Centers and Server 97
7.3.1 Virtualization of Servers 97
7.3.2 Data Cleansing 98
7.3.3 Data Compression 98
7.3.4 Reduction of Data Volume by Removing Redundant Information 99
7.3.5 Energy Efficiency in IoT Applications 99
7.3.6 Energy-Efficient Servers 99
7.3.7 Ventilation and Cooling of Servers and Data Centers 100
7.3.8 Applications and User Interfaces 100
7.3.9 Smart Homes and Home Automation 101
7.3.10 Smart Street Lighting System 103
7.3.11 Communication in the System 103
7.3.12 System Operation 104
7.3.13 Light Control 104
7.3.14 Remote Monitoring and Management 104
7.3.15 Smart Roads 104

7.3.16 Smart Parking Ecosystems 105
7.3.17 Sensor Parking and Management Tools for City Parking 105
7.3.18 Algorithm for Smart Parking System 105
7.4 Benefits of IoT in Smart Parking Systems 106
7.4.1 IoT Sensors for Smart City Monitoring 106
7.4.2 Useful Reports from IoT Sensors 106
7.4.3 Smart Offices 106
7.4.4 Smart Transportation 108
7.4.5 Some of the Solutions for Smart City Traffic Control 109
7.4.6 Smart Grids 110
7.4.7 IoT in Smart Grids 110
7.4.8 Processes in Smart Grids 110
7.4.9 Technologies in Smart Grids 111
7.4.10 Smart Energy Management Systems 111
7.4.11 Smart Waste 112
7.5 Layer of Communication Equipment and Sensors 113
7.5.1 Information Management Layer 115
7.5.2 Application Layer 116
7.6 Conclusion 117
References 118

8 Bio-Based Fuel Production, Environmental Impact, and Economic Assessment 123
Kavinkumar Ravikumar, Nicoletta Ravasio and Milind Shrinivas Dangate
8.1 Motivation 124
8.2 Methods 126
8.2.1 Cellulosic Ethanol (CE) by Fermentation 127
8.3 Results 127
8.4 Discussion and Outlook 130
8.5 Conclusions 131
8.6 Life Cycle Assessment 131
8.6.1 Abstract 131
8.6.2 Introduction 132
8.6.3 Methodology 132
8.6.3.1 Life Cycle Impact Categories 133
8.6.3.2 System Definition 133
8.6.3.3 Life Cycle Inventory (LCI) 134
8.6.4 Results and Discussion 138
8.6.4.1 Contribution Analyses 138

8.6.4.2 Energy Ratio 139
8.6.5 Conclusion 142
References 144

9 Improving Energy Consumption in Smart Homes Based on the Internet of Things 147
Sanjeevikumar Padmanaban, Mostafa Azimi Nasab, Amritansh Sagar, Mohammad Zand, Mohammad Ali Dashtaki and Morteza Azimi Nasab
9.1 Introduction 147
9.1.1 The Role of Objects in the Internet of Things 148
9.1.2 Bulk Data at the Heart of the Internet of Things Architecture 149
9.1.3 Comprehensive Framework Based on the Internet of Things for Smart Homes 150
9.2 Smart Home 150
9.2.1 Cloud 151
9.2.2 Beneficial Program 151
9.2.3 Third Party 151
9.2.4 User Interfaces 152
9.2.5 The Main Function of the Internet of Things 152
9.2.6 Security in the Internet of Things 152
9.2.7 Operational Structure for Personal Medical Services Based on the Internet of Things 154
9.3 Regulatory Applications and Their Use in Health Services 154
9.3.1 Challenges 154
9.3.2 Concerns About Healthcare 154
9.3.3 Architecture of Software-Oriented Networks 155
9.3.4 SDN for Cloud Service Providers 156
9.4 Previous Research 156
9.5 Conclusion 157
References 157

10 Performance Analysis of Supercapacitors in Renewable Energy System/Microgrid/Electric Vehicle, and Comparative Study of Various Converters With Their Applications 163
Moovendan Murugu, Arul Rajagopalan, V. T. Sreedevi, R. Sitharthan and R. Kanimozhi
10.1 Introduction 164
10.2 Supercapacitor Fundamentals 166
10.3 Applications and Functions of Supercapacitor 167

10.3.1 Energy Storage Technology 167
10.3.2 Uninterruptible Power Supplies (UPSs) 168
10.3.3 Power Electronic Device 170
10.3.4 Renewable Integration 170
10.3.5 Stabilization and Traction 170
10.4 Power Converter Topologies for Energy Storage 172
10.4.1 Passive Power Converter Topology 172
10.4.2 Semi-Active Power Converter Topology 173
10.4.3 Active Power Converter Topology 173
10.5 Comparative Study of Topologies 174
10.6 Converter Technologies in the Power Electronic Field 174
10.6.1 Converters for AC Source DC Applications 175
10.6.2 Converters for DC Source AC Applications 177
10.6.2.1 Neutral Point Clamped Topologies 178
10.6.2.2 Inverter Topologies With Multiple Levels 178
10.6.3 Converters for DC Source DC Applications 179
10.6.3.1 Non-Isolated Converters 179
10.6.3.2 Isolated Converters 181
10.6.4 Converters for AC Source AC Applications 183
10.7 Conclusion and Future Scope 184
Acknowledgments 186
Conflict of Interest 186
References 186

11 Diesel Engines and Related Environmental Pollution Issues 197
Sathya Narayanan, Jayendra Kasture and Milind Shrinivas Dangate
11.1 Introduction 198
11.2 Background 198
11.3 Diesel Engine Emissions 199
11.3.1 Diesel Engine Emission Control Technologies 199
11.3.2 Objectives 200
11.4 Particulates 201
11.4.1 Diesel Particulate Filter (DPF) 201
11.4.2 Exhaust Gas Recirculation (EGR) 202
11.4.3 Alternative Fuels for Diesel Engines 203
11.5 Biodiesel 204
11.5.1 Gas-to-Liquid (GTL) Diesel 204
11.5.2 Combined Alternative Fuels with New Injection Strategies 205

11.6 Summary 206
References 206

12 **Production of Bio-Diesel and By-Product Processing** **209**
Sathya Narayanan, Jayendra Kasture and Milind Shrinivas Dangate
12.1 Introduction 210
12.2 Fossil Fuel Resources 210
12.3 Environmental Issues 211
12.4 Fueling the Future: Vegetable Oil as an Alternative for Diesel Engines 211
12.5 Objectives of This Study 212
12.6 Requirements of Alternative Fuels to Diesel 212
12.6.1 Microbial Conversion of Glycerol to Alcohols: Enhancing Biodiesel Production and Sustainability 213
12.6.2 Methods of Utilizing Vegetable Oils 214
12.6.3 Transesterification 215
12.6.4 Catalysts 215
12.6.5 Acid-Catalyzed Transesterification 216
12.6.6 Base-Catalyzed Transesterification 216
12.7 Factors Affecting Kinetics 217
12.8 Legislation Governing Biodiesel 218
12.9 Conclusion 219
References 220

13 **Effective Use of Biofuel in a Diesel Engine** **221**
Sathya Narayanan, Nasrin I. Shaikh and Milind Shrinivas Dangate
13.1 Energy Resources 222
13.2 Rationale and Significance of Biofuels 223
13.3 Types of Biofuels and Their Characteristics 223
13.4 Advantages and Challenges of Biofuels 223
13.5 Ultracarbofluids: Pioneering Advances in Biodiesel Production 224
13.6 The Potential of Ultracarbofluids in Biodiesel Production 224
13.7 Incorporating Ultracarbofluids: Challenges and Future Prospects 225
13.8 Experimental Methodology 225
13.8.1 Performance Evaluation of Bio-Ultracarbofluid Blends 225

13.8.2 Bio-Ultracarbofluids: A Paradigm Shift in Biodiesel Evolution 226
13.8.3 The Potential of Bio-Ultracarbofluids in Biodiesel Production 226
13.8.4 Navigating Challenges and Future Prospects 226
13.8.5 Emissions Analysis and Environmental Impact 227
13.8.6 Influence of Charcoal Properties on Performance 227
13.8.7 Economic Considerations and Policy Implications 227
13.9 Conclusion and Future Directions 228
References 228

14 Enhancing Bio-Oil Stability and Fuel Quality Through Additive Integration 231
Sathya Narayanan, Nasrin I. Shaikh and Milind Shrinivas Dangate
14.1 Background Problems 232
14.2 Biomass Source Materials 233
14.3 Dedicated Energy Crops 233
14.4 Biomass Thermal Conversion 234
14.4.1 Slow Pyrolysis 234
14.4.2 Fast Pyrolysis 234
14.4.3 Flash Pyrolysis 235
14.4.4 Enhancing the Properties of Pyrolysis Bio-Oil through Physical Upgrading 235
14.4.5 Enhancing Pyrolysis Bio-Oil Through Chemical Upgrading 236
14.5 Employing Ultrasound for Bio-Oil and Diesel Emulsification 237
14.6 Emulsifying Pyrolysis Bio-Oil and Diesel 238
14.7 Utilizing RST 239
14.8 Conclusion 241
References 241

15 Biofuels from Microbes and Waste Resources 243
Sathya Narayanan, Nasrin I. Shaikh and Milind Shrinivas Dangate
15.1 Background 244
15.2 Biodiesel 245
15.3 Research Objectives 245
15.4 Biodiesel From Coffee Ground Waste 246
15.5 Pyrolysis for Controlled Fuel Properties 247

15.6 Cross Metathesis 248
15.7 Toward Enhanced Sustainable Liquid Biofuel Production 248
15.8 Conclusion 251
References 252

16 Optimal Scheduling of a Microgrid Incorporating Renewables and Demand Response Using a New Heuristic Optimization Technique 255
Karthik Nagarajan, Arul Rajagopalan, Krishna Kumba, R. Sitharthan, Sowmmiya Uthayakumar, Rajkumar Murugesan and R. Kanimozhi
List of Symbols and Abbreviations 256
16.1 Introduction 258
16.2 Problem Formulation 260
16.2.1 Grid-Connected Microgrid 261
16.2.2 Mathematical Modeling of Demand Response 263
16.2.2.1 Customer Cost Function 264
16.2.3 Combined Grid-Connected Microgrid with Demand Response Model 267
16.3 Fuzzy Logic-Based Assortment of Finest Compromise Solution 267
16.4 Enhanced Pelican Optimization Algorithm 268
16.4.1 Phase 1: Move to Prey (Exploration Phase) 270
16.4.2 Phase 2: Flying Over Water (Development Phase) 271
16.4.2.1 Disturbance Suppressor 272
16.5 Results and Discussion 274
16.6 Conclusion 281
References 282

17 Dual Battery Management System for Hybridization of Photovoltaic Systems with Batteries 285
P. Devadharshini, S. Arun Kumar, S. Rishwanth and G. Kanimozhi
17.1 Introduction 286
17.2 Battery Charging Algorithms 287
17.2.1 Battery Switching Technique 287
17.2.2 Schiffer Model 289
17.2.3 Hysteric Control Algorithm 290
17.2.4 Fuzzy Logic Technique 291
17.2.5 Coulomb Counting Method 292

17.2.6 PWM Control Method (Perturb and Observe Method) 292
17.3 DBMS Algorithms 293
17.3.1 Dual-Balancing Mechanism 293
17.3.2 Unitized Charging and Discharging (UCD) BMS for Rechargeable Batteries 293
17.3.3 Percentage of Discharge (POD) 295
17.3.4 Deep-Discharge Method 295
17.3.5 SOC Estimation 296
17.3.6 State of Health (SOH) Estimation 297
17.3.7 Thermal Management 297
17.3.8 Cell Balancing 297
17.3.9 Electrostatic Discharge Algorithm (ESDA) 299
17.4 DBMS Algorithm Comparison 300
17.5 Future Scope 301
17.6 Conclusion 302
References 303

18 Advancing Smart City Energy Management: Very Short-Term Photovoltaic Power Generation Forecasting Using Multi-Scale Long Short-Term Memory Deep Learning 305
Battula Srinivasa Rao, Mudiyala Aparna and Mallela Siva Naga Raju
18.1 Introduction 306
18.1.1 The Role of Deep Learning in Smart City Energy Management 306
18.1.2 Objectives and Structure of the Chapter 307
18.2 Methodology: Multiscale LSTM-Based Deep Learning Approach 307
18.2.1 Long Short-Term Memory (LSTM) Networks 308
18.2.2 Incorporating Multiscale Decomposition 309
18.2.3 Data Collection and Preprocessing 309
18.2.4 Training and Validation Strategies 309
18.2.5 Evaluation and Testing 309
18.3 Comparative Analysis with Traditional Forecasting Methods 310
18.3.1 Quantitative Metrics: Mean Absolute Error and Root Mean Squared Error 310
18.3.2 Interpretation of Comparative Results 311
18.4 Benefits and Implications for Smart City Energy Management 311

18.4.1 Challenges and Future Directions 313
18.5 Conclusion 314
References 315

19 Artificial Intelligence Applied to the Management and Operation of Solar Systems 317
Neel Ghoshal and B.K. Tripathy
19.1 Introduction 318
19.2 Background 319
19.3 AI for Solar System Management 320
19.3.1 Maintenance 320
19.3.2 Market-Based Strategy Optimization 323
19.3.3 Energy Storage Management and Optimization 324
19.3.4 Remote Monitoring and Control 326
19.4 AI for Solar System Operation 327
19.4.1 Solar Energy Optimization 328
19.4.2 Demand Response 329
19.4.3 Solar Power Forecasting 330
19.4.4 Grid Integration 331
19.5 Challenges and Future Scope 332
References 333

20 Robust Rearrangement of Interconnected Microgrids to Reduce the Effects of Physical Cyberattacks on Intelligent Distribution Networks 339
Sanjeevikumar Padmanaban, Mostafa Azimi Nasab, Omid Halimi Milani, Mohammad Zand, Mohsen Hatami, Morteza Azimi Nasab and Mohsen Khalili
20.1 Smart Grid 340
20.2 The Benefits of Smart Grids 340
20.3 Reduce Subscriber Shutdown Time 341
20.4 The Benefits of the Smart Distribution Network for Subscribers 341
20.5 Capital Optimization 343
20.6 Concepts of Security in Telecommunication Systems 343
20.7 Information Security 344
20.8 Information Technology Security 344
20.9 Information Security Management System 1 (ISMS) 345
20.9.1 Information Security Management Standard 345
20.9.2 Security in Smart Grid Telecommunication Systems 346

20.9.3 Threats in the Smart Power Grid 347
20.9.4 Comparison of Security of Telecommunication Systems 348
20.9.5 Attack on the Architecture of the Intelligent Measurement System 348
20.9.6 Internet of Things 350
20.9.7 Cyber Security 351
20.9.8 Physical Security 354
20.10 Overview of References 356
20.11 Summary and Conclusion 358
References 359

21 Reinforcement Learning Method Optimizes the Planning and Design of the Solar Energy Systems 363
R. Dhanasekaran, S. Sreenatha Reddy, K. Santhi, Ch. Nagaraja Kumari and Saba Noureen
21.1 Introduction 364
21.2 Objectives and Methodology 365
21.3 Solar Power Plant Operations Using RL 365
21.3.1 Solar Energy Production with RL 365
21.3.2 Improving Solar Panel Efficiency Using RL 366
21.3.3 Energy Control in Microgrid Using RL 366
21.3.4 Solar Irradiance Prediction Using RL 367
21.4 Future Scopes of AI in Solar Power 367
21.5 Conclusion 368
References 368

22 Photovoltaic-Thermal Solar-Assisted Heat Pump Systems for Building Applications: Integration and Design Methods 371
Krishna Kumba, Venkateswarlu Gundu and R. Arul
22.1 Introduction 371
22.2 Functioning of Photovoltaic and Thermal Heat Pump Systems 372
22.3 Mathematical Formation of PV-Thermal–Assisted System 373
22.3.1 PV and Thermal Efficiency 373
22.4 Heat Pump Systems with Solar PV and Thermal for Building Applications 376
22.4.1 Direct-Expansion Systems 376
22.4.2 Single-Source Systems 376
22.4.3 Dual-Source Systems 376
22.4.4 Indirect-Expansion (IDX) Systems 377

22.4.5 Single-Source Indirect-Expansion Systems 378
22.4.6 Dual-Source Indirect-Expansion Systems 378
22.5 Challenges of PV-Thermal–Assisted System 380
22.6 Conclusion 381
References 381

23 Critical Review: IoT-Based Data Logger for SAPV System 383
Kshitij Tiwari, Anil Kumar Jha, Manish Kumar, Ayush Shankar, Aditi Verma and O.V. Gnana Swathika
23.1 Introduction 383
23.2 Grid Connected PV System Opal RT 386
23.3 IoT Advancements and Implementation in SAPV Monitoring 396
23.4 Photovoltaic System and IoT Innovations 400
23.5 IoT in Hybrid Systems 405
23.6 IoT in SAPV and Its Data Logging (An Analysis) 408
23.7 Other Works in the Field 416
23.8 Conclusion 418
References 419

Index 427

Preface

The focus on identifying sustainable power generation is gearing up to meet the growing power demand of smart cities. This book discusses two energy sources: biomass and solar. Biomass is organic and it can be burned to create heat (direct), converted into electricity (direct), or processed into biofuel (indirect). Biomass is the only renewable energy source that can be converted into liquid biofuels such as ethanol and biodiesel. Biofuel is used to power vehicles. Biomass is rich in hydrogen, which can be chemically extracted and used to generate power and to fuel vehicles. Stationary fuel cells are used to generate electricity in remote locations, such as spacecraft and wilderness areas. Biomass is a clean, renewable energy source. Its initial energy comes from the sun, and plants or algae biomass can regrow in a relatively short amount of time. Trees, crops, and municipal solid waste are consistently available and can be managed sustainably. If trees and crops are sustainably farmed, they can offset carbon emissions when they absorb carbon dioxide through respiration. Unlike other renewable energy sources, such as wind or solar, biomass energy is stored within the organism, and can be harvested when it is needed.

This book will additionally present the reality of renewable energy sources, specifically those based on solar energy. This reality focuses on the integration of solar energy in cities (smart cities), buildings (smart buildings), microgrids, smart rural grids, and other similar environments. The applicability of solar energy is fundamental for the sustainable development of previous infrastructures.

Globally the focus is on practical large scale implementation of solar energy technologies in combination with new build and retrofit low energy building in the cities. This book will benefit those in a situation where many cities aim at a "Digital City" development, but without clear policies of how to achieve that in practice. This book will indicate how solar energy as the worlds number one energy source can play a major role here and how this can be combined with energy efficiency policies, use of district heating and combined heat and power. This book aims to provide a forum for researchers and practitioners to exchange ideas and achieve progress in smart city applications, advances, and research challenges. It is highly recommended to be used as a reference book for graduate and post-graduate level university programs. Case studies range across all the chapters.

1

Additives in Bio-Oil Components for Increased Stability, Quality, and Legal Status

Abhinav Koushik[1], Jayendra Kasture[2*] and Milind Shrinivas Dangate[1†]

[1]Chemistry Division, School of Advanced Sciences, Vellore Institute of Technology, Chennai, Tamil Nadu, India
[2]VIT School of Law, Vellore Institute of Technology, Chennai, India

Abstract
Malaysia, one of the biggest producers of palm oil worldwide, faced a significant disposal issue due to the enormous amount of oil palm biomass wastes generated. Fast pyrolysis technology can be used to liquefy these biomass wastes into biofuel in order to solve this issue. To get rid of its unwanted qualities, fast pyrolysis bio-oil has to be upgraded further via direct solvent addition. To establish competitiveness with regular diesel fuel, the envisioned blend of solvent and oil must align with benchmarks for both fuel efficiency and cost-effectiveness. Chemical signature descriptors were harnessed to correlate predictions across multiple classes of properties within the design challenge. Nonetheless, the intricacy inherent in the computer-aided molecular design (CAMD) challenge escalates as the length of the signature expands, owing as made to introduce a consistency rule, aimed at diminishing the scale and complexity of the CAMD undertaking. Then, to expand the CAMD problem to incorporate the economic components, a fuzzy optimization with multiple objectives approach was applied. Then, a rough set–based artificial intelligence model has been constructed to connect the feedstock description and pyro condition with the decomposition bio-oil properties. The generated decisions were carefully investigated in order to identify the deeper trends and to confirm that the created choice rules made sense. To manufacture pyrolysis bio-oil with desired fuel qualities, the best feedstock composition and pyrolysis conditions can be chosen using the decision rules developed. Next, experimental research was

**Corresponding author*: dr.jakasture@gmail.com
†*Corresponding author*: milind.shrinivas@vit.ac.in

O.V. Gnana Swathika, K. Karthikeyan, Milind Shrinivas Dangate and Nicoletta Ravasio (eds.)
Biomass and Solar-Powered Sustainable Digital Cities, (1–12) © 2024 Scrivener Publishing LLC

used to confirm the conclusions of the computational methodologies. To sum up, this chapter discussed how to improve the fuel qualities of pyrolysis bio-oil using a combination of computational and experimental methods.

Keywords: Palm oil, bio-oil, pyrolysis, emulsification, additives

1.1 Introduction

During the past century, burning fossil fuels has produced practically all of the energy needed for daily life. As of this writing, fossil fuels still meet roughly 80% of the global daily energy needs [1]. According to reports, 98.8 million kilograms of liquid gasoline and petroleum were utilized each day in the world in July 2022. It was anticipated that, by 2023, this amount will increase even further, averaging 101.5 million barrels of oil per day [2]. For the thermochemical conversion of palm biomass, a wide range of biofuels could be created through methods like liquefaction, gasification, and pyrolysis. The advantage of the pyrolysis process over other biomass conversion technologies is that it can produce bio-oil, charcoal, and gaseous byproducts as a result of the conversion of solid palm biomass feedstock. On the contrary, pyrolysis bio-oil presents an intricate mixture primarily consisting. According to a study by Khosravanipour Mostafazadeh *et al.* [3], efforts to directly utilize pyrolysis bio-oil in gas turbines or diesel engines have encountered difficulties due to its limited compatibility with traditional petroleum fuels. To overcome these challenges, various techniques for upgrading pyrolysis bio-oil have been proposed and explored, involving both physical and chemical approaches of the bio-oil produced through pyrolysis can be enhanced by incorporating solvents [3].

1.2 Background Problems

For oil and solvents, earlier research primarily focused on mechanically blending decomposition bio-oil with various solvents, selected based on their general characteristics. The conventional approach to selecting and pinpointing solvents typically involved a time-consuming experimental process of trial and error from a wide array of possibilities. This exhaustive testing of all potential solvents could be impractical, leading to an extended timeframe for assessment [4] frequently employed to develop viable solvent options that fulfill specific property criteria and constraints [5]. Computer-aided molecular design (CAMD), built around predetermined target properties, operates as a method of reverse engineering, predicting and approximating suitable molecules while assembling them from a set of molecular building blocks. Previous endeavors have been undertaken to create solvents appropriate for bio-oil applications using the CAMD approach.

1.3 Pyrolysis Bio-Oil

When pyrolysis process happens, biomass source material is thermally broken to produce syngas, biochar, and bio-oil. The pyrolysis process is akin to the well-known way of creating coal [6]. Operational variables including humidity, vapor residency time, heating, and feedstock have an impact on how pyrolysis products are distributed at the end. The pyrolysis product's ultimate composition is most strongly influenced by temperature, though. On the other hand, the faster heating and higher temperature (400°C to 500°C) stimulate the creation of levoglucosan, the catalyst for the development of the substance that causes smaller molecules to form, giving rise to liquid products. A few examples of the gaseous products produced by the decomposition of biomass at temperatures exceeding 650°C are carbon monoxide, hydrogen, and methane [7]. The three types of pyrolysis reactions—slow, rapid, and fast pyrolysis—can be classified depending on the operating parameters.

1.3.1 Fast Pyrolysis

Fast pyrolysis is a process that swiftly elevates the temperature of biomass to a high degree while impeding the ingress of oxygen. However, it is important to note that fast pyrolysis does not achieve the same rapid temperature increase as flash pyrolysis. According to findings by [8], the output of high-quality bio-oil is typically associated with fast pyrolysis. For instance, in the temperature range of 400°C to 700°C, swift pyrolysis of teak sawdust led to notable results. At 600°C, a peak bio-oil production of 48.8 was achieved [9]. Furthermore, a maximum yield of 44.16 weight percent of bio-oil from wood sawdust pyrolysis was attained through pre-treatment of empty fruit bunches using diluted nitric acid [9]. Additional advancements have been observed by employing diluted nitric acid through acid washing in order to enhance liquid output.

In a separate study by Solikhah *et al.* [10], the quick pyrolysis technique was employed to generate pyrolysis bio-oil using oil palm fronds and empty fruit bunches from palm oil.

Bio-oils from empty fruit bunches and oil palm fronds were shown to have higher heating values, with respective values of 12.19 MJ/kg and 26.49 MJ/kg. To quickly pyrolyze biomass for bio-oil, a variety of reactor configurations are suitable [11]. These configurations include fluidized bed, entrained flow, wire mesh, vacuum furnace, vortex, and rotating and circulating fluidized bed reactors. The fluidized bed reactor is the most promising of the aforementioned reactors for fast pyrolysis because it provides for high heating rates, quick de-volatilization, simple control, and inexpensive costs.

1.3.2 Chemical Improvement of Bio-Oil From Pyrolysis

Prior to gasification with synthesis gas, pyrolysis can function as a thermal pre-treatment, as pointed out [7]. The modest energy penalties such as reduced energy effectiveness of pyrolysis, increased transportation energy, and the need for an additional step of bio-oil gasification, as outlined, can be viewed in the context of this process.

1.4 Using the Characterization and Pyrolysis Conditions of the Biomass Feedstock, One May the Qualities of the Bio-Oil Produced by Fast Pyrolysis

A driven by information rough set–based machine learning (ML) model was suggested for the final subregion in order to estimate pyrolysis biological oil qualities based on the temperature of pyrolysis and feedstock parameters. On the basis of peer-reviewed studies and experiments, a database containing preliminary and thorough evaluations of the wood substance, pyrolysis the temperature, the pH value of the bio-oil, and the hydrophilic strength of the bio-oil is initially created. A research scientific study is carried out to identify fundamental trends or patterns in order to check that the generated rules of decision-making make sense and work for potential usage. Depending on the mechanical validity and distinctive viewpoints of the choice rules, the best models of prediction and choices are selected.

1.4.1 CMD Stands for Computer-Aided Molecular Design

In previous instances, the design of solvents has often relied on heuristic rule-based or experimental approaches, resulting in the development of product designs that are reliable and safe, as highlighted by [12]. However, these conventional methodologies for solvent design have exhibited drawbacks such as being laborious, costly, and time-intensive. Moreover, it is practically infeasible to exhaustively explore all potential options or identify the optimal solution can replace the conventional trial-and-error method. CAMD stands as a viable substitute for the conventional trial-and-error technique, offering a model-based design process that enhances solvent design efficacy. A kind of property forecast model–based reverse engineering technique is the CAMD process. With a certain set of elements and targeted properties, the ideal atomic structure and/or structure can be found using the CAMD technique [13]. In other words, the design purpose was defined in terms of the targeted qualities and performance in order to create the molecular structure with the best performance [14].

1.5 Multi-Stage Computational Framework for Improving Pyrolysis Bio-Oil Through Solvent Design

Utilizing CAMD technologies, it becomes possible to develop a solvent possessing favorable physical attributes and minimal ecological impact. While Group Contribution (GC) techniques have traditionally found extensive application in estimating properties tied to thermodynamics, the current landscape predominantly employs Transferable Interaction (TI) models for predicting environmentally related properties. The challenge arises from the mathematical intricacy of simultaneously integrating and solving diverse sets of prediction models, each employing distinct mathematical formulations. To address this, the proposal to introduce molecular signature descriptors as a shared foundation for structuring the design problem, incorporating a variety of property prediction models, holds merit. Nonetheless, to manage a workable scale for the CAMD problem, the development of a multi-stage framework is imperative. This framework systematically eliminates impractical signatures across various levels, ensuring the maintenance of a manageable CAMD problem size.

1.5.1 Solvent Design Taking Into Account Economics

In addition to low gas quality, another barrier to commercialization of biofuels is the huge cost of producing pyrolysis bio-oil. Solvents are frequently needed to improve the qualities of the biofuel. In some circumstances, the chosen solvent may be produced as a specialty chemical, which is typically expensive. This drives up the price of producing biofuels even more. The evaluation of the cost involved in upgrading bio-oil is further motivated by the context of current and evolving biofuel legislation.

Economic considerations, such as product pricing, revenue, share of the market, and operating costs, were implemented into the manufacturing process in a number other recent contributions. When creating a chemical-based design framework for an ancient Chinese medicinal supplement, customer preferences for the product's quality and economic factors were taken into account. In terms of an interdisciplinary hierarchical methodology that incorporated design issues including management of projects, market research, economic assessment, product layout, and process design, this framework provided an overview of the design of chemical products. Additionally, a facility for producing activated carbon from industrial waste the nutshells was suggested.

1.5.2 Problem of Multi-Objective Optimization

The majority of chemical product design frameworks in use today only optimize one target feature. The term "single-objective optimization problem" is frequently used to describe this kind of design problem. However, to develop an ideal product, numerous product characteristics, it is necessary to take into account elements including the thermal properties, health, security, and environmental aspects, as well as financial considerations. When there are numerous objective functions that need to be optimized, the CAMD layout problem must be approached as a broad optimization (MOO) problem. For a non-trivial MOO problem, there exists a nondominated set rather than a single answer that can concurrently optimize all the goals.

1.6 Rough Set Theory

Rough Set Theory (RST) stands out as a groundbreaking mathematical approach designed to tackle the challenges posed by confusion, imprecision, inconsistencies, and uncertainty within knowledge and information. RST builds upon set theory, offering a framework to manage data where boundaries are not well-defined. At its core, RST assumes that each element in a given domain is associated with some relevant data. This distinctive characteristic enables RST to exhibit robustness, making it capable of handling information of subpar or conflicting quality.

RST serves as the bedrock RSML, boasting the advantage of simplicity by only needed a dataset, without necessitating additional factors like likelihood in economics or grade of membership in fuzzy collection theory. Its practicality lies in its ability to unveil latent patterns within data. The versatility of RST extends beyond its robustness and simplicity. It also excels in concurrent processing, data reduction (identifying minimal data sets), evaluating significance, autonomously generating sets of decision rules from data, and providing clear interpretations of outcomes.

Moreover, RST finds extensive application in artificial intelligence (AI) and cognitive sciences, particularly in machine learning (ML) and knowledge discovery (KD). A key strength of RSML lies in handling sparse datasets, prevalent in various engineering applications. The methodology that it offers proves highly beneficial in these scenarios. In essence, RST and RSML provide a powerful toolkit for navigating the complexities of uncertain and incomplete information, advancing both the theoretical foundations and practical applications of various disciplines.

1.7 Verification

Verification is necessary to make sure that the compounds produced in earlier processes are possible and useful. The produced molecules were checked in this step using database searches from a variety of platforms, including ChemSpider and PubChem. Modifying the property characteristics and limitations allowed the preceding phase to be repeated for compounds that were not found in the database or were deemed to be impractical.

1.8 Enhancing the Fuel Qualities of Pyrolysis Bio-Oil Through a Fusion of Computational and Experimental Methods Involves a Systematic Approach

1.8.1 Feedstock Selection and Pyrolysis Condition Optimization

Start by selecting appropriate feedstocks for pyrolysis that yield bio-oil with desirable properties. Employ computational modeling to predict the outcome of various pyrolysis conditions (temperature, residence time, etc.)

on bio-oil composition. Optimize these conditions to maximize the desired fuel properties.

1.8.2 Characterization of Bio-Oil

Utilize experimental techniques to thoroughly characterize the chemical composition of the produced bio-oil. This includes analyzing its physical properties (density, viscosity, etc.) and chemical composition (elemental analysis, functional groups, etc.).

1.8.3 Blending and Additive Selection

Use computational models to screen and select potential additives or blending agents that can enhance the fuel qualities of bio-oil. These agents might include catalysts, stabilizers, or other bio-based materials that can modify viscosity, improve combustion efficiency, and reduce undesirable components.

By combining computational predictions with targeted experimental validations, this integrated approach can effectively lead to the development of bio-oil blends with enhanced fuel properties, improved stability, and greater potential for various applications in the energy sector.

1.9 Conclusion

The influential role of predicted fuel properties, attainable through GC property prediction algorithms, in shaping the composition of the current bio-oil is evident. However, the holistic CAMD as shown in Figure 1.1 framework must not only consider these factors but also incorporate environmental and financial considerations to ensure the resultant oil competes effectively with diesel fuel. Within this chapter, various raw biomass substrates and pyrolysis techniques are also covered. Notably, comprehensive information about feedstock and pyro operating conditions, necessary for assessing pyrolysis bio-oil fuel qualities, is not readily available. To bridge this gap, the implementation of the RSML technique, serving as a preliminary modeling tool, proves advantageous.

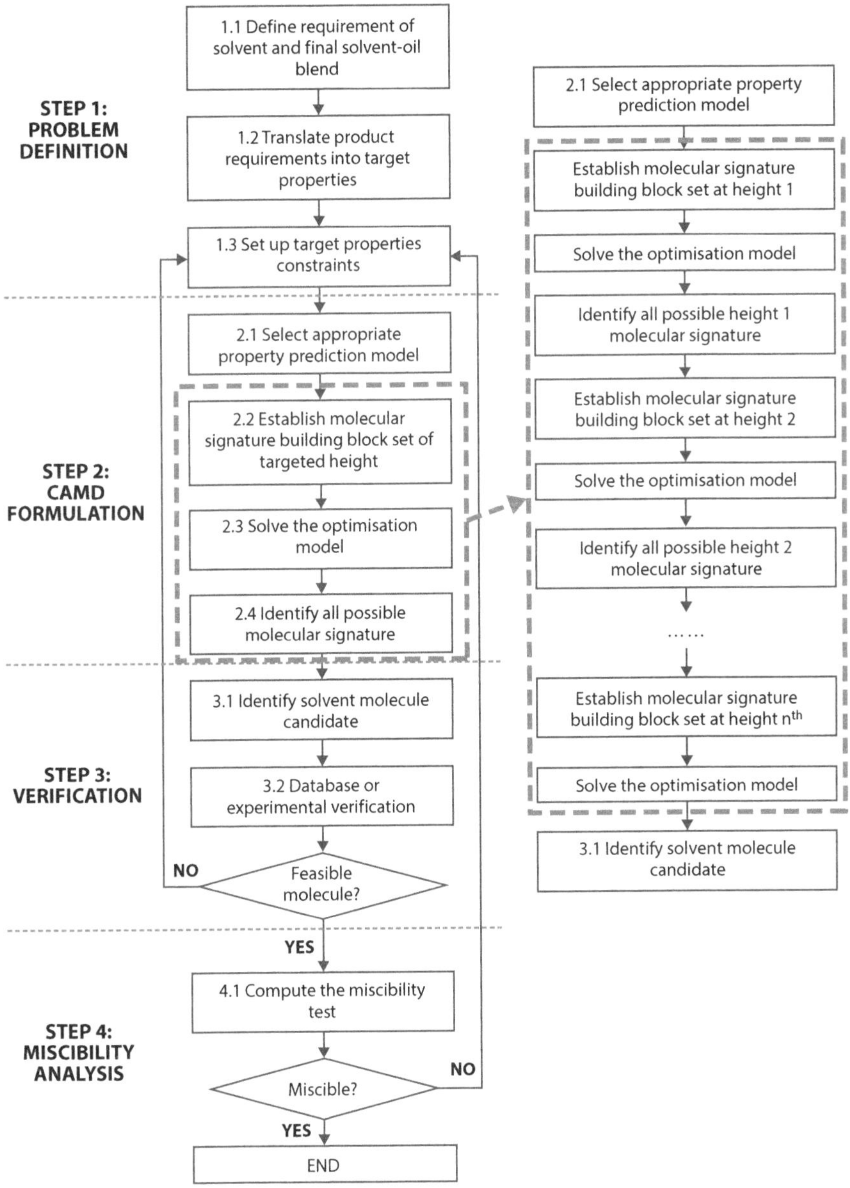

Figure 1.1 Restructuring the foundation for formulating a computer-aided molecular design (CAMD) model targeted at devising solvent additive designs.

Emulsification emerges as a swift, cost-effective, and proficient physical approach to enhance pyrolysis bio-oil characteristics. Nevertheless, the intricacies within the CAMD framework necessitate the development of a solvent capable of mitigating all associated drawbacks linked to pyrolysis bio-oil. The complexity of the CAMD challenge escalates in tandem with the molecular signature descriptor's complexity, which inherently presents a combinatorial nature.

Pyrolysis bio-oil production and characterization may need complicated, time-consuming, and expensive procedures. Finding the best pyrolysis conditions and biomass feedstock properties for the intended bio-oil results frequently requires using trial-and-error methods. In response, an RSML-based rule-based prediction model has been developed. This model generates visible and understandable decision rules by computing pyrolysis bio-oil properties based on feedstock biomass parameters and pyrolysis operating circumstances.

With completed case studies as a cornerstone reference, stringent selection criteria have been devised, grounded in scientific reasoning. The utilization of the suggested methodology has led to the development of bio-oil/diesel blends that demonstrate stability lasting beyond 15 days. Additionally, this thesis delves into innovative computational and experimental techniques designed to improve the fuel and environmental characteristics of pyrolysis bio-oil, all the while maximizing its economic feasibility.

References

1. EESI, 2021. Fossil Fuels [WWW Document]. *Environmental and Energy Study Institute.* URL https://www.eesi.org/topics/fossil-fuels/description (accessed 9.3.22).
2. EIA, 2022a. *Short-Term Energy Outlook.* Washington, D.C.
3. Khosravanipour Mostafazadeh, A., Solomatnikova, O., Drogui, P., Tyagi, R.D., 2018. A review of recent research and developments in fast pyrolysis and bio-oil upgrading. *Biomass Convers. Biorefin.* 8, 739–773. https://doi.org/10.1007/s13399-018-0320-z.
4. Venkatasubramanian, V., Chan, K., Caruthers, J.M., 1994. Computer-aided molecular design using genetic algorithms. *Comput. Chem. Eng.* 18, 833–844. https://doi.org/10.1016/0098-1354(93)E0023-3.
5. Achenie, L.E.K., Gani, R., Venkatasubramanian, V., 2003. *Computer Aided Molecular Design: Theory and Practice,* 1st ed. Elsevier B.V., Amsterdam.

6. Jenkins, R.W., Sutton, A.D., Robichaud, D.J., 2016. Pyrolysis of Biomass for Aviation Fuel, in: *Biofuels for Aviation*. Elsevier, pp. 191–215. https://doi.org/10.1016/B978-0-12-804568-8.00008-1.
7. Müller-Langer, F., Klemm, M., Schneider, J., 2017. Biofuels Production Processess and Technologies, in: Riazi, M.R., Chiaramonti, D. (Eds.), *Biofuels Production and Processing Technology*. CRC Press, Boca Raton, pp. 153–182.
8. Canabarro, N., Soares, J.F., Anchieta, C.G., Kelling, C.S., Mazutti, M.A., 2013. Thermochemical processes for biofuels production from biomass. *Sustainable Chem. Processes* 1, 22. https://doi.org/10.1186/2043-7129-1-22.
9. Varma, A.K., Thakur, L.S., Shankar, R., Mondal, P., 2019. Pyrolysis of wood sawdust: Effects of process parameters on products yield and characterization of products. *Waste Manage.* 89, 224–235. https://doi.org/10.1016/j.wasman.2019.04.016.
10. Solikhah, M.D., Pratiwi, F.T., Heryana, Y., Wimada, A.R., Karuana, F., Raksodewanto, A., Kismanto, A., 2018. Characterization of Bio-Oil from Fast Pyrolysis of Palm Frond and Empty Fruit Bunch. *IOP Conf. Ser. Mater. Sci. Eng.* 349, 012035. https://doi.org/10.1088/1757-899X/349/1/012035.
11. Goyal, H.B., Seal, D., Saxena, R.C., 2008. Bio-fuels from thermochemical conversion of renewable resources: A review. *Renew. Sustain. Energy Rev.* 12, 504–517. https://doi.org/10.1016/j.rser.2006.07.014.
12. Zhang, S., Yang, X., Zhang, H., Chu, C., Zheng, K., Ju, M., Liu, L., 2019. Liquefaction of Biomass and Upgrading of Bio-Oil: A Review. *Molecules* 24, 2250. https://doi.org/10.3390/molecules24122250.
13. Gani, R., Achenie, L.E.K., Venkatasubramanian, V., 2003. *Introduction to CAMD*. pp. 3–21. https://doi.org/10.1016/S1570-7946(03)80003-2.
14. Gertig, C., Leonhard, K., Bardow, A., 2020. Computer-aided molecular and processes design based on quantum chemistry: current status and future prospects. *Curr. Opin. Chem. Eng.* 27, 89–97. https://doi.org/10.1016/j.coche.2019.11.007.

2

Potential Source of Energy: Microalgae and Legal Commercialization

Abhinav Koushik[1], Jayendra Kasture[2†] and Milind Shrinivas Dangate[1*]

[1]*Chemistry Division, School of Advanced Sciences, Vellore Institute of Technology, Chennai, Tamil Nadu, India*
[2]*VIT School of Law, Vellore Institute of Technology, Chennai, India*

Abstract

This chapter's goal was of microalgae and the catalytic improvement of the resulting bio-crudes from hydrothermal liquefaction (HTL). Three microalgae species were found to be suitable candidates for hydrothermal processing from the selected feedstock. Results from characterization studies serve as a foundation for reaction pathways, feedstock screening, and the creation of bio-crude. A larger-scale treatment for the model components and algae was then performed in a 300-ml batch machine at 200°C to 350°C after the initial HTL testing for macro plus microalgae species reactor. Comparatively speaking, macroalgae produced less biocrude than microalgae. The maximum bio-crude yield, 51.6 weight percent (wt%), was attained during liquefaction, and bio-crude yields increased with higher temperature HTL processing. It was discovered that the bio-crudes have large quantities of nitrogen and oxygen, which lessens their usefulness as a fuel for transportation. The use of these cutting-edge methods made it possible to characterize the entire batch of bio-crude and upgraded liquid samples, as opposed to simply the portion that was able to elute from a gas chromatography (GC) column before mass spectrometry (MS) detection. Supercritical water upgrading at 450°C was examined as a method of upgrading the oils. The upgraded oils exhibited decreased nitrogen content, ranging from 1.9% to 2.6% by weight. Remarkably, the oxygen content underwent a significant reduction, dropping from 19.9% to values within the range of 1.3% to 4.8% by weight. Analysis of the data underscores these trends. The upgraded oil revealed that the content was predominantly composed of C6-C16 alkanes and substituted aromatic compounds. The oxygen content dropped sharply, from 19.9 wt% to

**Corresponding author*: milind.shrinivas@vit.ac.in
†*Corresponding author*: dr.jakasture@gmail.com

O.V. Gnana Swathika, K. Karthikeyan, Milind Shrinivas Dangate and Nicoletta Ravasio (eds.) Biomass and Solar-Powered Sustainable Digital Cities, (13–20) © 2024 Scrivener Publishing LLC

values between 1.3 wt% and 4.8 wt%. The enhanced oil's composition was mostly made up of substituted aromatic compounds and C6-C16 alkanes, according to analysis. Despite being technically possible, the transforming energy access (TEA)'s economic findings suggest an MFSP of £2.64 per liter, showing that algae-derived biofuels are not at all competitive in this scenario.

Keywords: Macroalgae, bio-crude yield, NMR, MS, oil

2.1 Introduction

Currently, discussions surrounding fossil-derived fuels and energy are pervasive, with numerous experts predicting the depletion of natural reserves by the latter part of this century. The consensus is that we stand at a pivotal juncture, marking the initiation of the decline of easily accessible and affordable fossil fuels. These include diminished productivity in previously, escalated, and complexities. These are now compelled to employ intricate and consequently pricier to meet the demands of a growing and more prosperous global populace.

The necessity to control emissions—and, ideally, to reduce them—in order to address the problems associated with warming temperatures and climate change has received substantial attention on a global scale. Climate change worries are present on a daily basis, whether via news broadcasts, consumer goods, or personal choices. Given our influence on climate change, many people believe that it is morally required that they reduce their impact on the ecosystems and atmosphere of the planet. Global warming is the term used to describe how human activities are affecting climate change. Since the mid-20th century, meteorological and oceanic warming have been detected, and it is commonly believed that human activity is the primary reason (90%–100% certainty) behind these changes [1]. Greenhouse gases, including CO_2 (carbon dioxide), CH_4 (methane), N_2O (dinitrogen monoxide), and fluorinated gases, contribute significantly to this warming. A significant milestone was reached with the adoption of the 2015 Paris Agreement. Nearly every nation on the planet endorsed this international accord, aiming to combat adverse consequences. This establishes a framework for global collaborative action on climate change mitigation, adaptation, and financial support. The UK took an initial step toward climate action with its 2008 domestic Climate Change Act, legally mandating a reduction of greenhouse gas emissions by at least 80% by 2050 compared to 1990 levels, according to [2].

Previous commercialization of microalgae-based biofuels relied on growing the algae for their lipid content, followed by their extraction

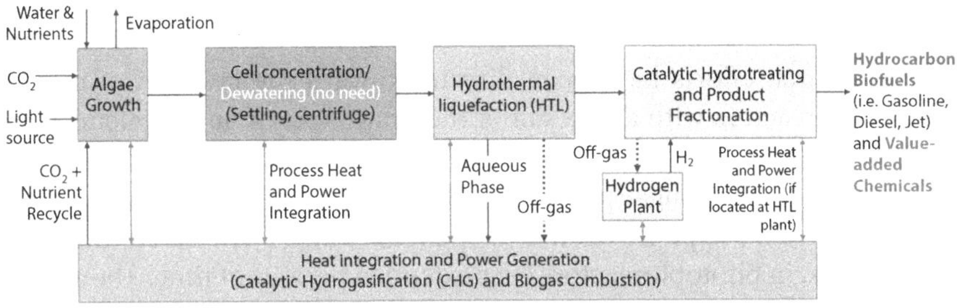

Figure 2.1 Diagrammatic description of the HTL method for microalgae.

and transesterification into biodiesel. Due to the significant expenditures involved in carefully controlled algae production, this approach is often considered economically challenging [3].

2.2 Algae Cultivation

These algae are distinct aquatic plant types. Macroalgae, also known as seaweeds, are multicellular aquatic organisms resembling plants. In comparison to terrestrial biomass sources, they exhibit significantly higher photosynthetic activities, measuring at 6%–8% for macroalgae (Figure 2.1) and 1.8%–2.2% for terrestrial sources [4]. Their relatively straightforward harvesting cycle occurs every, and they contain elevated levels of carbohydrates. On the other hand, microalgae are primitive plants characterized by their lack of roots, leaves, and stems. Predominantly unicellular, they typically have a diameter of 0.4 mm or smaller [5].

2.3 Open-Air Systems

The bulk of large-scale commercial systems are built on what are known as "open pond systems," which are open-air arrangements. These can be divided into four main groups: raceway ponds, tanks, big shallow ponds, and circular ponds. The option of a particular open pond system depends on a number of factors, including the cost of the land, the local climate, the intended algae's natural qualities, and the cost of the water [6]. Raceway ponds are the most popular option among these because they stand out as an established technique. Since the 1950s, these ponds have been used for extensive microalgae farming [5].

2.3.1 Shutdown Systems

Many research efforts have been directed toward the creation of closed tubular bioreactors with as a result of the drawbacks of open systems outlined in the previous part [6]. They have been utilized for a while, but, until recently, commercial algae production was not carried out in them because of a higher capital expense relative to open raceways [7]. Monocultures can be produced in photobioreactors over extended periods of time. The ability to generate algae free from contaminants and pollutants is crucial if the algae are delicate to contamination by different organisms or if it may be helpful in byproducts like polyunsaturated fatty acids (PUFAs) and colors.

2.4 Chemical Composition

Algae have a chemical composition that is noticeably different from traditional biomass sources like terrestrial plant species. Traditional plants often contain cellulose, hemicellulose, and lignin, which together can make up more than 95% (Figure 2.2) of their total weight in some cases [8]. Proteins, lipids, waxes, and carbohydrates are the main components of algae, in contrast to other useful substances such vitamins [9, 10] and colors [11], though in smaller amounts.

The chemical composition can significantly differ from one algae species to another. While some algae may possess high protein content, others might exhibit elevated lipid concentrations. These inherent variances in composition result in certain species being better suited for specific

Figure 2.2 A medium-sized tubular photobioreactor at the USA site of Varicon Aqua.

applications. For instance, algae rich in lipids are well suited for biofuel conversion, whereas algae abundant in protein and vitamins find utility in applications like nutritional supplements and animal feeds.

2.4.1 Chlorophyll

Algae have a wide range of pigments, some of which are unique for them while others have commonality with organisms such as bacteria, plants, and yeasts. Certain pigments generated by algae have great commercial importance and value. Chlorophyll is one of these pigments that is found in all algal species. Within the nine kingdoms of algae, five distinct forms of photosynthesis have been recognized. Notably, chlorophyll-a is present in all algae species and solely found in cyanobacteria like Spirulina (Arthrospira). In Spirulina, which includes A. maxima and A. platensis, chlorophyll-a constitutes around 0.8%–1.5% of the dry weight [12].

2.5 Microalgae

Microalgae, being rudimentary plants, lack conventional structures like leaf. Most of the microalgae are single-celled and smaller. These microorganisms primarily function as energy converters; they capture sunlight, absorb carbon dioxide from the air, and extract nutrition from their environment. Notably, microalgae are prolific producers of lipids, particularly advantageous for bio-diesel production. Microalgae exhibit a high lipid content compared to macroalgae, typically ranging from 20% to 50% of their dry weight. In a noteworthy environmental aspect, 1 kg of algae can sequester 1.83 kg of CO_2, proving especially relevant for mitigating air pollution from heavy industries. Moreover, owing to their ability for continuous cultivation, microalgae surpass the oil productivity biodiesel yields from microalgae are strikingly higher, reaching 12,000 liters per hectare, in contrast to rapeseed's 1,190 liters per hectare [6].

2.6 HTL Bio-Crude Upgrading

Through hydrothermal liquefaction (HTL), biomass can be processed into energy-rich bio-crude. Microalgae, due to their inherent high moisture content, serve as an ideal feedstock for this method. The elevated moisture content need not be reduced prior to processing, as thermal energy can be applied to vaporize some of the moisture during HTL. This characteristic of

HTL to accommodate high moisture feedstock leads to significant energy savings in comparison to technologies demanding low moisture content feedstock [13].

2.6.1 Analytical Pyrolysis (Py-GC-MS)

Model chemicals and samples of algae were subjected to pyrolysis that is elution of gasses and mass spectra (Py-GC-MS) examination. For the analysis, Chemistry Data Systems' CDS Analytic 5200 Pyroprobe from Oxford, Pennsylvania, along with the CDS 5200 HPR framework, was employed. Both a Clarus 600S class mass spectrometer and a Clarus 680 series liquid chromatograph, both manufactured by PerkinElmer, were connected to this system. As part of the process and the temperature, the specimens, each weighed 5 mg and which were housed in quartz sample tubes, underwent pyrolysis at 500°C for 30 s of holding it [14].

2.7 One- and Two-Dimensional NMR Spectroscopy

After consulting with them on this PhD research, Demirbas and Demirbas [15] provided precise measurement methods that were used to determine the composition of both HTL-derived bioliquids and improved liquids using NMR analysis. The residual solvent signals from the Bruker

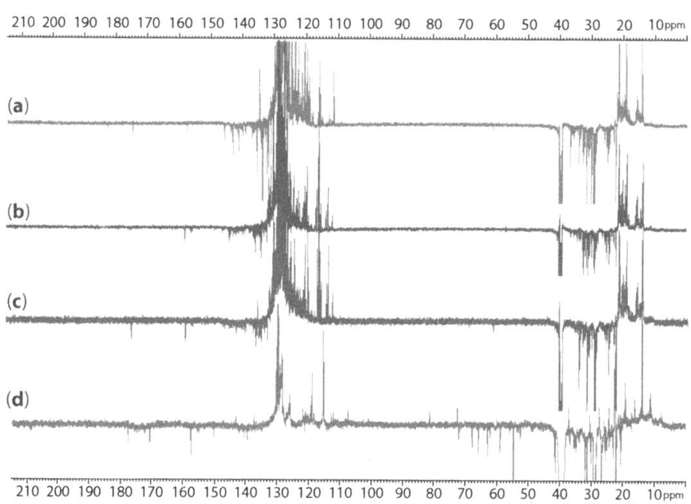

Figure 2.3 PENDANT ^{13}C NMR.

Avance-300 spectrometer were used as a reference for the nuclear magnetic resonance (NMR) (Figure 2.3) study, which was carried out at room temperature.

2.8 Two-Dimensional NMR

Two-dimensional (2D) ^{1}H-^{13}C Heteronuclear Single Quantum Coherence (HSQC) spectra were obtained, facilitating the correlation of ^{1}H and ^{13}C chemical shifts through a one-bond heteronuclear scalar coupling (1JCH). The cross peaks observed in the ^{1}H-^{13}C HSQC spectra exhibited ^{1}H chemical shifts on the horizontal axis, correlating with ^{13}C (on the vertical axis). These correlated pairs of chemical shifts belonged to other. To enhance the overall sensitivity of the experiment, the pulse sequence incorporated several polarization transfer steps.

2.9 Conclusions

The goal of analyzing micro- and macroalgae was to learn more about and possible appropriateness such as HTL. The aim was to have a better knowledge of their potential to form bio-crudes during HTL. The research included the examination of five algae and three microalgae types. In comparison to macroalgae, microalgae demonstrated more beneficial properties, such as increased volatile content, lower ash level, and enhanced lipid content. Microalgae, on the other hand, demonstrated which can contribute to elevated levels of nitrogen in resulting HTL bio-crudes, necessitating extensive denitrogenation during upgrading processes.

Initially performed on a smaller scale, HTL experiments were conducted with two microalgae and two macroalgae species. Notably, microalgal species showcased notably higher average bio-crude yields ranging from 30.1% to 48.6%, whereas macroalgae exhibited yields between 16.6% and 18.5%. The resultant bio-crudes from both categories contained considerable nitrogen (3.8–8.8 wt%) and oxygen (12.6–18.6 wt%), and their higher heating values significantly increased, signifying an enhanced energy density in the bio-crude products.

In a larger-scale study of the effect of reaction temperature, the HTL of the microalgae a plant called Nan sp. revealed optimum bio-crude yields of 51.6 wt% at 300°C. These yields exceeded the amount of lipids of the algae, confirming HTL's ability to process complete algal cells while improving energy recovery. The elevated yields and molecular composition of microalgal HTL bio-crudes point to their potential for additional processing to improve fuel attributes and quality.

Looking toward commercial applications, a commercial-scale HTL processing plant was considered with a focus on microalgae as feedstock.

References

1. IPCC, Climate Change 2021: *The Physical Science Basis*. 2021.
2. Government, U., *Climate Change Act 2008*, 2050 amendment. 2019.
3. Rodolfi, L., *et al.*, Microalgae for oil: strain selection, induction of lipid synthesis and outdoor mass cultivation in a low-cost photobioreactor. *Biotechnol. Bioeng.*, 2009. **102**(1): p. 100-12.
4. Ziegler, M.S. and J.E. Trancik, Re-examining rates of lithium-ion battery technology improvement and cost decline. *Energy Environ. Sci.*, 2021. **14**(4): p. 1635-1651.
5. Chisti, Y., Biodiesel from microalgae. *Biotechnol. Adv.*, 2007. **25**(3): p. 294-306.
6. Brennan, L. and P. Owende, Biofuels from microalgae—A review of technologies for production, processing, and extractions of biofuels and co-products. *Renewable Sustainable Energy Rev.*, 2010. **14**(2): p. 557-577.
7. Borowitzka, M.A., Commercial production of microalgae: ponds, tanks, tubes and fermenters. *J. Biotechnol.*, 1999. **70**(1-3): p. 313-321.
8. Guo, Y., *et al.*, A review of bio-oil production from hydrothermal liquefaction of algae. *Renewable Sustainable Energy Rev.*, 2015. **48**: p. 776-790.
9. Fabregas, J. and C. Herrero, Vitamin content of four marine microalgae. Potential use as source of vitamins in nutrition. *J. Ind. Microbiol. Biotechnol.* 1990. **5**(4): p. 259-263.
10. Lubián, L.M., *et al.*, Nannochloropsis (Eustigmatophyceae) as source of commercially valuable pigments. *J. Appl. Phycol.*, 2000. **12**(3): p. 249-255.
11. Guschina, I.A. and J.L. Harwood, Lipids and lipid metabolism in eukaryotic algae. *Prog. Lipid Res.*, 2006. **45**(2): p. 160-86.
12. Ciferri, O., Spirulina, the edible microorganism. *Microbiol. Rev.*, 1983. **47**(4): p. 551-578.
13. Appenteng, M.K., *et al.*, Multi-element analysis of Ghanaian crude oils by instrumental neutron activation analysis. *J. Radioanal. Nucl. Chem.*, 2012. **292**(3): p. 1197-1206.
14. Ramamoorthy, S. and N. Sulochana, Enhancement of biogas production using algae. *Curr. Sci.*, 1989. **58**(11): p. 646-647.
15. Demirbas, A. and M. Fatih Demirbas, Importance of algae oil as a source of biodiesel. *Energy Convers. Manage.*, 2011. **52**(1): p. 163-170.

3

Renewable Energy Integration for Rural Electrification Legally

Abhinav Koushik[1], Nasrin I. Shaikh[2*] and Milind Shrinivas Dangate[1†]

[1]*Chemistry Division, School of Advanced Sciences, Vellore Institute of Technology, Chennai, India*

[2]*Department of Chemistry, College of Military Engineering, Old Mumbai Pune Hwy, Dapodi, Pimpri-Chinchwad, Maharashtra, India*

Abstract

Renewable energy technology is utilized to create pure current, quickly electrify rural areas, and provide dependable electricity at an affordable price. The study will encompass the examination and modeling of load profiles specific to rural households. In the process of determining the optimal configuration of components within the design scope of the hybrid renewable energy system (HRES), factors such as biomass availability and solar radiation statistics from the targeted region are carefully considered. Leveraging survey information encompassing variables like the number of beds, household size and classification, occupant behavioral patterns, and appliance ownership, we can gain comprehensive insights into space utilization and energy consumption patterns. An analysis of simulated load profiles indicates an average daily energy consumption of 3.13 kWh at the study location. In the context of estimating solar radiation at the studied location, the evaluation underscores the significance of temperature as a crucial meteorological characteristic. This evaluation underscores the importance of temperature as a key variable when constructing models aimed at estimating solar radiation within the examined area. While the lowest amount of time needed to measure data to estimate past solar radiation indicates that more than 2 years of recent data are needed to obtain R2 values greater than 0.75 and more than 5 years are needed to achieve R2 values larger than 0.9. The study location has a restricted supply of recoverable household and animal biowaste, according to the assessment of biomass availability. Power flow analysis of the effects of aggregating five regional

**Corresponding author*: dr.nasrinshaikh@yahoo.com
†*Corresponding author*: milind.shrinivas@vit.ac.in

O.V. Gnana Swathika, K. Karthikeyan, Milind Shrinivas Dangate and Nicoletta Ravasio (eds.) Biomass and Solar-Powered Sustainable Digital Cities, (21–30) © 2024 Scrivener Publishing LLC

loads reveals that doing so lowered the peak load in each of the five regions by 23%. The regional grid's power flow research also demonstrates that installing generators that are dispersed according to the apparent power consumed by each load bus on the regional grid will minimize power losses. The creation of an electricity pricing strategy that permits the shifting of peak loads as well as research into how the strategy affects LCOE would be beneficial for work to come.

Keywords: Rural electrifications, biowaste, efficient combination, biomass

3.1 Introduction

Energy plays a crucial role in elevating living standards and cultivating economic sustainability. In this context, the adaptable communal grid system presents advantages over decentralized technologies such as solar household systems. Communal grids can be seamlessly integrated with national grids or interconnected to construct expansive electrical networks. Operating in an islanded mode while interconnected allows for more effective control of generated electricity and enhances the energy system's security. In contrast to the limited impact of decentralized technologies like solar household systems, communal grid energy systems have the potential to significantly drive sustainable development in rural areas [1].

Several developing nations, including India, advocate for communal grid systems as a viable solution for rural electrification, owing to their inherent benefits and the positive impact that they can bring to underserved regions [2].

3.1.1 Grid and Environment for Communities

Human activities, especially the burning of fossil fuels, which raise atmospheric CO_2 and GHG levels, are of global concern. The majority of GHG emissions are caused by the energy sector, with world current CO_2 emissions leveling a record high of 33.1 Gt in 2018. Sixty-seven percent [3] of the carbon-dioxide emissions were caused by the power sector.

3.1.2 Configuration of a Hybrid Renewable Energy System

The best possible combination of renewable energy technologies is needed to reduce supply and load mismatch. There are three basic configurations for HRES: DC-coupled, AC-coupled, and hybrid-coupled. These configurations are essential for AC load supply and DC battery charging.

3.1.3 Configuration of an AC-Coupled HRES

The design option for off-grid renewable energy applications uses an AC bus to link storage and renewable energy systems to the integrated shared AC load. Compared to DC-coupled solutions, this design offers greater energy security. However, it has trouble synchronizing power produced by renewable energy sources with the AC bus and has problems with harmonics and reactive power in the power supply [4].

3.1.4 Coupled Hybrid HRES Configuration

An off-grid hybrid-coupled configuration option for renewable energy applications links DC and AC buses using a bidirectional converter. By linking AC and DC energy systems individually, this arrangement increases system efficiency while lowering energy conversion losses and unit cost. Recent advancements in power converter technology can reduce complexity even when control and energy management are complex. The HRES setup in this study is this hybrid-coupled design option [5].

3.1.5 Components of Hybrid Renewable Energy Systems

The photovoltaic (PV), the biomass generator, and the battery are the three basic elements of hydrogen-powered energy systems (HRESs) [6]. This lengthens battery life and decreases energy losses. A biogas generator fills the load in the absence of PV systems, and the battery is employed to reduce peak power. Local biowaste collection promotes community involvement in the production of biogas energy.

3.2 Solar Energy Technology

P-n junctions are formed by PV semiconductor materials, absorbing incident photons and producing electron-hole pairs. Excited electrons move through closed circuits when linked, affecting a PV system's electrical output [7].

3.3 Technique for Solar Radiation Modeling

Information on solar radiation is essential for developing and assessing solar energy technology. Because of the Earth's movement and the erratic atmosphere, measurement is difficult. Solar radiation is measured with group equipment or satellite image processing. Ground-measured data are frequently unavailable in poor nations like Nigeria. A different strategy is to empirically global sun radiation data using long-term meteorological characteristics. Based on input parameters such as astronomical, geographic, geometric, physical, and meteorological elements, solar radiation estimating models are created, offering insights into atmospheric and meteorological conditions.

3.4 Iterative Procedures

When specific selection criteria are satisfied, an iterative technique is a recursive process that chooses the best configuration of a hybrid energy system. Using choice factors that change linearly, it replicates the best possible outcome. On Kerkennah Island in Southern Tunisia, the approach was also utilized to estimate the lowest possible cost for a seawater desalination facility.

3.5 Methodology for Modeling Load Profiles

Drawing from surveys focused on household occupancy and employing stochastic modeling techniques to analyze the survey outcomes, a bottom-up modeling approach is adopted to construct a stochastic load profile for rural households. Acknowledging the substantial influence of occupant behavior on energy usage, this model is constructed around the schedule of residents' activities.

3.6 Data Collection for Surveys

When there is a lack of data and end users need to understand, be aware of, and accept the data, surveys are used to collect household energy consumption statistics. To validate the model, this study gathers information on household demographics, resident activity patterns, and appliance ownership. Instead of using data from the country's population and housing census. According to Power Holding Company of Nigeria (PHCN), unreported clients and unpaid electricity bills account for 36% of the electricity that has been distributed. Reducing the number of residential consumers connected to its distribution network who are not being captured is one of PHCN's biggest concerns.

3.7 System of Energy Modeling

To estimate the household population in the survey location, the created stochastic demand model employs survey data, utility company data, and data from the national housing and population census. The distribution of housing units among various household classes is based on national population and housing census statistics; occupant consumption patterns, appliance ownership, and appliance use are determined using survey data. The weights that were assigned were from the study area's national population and housing census data for Nigeria. The hourly stochastic modeling of occupant consumption behavior and the minutely detailed simulation of energy usage patterns provide a comprehensive analysis of building energy demands.

3.7.1 Modeling of Converter Systems

The hybrid-coupled topology converter, which acts as the link between the AC and DC buses, permits bidirectional power transmission. It serves as a rectifier for biomass generator energy, converting AC to DC power, and an inverter for photovoltaic system energy, converting DC to AC power. This makes it possible for battery systems to generate and store energy effectively [12].

3.7.2 Worldwide Solar Radiation Estimation

Evaluating the techno-economic viability of solar energy plants requires accurate forecasting of global sun radiation for a design location. Estimated global radiation data offer important insights into the solar energy output from the specified region. The energy output is influenced by the PV panel's orientation in regard to the sun. Estimates are more precise than satellite measurements because local meteorological stations measure site-specific features.

3.8 Technical Design Approach

In order to reduce peak demand shaving, technical design strategies for hybrid energy systems concentrate on reducing unmet load and aggregating electrical load. These tactics seek to reduce load diversity and boost effectiveness [13].

3.8.1 Technical Design Approach for Distribution of Power on Regional Grids

A power flow analysis is conducted to tell the most efficient technical design strategy for distributing electricity in a regional grid. Following the initialization of power network factors such as electric loads, generation requirements, and constraints, the study calculates essential characteristics including bridge admittance, conductance, and susceptibility. The problem is examined using the Newton–Raphson approach because it yields more accurate results faster than the Gauss–Siedel and Jacobi methods [11].

3.8.2 Analysis and Design of a Hybrid Energy System

Electricity is essential for raising the level of living, particularly in emerging nations like Nigeria. Rural areas frequently have unstable networks and limited access to electricity. A practical way to provide dependable electricity is through a communal-grid system that aggregates community loads [8].

3.8.3 System Powered by Biomass

It offers a method for calculating a PV panel's hourly solar insolation at the right tilt angle. The National Centre for Energy and Environment (NCEE) in Benin City measured the hourly solar insolation at 11.6°C and the ambient temperature, which were the climatic variables utilized to simulate PV-generated electricity [9].

The amount of biogas produced daily by the bio-digester/reactor and the biogas generator's rated capacity have a direct impact on the overall energy production. As mentioned before, the biogas generator operates at its rated capacity whenever the power demand is less than 50% (at a minimum load ratio). The biogas engine is turned off at this point [8].

Important decisions that need to be made during the planning stage for the biomass-powered power system can be guided by the results of the modeling of the methane energy system. Because constant use increases the size of an anaerobic digestion system resulting in a high need for waste that breaks down that exceeds the location's predicted capacity for producing it, a biogas generator should only be used during times of high energy consumption [9].

3.8.4 Techno-Economic Examination of Hybrid Energy Systems

To determine the best possible combinations for a steady supply of energy to the load profile, the cost of HRES components is covered in this section. Through yearly hourly simulations that took into account seasonal inputs like solar radiation and ambient temperature, accurate hybrid energy system modeling was made possible [10].

3.9 Conclusion

Electrifying rural residential electrical loads within a community or regional grid presents a pragmatic alternative to national grid electrification. This approach facilitates load aggregation and provides swift electrification to electricity-deprived communities, as well as reliable power to areas linked to an unstable national grid. To ascertain the cost-effectiveness of implementing a hybrid PV and biogas generator system for ensuring consistent electricity supply to residential loads within a community and to evaluate the impact of load aggregation on a broader regional grid, this chapter conducted two comprehensive studies: a techno-economic analysis and a power flow assessment.

The assessment of biomass availability in the examined locale revealed limited recoverable human and animal waste. Biogas generators are deployed only when the required energy surpasses 50% of the generator's rated capacity. In contrast, a meticulously designed hybrid renewable energy system (HRES) necessitates a notably smaller battery capacity compared to integrating a biogas generator and battery or a PV and battery system for rural electrification, as elucidated in a study investigating the influence of diverse combinations of PV, biogas generator, and battery setups on optimal battery sizing.

Notably, the LCOE of the HRES is a formidable rival to both the LCOE of a national grid and that of an exclusively diesel-powered energy system. Furthermore, HRES's loss of power supply probability (LPSP) of

3.7% shows a positive position in comparison to the LCOE assigned to the national grid. The HRES's LCOE may become even more competitive with the LCOE of the national grid with the incorporation of more residential loads into the regional grid. This consolidation resulted in a stunning 23% decrease in peak loads across the five regions, as shown by a power flow study examining the effects of combining five regional loads. Hybrid renewable energy system (HRES) stands as a strong competitor against both the LCOE of a national grid and that of an exclusive diesel-powered energy system. Moreover, the HRES's LCOE demonstrates a favorable position in comparison to the LCOE attributed to the national grid for a LPSP of 3.7%. The integration of additional residential loads into the regional grid has the potential to render the HRES's LCOE even more competitive with the national grid's LCOE. As evidenced by a power flow study investigating the consequences of aggregating five regional loads, this consolidation led to a remarkable 23% reduction in peak loads across the five regions.

In order to accurately assess the hourly variance and ensure that the weekly apparent power% loss stays within the authorized benchmark for the surveyed site, an accurate estimation of the maximum apparent electricity percentage loss becomes crucial.

References

1. BP, "*Energy outlook*," London, UK, 2018. [Online]. Available: https://www.bp.com/content/dam/bp-country/de_ch/PDF/Energy-Outlook-2018-edition-Booklet.pdf. [Accessed: Sep. 2, 2019].
2. IMF, "*World economic outlook: Less even expansion, rising trade tensions*," Washington, USA, 2018. [Online]. Available: https://www.imf.org/en/Publications/WEO/Issues/2018/07/02/world-economic-outlook-update-july-2018. [Accessed: Feb. 17, 2019].
3. IEA, "*Global energy & CO2 status report: The latest trends in energy and emissions*," Paris, France, 2018. [Online]. Available: https://webstore.iea.org/global-energy-co2-status-report-2018. [Accessed: Apr. 21, 2019].
4. IEA, "*Energy access outlook: From poverty to prosperity*," Paris, France, 2017. [Online]. Available: https://www.iea.org/energyaccess/database/. [Accessed: Mar. 6, 2018].
5. IEA, "*World energy outlook: Electricity database*," Paris, France, 2017. [Online]. Available: https://www.iea.org/energyaccess/database/. [Accessed: Oct. 14, 2018].
6. World Bank, "*World development indicators*," Washington, USA, 2019. [Online]. Available: https://data.worldbank.org/country/nigeria. [Accessed: Sep 2, 2019].

7. E. Cecelski, "*Enabling equitable access to rural electrification: Current thinking and major activities in energy, poverty and gender*," World Bank, Washington, USA, 2000. [Online]. Available: https://www.semanticscholar.org/paper/Enabling-equitable-access-to-rural-electrification. [Accessed: Nov 6, 2015].
8. R. A. Cabraal, D. F. Barnes, and S. G. Agarwal, "Productive uses of energy for rural development," *Annual Review of Environment and Resources*, vol. 30, pp. 117–144, 2005.
9. D. F. Barnes, *The challenges of rural electrification: Strategies for developing countries*. Michigan: Routledge, 2007.
10. M. Gustavsson, "Educational benefits from solar technology-Access to solar electric services and changes in children's study routines, experiences from eastern province Zambia," *Energy Policy*, vol. 35, pp. 1292–1299, 2007.
11. World Bank, "*The welfare impact of rural electrification: A reassessment of the costs and benefits*," World Bank, Washington, USA, 2008. [Online]. Available: https://siteresources.worldbank.org/EXTRURELECT/Resources/full_doc.pdf. [Accessed: Jan. 16, 2017].
12. J. Finucane, S. V. Bogach, and L. E. Garcia, "*Promoting productive uses of electricity in rural areas of Peru: Experience & lesons learned*," World Bank, Washington, USA, 2012. [Online]. Available: https://www.worldbank.org/en/results/2013/04/10/pomoting-poductive-uses-electricity-rural-areas-peru. [Accessed: May 7, 2018].
13. H. Ahlborg and L. Hammar, "Drivers and barriers to rural electrification in tanzania and mozambique - grid-extension, off-grid, and renewable energy technologies," *Renewable Energy*, vol. 61, pp. 117–124, 2014.

4

Fuels from Waste Biomass for the Transport Industry

Sakthi Ashok[1], Abhinav Koushik[1], Milind Shrinivas Dangate[1] and Nasrin I. Shaikh[2]*

[1]*Chemistry Division, School of Advanced Sciences, Vellore Institute of Technology, Chennai, India*
[2]*Department of Chemistry, College of Military Engineering, Old Mumbai Pune Hwy, Dapodi, Pimpri-Chinchwad, Maharashtra, India*

Abstract
This study looked at whether making fuel from burned plants can be a good idea for a long time. There are many new kinds of fuel that can help us use less oil. These fuels can be made in different ways, each with its own good and bad points. We looked at six ways to make this kind of fuel using special machines. One of the ways that we looked into was heating corn leftovers in a special tank. We also looked at other processes like making certain chemicals and improving the fuel. We found out that, when you look at all the pollution from making the fuel and using it, this kind of fuel can actually help reduce pollution. For example, one way of making this fuel might release much less pollution compared to making diesel fuel. We also checked how much this fuel might cost to make. It turns out it could be cheaper than we thought. We carefully compared the six ways to make this fuel using some methods called analytic hierarchy process and technique for order of preference. Then, we made a suggestion based on what we found out.

Keywords: Artificial fuel, pyrolysis, bioenergy, analytic hierarchy process, technique for order of preference by similarity to ideal solution

4.1 Introduction

This chapter talks about how climate change affects our daily lives and how using different kinds of fuels, especially for cars and trucks, can make a

**Corresponding author*: dr.nasrinshaikh@yahoo.com

O.V. Gnana Swathika, K. Karthikeyan, Milind Shrinivas Dangate and Nicoletta Ravasio (eds.) Biomass and Solar-Powered Sustainable Digital Cities, (31–42) © 2024 Scrivener Publishing LLC

difference. It also discusses the problems we face with these new fuels. The chapter explains what the thesis aims to do. It looks at turning plants into fuels through different methods. One part of the thesis focuses on making fuels for vehicles. It describes how we compared six different ways of making these fuels. We used a detailed method to see how these new fuels stack up against traditional ones like gasoline and diesel. The main goal is to figure out if these synthetic fuels are good for the environment in the long run. When dealing with complex problems like this, there is not always just one answer. That is where multi-criteria decision analysis or multi-criteria decision-making (MCDM) comes in handy. It helps us figure out the best options among many choices.

4.1.1 Background Problems

The majority of nations have significant rates of biomass production, which produces enormous amounts of garbage (such as municipal, industrial, and agricultural waste). As a large amount of this trash builds up, disposal challenges arise, such as when it is burned, which has detrimental environmental effects [1]. Researchers looked into whether using zeolites to improve bio-oil could make it more stable when heated and reduce the amount of oxygen in the oil. Zeolites are often used in factories that make chemicals from oil and in oil refineries.

4.1.2 Effects of Climate Change

It is believed that global warming is the primary source of many climate change repercussions. Extreme weather events like downpours and a drop in summer precipitation are just two examples of how the drought will likely get worse. Changes in how much rain and snow we get might happen because of shifts in weather patterns. Global warming can also make the oceans warmer and hold more water in the air. When ice at the North and South Poles melts, it can lead to stronger storms and the sea level going up. The effects of climate change are anticipated to be mixed, with some regions of the world benefiting, whereas, in other areas, the frequency of dryness and high rainfall may increase due to earlier spring weather. Increased forestry development was recommended as a possible treatment in earlier research, but the good effect was negated by forest fires linked to global warming.

Electric hybrid cars, which use smaller batteries that can be charged while driving, and fuel cell electric cars, which use hydrogen to power an electric engine, are both possible future technologies, allowing for the interchangeability of electricity and fossil fuels. Compared to internal combustion vehicles, fuel cell electric vehicles need a completely new vehicle technology and energy infrastructure.

Biofuels are a promising replacement that are currently used all over the world. They can be used in the same ways and distributed using the same infrastructure as present fuels. Certain biofuels might need changes to the infrastructure and vehicles they are used in. However, when we look at it, the technology for regular cars is already here, and these biofuels, especially for hydrogen and electric vehicles, do not need big changes to work with them, unlike regular fuel for traditional cars [2]. While energy can be produced using a variety of renewable resources Gasoline and diesel can be directly replaced by biofuels. Additionally, they are currently the only resources that can replace fossil fuels in the production of chemicals and commercial aviation fuels.

According to the International Energy Agency [2], oil now supplies the bulk of the world's total energy needs for transportation, followed by natural gas and other fuels at 3%, bio-fuels at 2%, and current at 1%. To reach the transportation goals for 2020, which aim for 10% improvement, important rules such as the renewable transport fuel obligation must be put into action by the UK Government.

4.2 The Issue with Fossil Fuels

Along with worries about climate change, the world's energy needs are growing quickly, and fossil fuel stocks are depleting due to the extraction of crude oil at a rate far quicker than its natural regeneration. As a result, discovering alternate sources of energy is of great importance.

The majority of the energy that powers our life originates from plants, either as food that we consume or as fossil fuels that date back to the Paleolithic period. When fossil fuels that were created by the fossilization of plant remains over hundreds of millions of years are burned, something that has been discreetly locked becomes freed and could have serious repercussions.

4.3 Existing Biofuel and Artificial Fuel Technologies

Different technological methods can be used to create liquid fuels that can replace conventional fossil fuels. Biofuels produced through fermentation and transesterification, such as bioethanol and biodiesel, are among the many options for feedstock types, as are artificial fuels produced using cutting-edge thermochemical treatment method. Fermentation uses microorganisms and enzymes to extract recoverable product from a substrate. that may be fermented [3].

When we burn biomass, we usually use it to make energy instead of fuels for transportation. But for processes like liquefaction using water, when water is added to the biomass, it can help make more biocrude. Through heat decomposition and exposure to various oxygen concentrations, biomass is chemically treated [4].

Syngas is created when biomass is gasified. This flammable gas mixture can then be converted into liquid artificial fuel using the Fischer–Tropsch process at temperatures of approximately 800°C to 900°C and low oxygen concentrations. Another way is by heating up biomass without oxygen. This process, called pyrolysis, can create various organic liquids that can later be turned into liquid fuels.

4.4 Pyrolysis Method

Charcoal has been created by pyrolysis, which is the heat decomposition of materials in an oxygen-free atmosphere, for thousands of years. It can be expressed using the straightforward equation shown below.

$$\textbf{Biofuel} \xrightarrow[\textbf{Inert gas}]{\Delta} \textbf{Coal} + \textbf{Liquid} + \textbf{Gas}$$

Slow pyrolysis, intermediate pyrolysis, fast pyrolysis, and flash pyrolysis are the four basic forms of pyrolysis. Char is the primary product gathered; hence, it cannot be used to produce high-quality bio-oil. Intermediate temperatures, solid and vapor residence times, and slow heating rates are the distinctive features of intermediate pyrolysis. It is a procedure that is frequently investigated and improved. Compared to quick pyrolysis, intermediate pyrolysis yields less oil. However, recent research has found that these materials work better with the fuels. There are two types of pyrolysis: fast and flash. They involve heating quickly, reaching high temperatures, and making gases and solids evaporate. Fast pyrolysis was developed during the oil crisis of the 1970s to produce more bio-oil. However, due to its flaws, such as its corrosiveness, thermal instability, and particle content, it is not ideal as a fuel for transportation. The immensely complex pyrolysis process includes both immediate and aftereffects. Organic substance can be heated using this technique without the use of oxygen or air. Components of biomass feedstock start to thermally breakdown between 350°C and 550°C and continue up to 700°C to 800°C by the primary pyrolysis method [6].

Cellulose, hemicellulose, and lignin are the three main natural substances that make up the majority of waste biomass. These, however, are present in various concentrations for every type of biomass and affect the pyrolytic process in various ways. Condensable vapor is the main byproduct of cellulose and hemicellulose, which can be either a liquid or a gas. Lignin, however, decomposes into gas, coal, and liquid products. Through volatilization or decomposition, extractives may have an impact on the liquid and gas products, and the ash will have the majority of the minerals found in feedstock [5].

4.4.1 Technique for Order of Preference by Similarity Method

For polymer electrolyte fuel cells investigated the technique for order of preference by similarity to ideal solution (TOPSIS) approach for choosing the metallic bipolar plate material. Using the entroy weighing technique, it was possible to compile a list of all plausible options, from the best to the worst. Using entropy weighting, Abbasi and Abbasi [5] employed the TOPSIS technique to evaluate the safety of coalmines. The four safety condition distinct mines were compared. Researchers studied how to choose the best shelter during an earthquake using a method called TOPSIS. They considered nine different factors, like safety and accessibility. Then, they used pair-wise regression analysis to decide how important each factor was. They used a few methods, like analytic hierarchy process (AHP) and entropy, to figure out the best shelter option. Another study used the TOPSIS method with entropy to make sure that materials were delivered on time by a power company [7].

4.4.2 Analytic Hierarchy Process Approach

The AHP approach has been employed as a processing tool in numerous fields. AHP for flexible manufacturing systems was employed by [8] as a decision-support tool for evaluating innovative technologies. Further sensitivity assessments were conducted after identifying the factors that were most important for accomplishing the aim.

AHP was also utilized [8] to select vendors based on metrics for both quality and quantity. They discovered that the AHP may be utilized to handle complex scenarios, and they came to the conclusion that the sensitivity analysis is crucial for making sensible selections because it might help prioritize the criteria. It all depends on the information and how it affects the whole system. After looking at the results, they found that using AHP was the most accurate way to divide things up in a manufacturing process [8].

Nixon *et al.* (2013) evaluated the machine using HANP (hierarchical analytical network process) and various methods for producing power in India from municipal solid trash. The accuracy of the model is increased by ANP, an extension of AHP that takes into account the interdependencies of the criteria [9]. Anaerobic digestion was determined to be the preferred technology among the combustion, gasification, and anaerobic digestion. The most effective energy recovery techniques in India and the UK were identified by Yap using AHP. Mass burn incineration, gas recovery, anaerobic digestion, gasification, and refuse-derived fuel incineration were among the technologies taken into account. Because the technological and economical standards in India and the United States are different, anaerobic digestion was found to be the best option. The best choice for the UK was gasification. AHP was employed as a flexible manufacturing decision-making tool. To meet client requests, they sought to implement tactics that were both successful and efficient. The authors came to the conclusion that AHP is a useful technique for multiple criteria analysis.

4.5 Esterification

The temperature for esterification usually falls between 70°C and 170°C. When using a zeolite catalyst at around 100°C, about 62% of the bio-oil can be improved. For esterification, the best ratios of ethanol (C_2H_5OH) to oil are between 3:1 and 5:1. But, if you use a small amount of sulfuric acid (H_2SO_4), then you can use lower ratios of ethanol to oil, like between 1:1 and 3:1, and still get around 60% improved bio-oil. Bioethanol can come from different plants like sugarcane, corn, and beet. The information for making bioethanol from corn comes from a database called E-coinvent. It is not a good idea to esterify bio-oil at very high temperatures, like between 250°C and 300°C, because it can change the oil's makeup. The energy needed for the process can range from 0.027 kWh/kg to 0.095 kWh/kg of bio-oil, depending on the starting temperature and how hot you need to make it during the process. The things you put in and get out of the esterification process are listed in Table 4.1.

4.6 Ketonic Decarboxylation

The light source oxygenate the portion of the pyrolysis oil can undergo ketonic decarboxylation at temperatures of about 400°C. The esterification

Table 4.1 Bio-oil esterification.

Products	Minutes	Exp	Maximum	Units
Source \current	0.027	0.061	0.095	kWh/kg of raw bio-oil
H_2SO_4	-	2	-	weight%
C_2H_5OH	1	2	3	kg/kg of raw bio-oil
Outputs Bio-fuel produced	0.55	0.62	0.65	%

process, which uses 0.25 kWh/kg of light oxygenates, is similar of electricity is estimated to be required for the process.

While the catalyst, reaction temperature, and time all affect the yield of processed bio-oil produced by ketonic decarboxylation, its anticipated to reach 54% when a TiO_2&C catalyst is utilized at a 5 weight % concentration The chemical reaction's byproduct, carbon dioxide, has not been added. Table 4.2 displays the inputs and outputs of ketonic decarboxylation.

When you heat up biomass to 300°C, it can split into different parts like CH_3COOH and $C_3H_6O_2$. These parts come from light oxygenates, sugars, and lignin. If you heat it more to 400°C, then you get furfurals, and, at 550°C, you get phenolics. Separating these parts needs extra energy, but that has not been considered yet. It is estimated that, after separation, you might get 10% from light oxygenates, 30% from sugars, and 60% from lignin. At 120°C, aldol condensation occurs; acetone is produced from the ketonic decarboxylation (Alibaba.com, 2018) of light oxygenates, and oils generated from sugar using a Esters made with a 5 weight % Pd/MgO-ZrO_2 catalyst have been generated [11].

Table 4.2 Ketonic decarboxylation of bio-oil.

Products	Minutes	Exp	Maximum	Unit
Inputs				
Current		0.25		kWh/kg of light
Oxygenates				
TiO_2/C catalyst		5		Weight %
Outputs				
Bio fuel produced	-	0.46	-	%

4.7 Hydro-Processing

Different levels of deoxygenation can be reached depending on how much hydrotreating is done.

4.7.1 Hydrotreating in Single Stage

Usually, special metals like ruthenium (Ru/C) and palladium (Pd/C) are used as catalysts in a process called single-stage hydrotreating. This process happens at high temperatures and pressures, up to 200 bar and 400°C, and lasts for about 4 h [12]. Ruthenium has been used to hydrotreat fast pyrolysis oils, reducing the oxygen concentration from about 40 to 18 weight % to 27 weight %. Improved bio-oil hydrotreating yields have been documented; however, they are more erratic, ranging from 30% to 65% with the best yields being attained when employing 5 weight % Ru/C. Using a Ru/C catalyst to hydrotreat corn stover pyrolysis oils can produce yields between 54% and 67% and has a product with 25 to 26 weight % oxygen [10]. Research indicates that poplar and mixed wood have demonstrated.

4.7.2 Hydrotreating in Two Stages

According to Jones *et al.* (2013), two-stage hydrotreating entails light hydrotreating at 150°C to 270°C and followed by light processing at 80 bar to 100 bar, 350°C to 425°C, and 140 bar to 200 bar. In total, the residence times for two-stage hydrotreating typically range from 2 h to 4 h. In the first step, Ru/C is used, and the second usually uses a NiMo catalyst. With lives of 700 h to 1,752 h, catalyst amounts are typically in the range of 3 to 5 weight % [12]. Therefore, 0.1 g/kg to 0.3 g/kg of bio-oil based on a 4-h residence period is the predicted catalyst requirement. According to some accounts, two-stage hydrotreating can reduce hydrogen usage by 13%, whereas, according to the other studies, the consumption of hydrogen remains proportional to the degree of deoxygenation. Deoxygenation levels typically fall between 6 to 10 weight %, although, sometimes, it can range from 6% to 11% according to Ahmad *et al.* (2010). The predicted consumption of hydrogen for hydrotreated biofuel is between 58 g/kg and 112 g/kg. According to other authors, 69 g/kg was reported. The usual steam reforming of natural gas is assumed to be the source of hydrogen; however, make-up hydrogen could be collected from the pyrolysis and hydro-processing stage off-gases. The overall yield of deoxygenated bio-oil for two-stage hydrotreating is anticipated to vary between 30% and 44%.

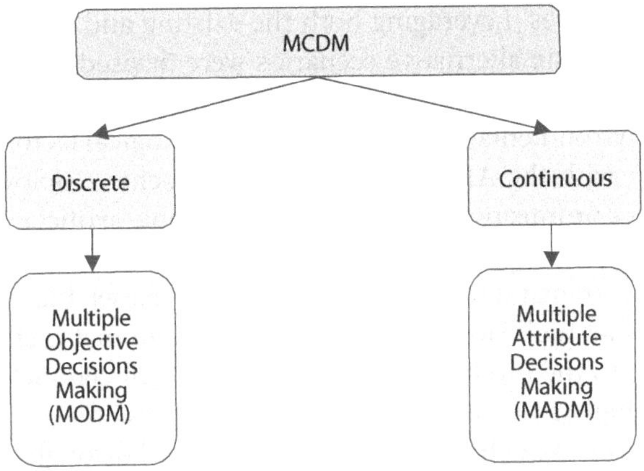

Figure 4.1 Based on data processing, the MCDM category is shown by Hwang and Masud (2012).

4.8 MCDM Categories

Understanding the MCDM classification is crucial for choosing the best decision-making approach for any kind of challenge. Figure 4.1 displays a generic classification of MCDM techniques.

The MADM category results must be picked. The text has a restricted range of potential answers. Because of this limitation, each criterion needs to consider a specific set of sub-criteria. Several possibilities in MADM must be measured against a set of standards, which are frequently elusive to define. The best alternative is chosen by contrasting each alternative with regard to each criterion.

4.9 Conclusion

Numerous methods exist for transforming biomass into biooil and further refining it into synthetic transportation fuels, a topic that has garnered significant attention in scholarly circles. Evaluating the most effective conversion techniques and considering the limited information available in this domain led to the identification of six potential scenarios warranting further investigation. Among these, pyrolysis oil emerges as a viable option for producing sustainable transportation fuels that meet essential standards, as suggested by Balat *et al.* (2009) and Chiaramonti *et al.* The methodology section details the approach employed to explore the sustainable utilization of waste biomass for artificial fuel production through various technologies. Leveraging both the existing and proposed technological synergies, six alternative scenarios were devised, which were then subjected to evaluation using the MCDM method.

Critical environmental, economic, and technological factors were taken into account, with the AHP and TOPSIS approaches employed to determine the most promising alternative for sustainable artificial fuel production. This analysis highlights the potential reduction in CO_2e emissions achievable by transitioning from conventional diesel fuel to pyrolysis-derived artificial fuel. However, it is crucial to note that inefficient methods may lead to the production of poor-quality artificial fuel, resulting in significantly higher CO_2 emissions compared to fossil fuels.

Moreover, the use of high-quality fuels generated through esterification, two-stage hydrotreating, and hydrocracking or through a combination of esterification, ketonic decarboxylation, adol condensation, two-stage hydrotreating, and hydrocracking is projected to have various environmental implications. One or two stages of hydrotreating and esterification appear to offer a balanced compromise between potential environmental benefits and final product quality.

It is worth noting that internal hydrogen generation results in a 53% increase in power usage, whereas external hydrogen production leads to a 50% to 62% rise, making it a significant contributor to CO_2 emissions. Therefore, future research may explore the feasibility of powering pyrolysis machinery with green energy sources to mitigate environmental impact.

Although the study and previous research have not considered the byproducts, it remains uncertain whether byproduction could offer environmental benefits. Further research is warranted to explore this aspect comprehensively.

References

1. Corma, A. (2003) State of the Art and Future Challenges of Zeolites as Catalysts. *J. Catal.*, 216, 298-312. https://doi.org/10.1016/S0021-9517(02)00132-X
2. https://www.iea.org/reports/world-energy-outlook-2012.
3. Cherubini, F. (2010) The Biorefinery Concept: Using Biomass Instead of Oil for Producing Energy and Chemicals. *Energy Convers. Manage.*, 51, 1412-1421. http://dx.doi.org/10.1016/j.enconman.2010.01.015
4. Bain, R. (2004) *An introduction to biomass thermochemical conversion, DOE/NASLUGC Biomass and Solar Energy Workshop, Golden.*

5. Abbasi, T. and Abbasi, S. A. (2010) 'Biomass energy and the environmental impacts associated with its production and utilization', *Renew. Sustain. Energy Rev.*, 14 (3), pp. 919-937.
6. Abdul Aziz, S. M., Wahi, R., Ngaini, Z., Hamdan, S. and Yahaya, S. A. (2017) 'Esterification of Microwave Pyrolytic Oil from Palm Oil Kernel Shell', *J. Chem.*, Vol. 2017 (ID 8359238).
7. Zhang, J. H. (2015). Research on Customer Purchase Intention Based on Wechat Marketing. Ph.D. Thesis, Shanghai University of Engineering and Technology.
8. Bayazit, V., Cayci, M.K. and Khan, K.M. (2005) Purification and Kinetic Properties of Glucose-6-Phosphate Dehydrogenase from Goat (Capra aegagrus hicrus) Erythrocytes. *J. Chem. Soc. Pak.*, 27, 518-526.
9. T. L. Saaty and L. G. Vargas, *The Logic of Priorities, Applications in Bu,sines.., energy, Health, Transportation*. KluwerNijhoff. The Hague (1981).
10. Capareda, S.C. (2014) *Introduction to Biomass Energy Conversions*. CRC Press, Taylor and Francis Group, Boca Raton. https://doi.org/10.1201/b15089.
11. Barrett, C., Chheda, J., Huber, G. and Dumesic, J. (2006) 'Single-reactor process for sequential aldol-condensation and hydrogenation of biomass-derived compounds in water', *Appl. Catal. B: Environ.*, 66 (1), pp. 111-118.
12. Wildschut, J., Mahfud, F.H., Venderbosch, R.H. and Heeres, H.J. (2009) Hydrotreatment of Fast Pyrolysis Oil Using Heterogeneous Noble-Metal Catalysts. *Ind. Eng. Chem. Res.*, 48, 10324-10334. http://dx.doi.org/10.1021/ie9006003

5

The Potential Use of Biodiesel in Diesel Engine and Legal Environmental Laws

Abhinav Koushik[1], Jayendra Kasture[2]* and Milind Shrinivas Dangate[1]

[1]Chemistry Division, School of Advanced Sciences, Vellore Institute of Technology, Chennai, India
[2]VIT School of Law, Vellore Institute of Technology, Chennai, India

Abstract
The replacement of diesel fuel with ultra carbo-fluids was thought to have the ability to reduce pollutants emissions such as smoke, hydrocarbons (HCs), carbon dioxide, and carbon monoxide. A minimal amount of stabilizing chemicals, along with an interruption of coal in water and fuel oil—often in a 5:3:2 ratio—make up these ultra carbo-fluids. The production of fuel oil and coal, as well as the properties of vegetable oils and charcoal and their economics, have all been deeply analyzed in the literature. The production of fuel oil and coal as well as the quality and economics of vegetable and animal oils and charcoal have all been severely analyzed in the literature. Charcoal and vegetable oils may not be able to completely replace coal and fuel oil. New bio-ultra carbo-fluid compositions were used and put to the test. These were unlike ultra carbo-fluid as they used vegetable oil and biodegradable charcoal instead of fuel oil and coal. In the studies, a four-stroke diesel engine with Mexa-32 1E, a Lister-Petter 600cc 2-cylinder, and Mexa-21 1E analyzers for analyzing a Heenan-Froude DPX I water brake dynamometer and exhaust pollutants to measure brake power output were both employed. Engine braking carbon dioxide, power output, carbon monoxide, smoke, and hydrocarbon emissions were measured at intervals of 200 rpm between 1,000 and 3,000 rpm. These results were compared to a widely used diesel fuel. The brake power outputs of all the bio-ultra carbo-fluid formulations ranged from 5.6% to 20.7% less, whereas the exhaust emissions of CO_2, HCs, and smoke were significantly reduced. The key influencing elements in the formulation were revealed by the quantity of charcoal to the type; charcoal making up 30% of mass and having a high volatile content of 27.2% shows the best results, resulting in only slightly

**Corresponding author*: dr.jakasture@gmail.com

O.V. Gnana Swathika, K. Karthikeyan, Milind Shrinivas Dangate and Nicoletta Ravasio (eds.)
Biomass and Solar-Powered Sustainable Digital Cities, (43–56) © 2024 Scrivener Publishing LLC

decreased braking substantially fewer exhaust pollutants and power output. It is claimed that bio-ultra carbo-fluids show great promise as a diesel fuel substitute because their harmful exhaust emissions are far lower than those of regular diesel fuel. The findings offer a starting point for defining a bio-ultra carbo-fluid's specifications in order to achieve peak performance. There are other areas that could benefit from more investigation, as well as potential engine upgrades. However, Long-term testing is required to examine the influence of bio-ultra carbo-fluids on engine parts. The current situation of the economy is unfavorable, although it is influenced by both fluctuations in crude oil prices and political variables like EU environmental policy and economics.

Keywords: Bio-ultra carbo-fluids, diesel fuel, hydrocarbon, renewable, biomass

5.1 Introduction

The non-renewable resources of the planet were diminishing quickly and might be depleted within the next few decades; it became apparent in the 1970s. The technologically advanced nations started to pay more attention to renewable energy sources' response to this unsettling prospect; this increased worries about the release of dangerous pollutants into the environment from power plants and car exhausts, especially greenhouse gases like CO_2 and gases that cause acid rain like NOX and S02. As a result, considerable technological advancements have been made in fields [1].

The enormous pollution that they produce when burned is the driving force behind them, whether completely or in part. However, using fuels made from biomass, namely, bio-ultra carbo-fluids, may also have strategic and financial benefits.

Each year, the energy required to support the world's population is equal to 80 billion tons of oil. In actuality, only two-fifth of this energy is created; the remainder is mostly produced by the combustion of coal, nuclear energy, and natural gas. In fact, a detailed examination reveals that over 90% of that energy is derived from fossil fuels, and the sources are limited and will soon run out.

World's population depends on lighting and transportation. The number of registered passenger cars has doubled over the last 20 years, and there are currently 400 million of them in the world. There are 50 automobiles per hundred people in North America, 40 cars per hundred people in Europe, and 30 cars per hundred people in Japan. The global average for the number of autos per 100 people is 10. In countries like China and India, hardly 1 out of 100 people own a car. Fossil fuel supply and demand are currently more than equal. Resources of about one trillion (1012) tons are presently known, compared to little under half of this twenty years ago [2].

5.1.1 Background Problem

Using a non-renewable energy source and the enormous pollution that they produce when burned are the driving forces behind the substitution of diesel fuels made from petroleum, whether completely or in part. However, using fuels made from biomass, namely, bio-ultra carbo-fluids, may also have strategic and financial benefits. Most of the world's energy resources exist as solids in nature, such as other types of biomass, coal, and waste products. Biofuels such as wood, charcoal, and agricultural waste can meet up to 14% of the world's current energy needs in many developing countries [2]. This is the same as what OPEC is currently producing, which is about 25 million barrels of oil per day. There is a sizable amount of arable land that might be used to cultivate energy crops. However, liquid fuels are the foundation of the majority of modern transportation. This reliance on liquid fuels is partially due to the abundance of petroleum, which can be easily extracted from the natural world, moved, and changed into a variety of liquid fuels due to its composition as a mixture of hydrocarbon liquids. However, gaseous and liquid fuels do have an edge.

5.1.2 Coal

About 25% of the world's energy demands are satisfied due to coal, despite the fact that it has slipped from the top spot to the second-most popular fuel. It continues to be the most accessible fuel. It will likely endure for 200 years if consumption continues at the current rate [3].

5.1.3 The Decline in Coal Consumption

Nowadays, electricity is largely produced by burning coal. The use of coal has decreased in formerly heavy industries like steel production. This decline is due to the greater price of coal compared to gas, and oil as well as the substantial amount of S0 [2] that is created during combustion.

5.1.4 Vegetable Oil Derivatives as Fuels

Vegetable oil was initially utilized as fuel for an engine by Rudolph Diesel in 1899. However, inexpensive and readily available petroleum-derived fuel surpassed vegetable fuel to take over as the main source. All animal fats and vegetable oils are composed of triacyl glyceride molecules, which are made up of three long fatty acid chains joined to a glycerol molecule. Oils and fats are produced by plants and animals because the long CH2 chain chains serve as an effective energy storage mechanism [3]. Vegetable oils can only be utilized with difficulty in pre-chamber diesel engines due to the fact that they degrade lubricants and create severe engine deposits, ring sticking, and injector cooking. Cooking is the buildup of carbon deposits on surfaces in the combustion chamber, such as the injector orifices, that can eventually obstruct the spray that is issued from the orifices, inhibit valve seating, and result in ring sticking and engine failure. By employing alkali and a trans-esterification procedure, the fatty acids are released from the glycerol to lessen these issues. Methanol (CH30H), the resultant acid, changes it into methyl ester [4].

By using vegetable oil methyl esters, carbon deposits are prevented from forming. However, issues like lubricating oil dilution and sludge formation also occur here. There are significant efforts being made to solve these issues. In some specialized industries, such as fueling, where it is needed to be easily biodegradable to maintain the delicate marine environment, the usage of vegetable oil can be quite advantageous. Plants that can be used to make vegetable oils include soybean, peanut, sunflower, cotton seed, rapeseed, coconut, and canola. Castor oil and linseed are also often used. Diesel fuel is made of saturated hydrocarbons, whereas vegetable oils are tri-glycerides composed of glycerol esters. In this regard, vegetable oils differ from diesel fuel [4].

Table 5.1 provides a summary of the fatty acid contents of the sample oils. Due to their peak molecular weights, peak oxygen concentrations, and peak unsaturation levels, some characteristics of vegetable oil are considerably different from conventional fuels in general. Because they produce fewer issues with the engine, vegetable species with ICE (internal combustion engine), and oleic acids (18:2), little to huge concentrations of stearic (18:1), and no linolenic acid (18:3) are preferred.

5.2 Properties of Vegetable Oils

Goering *et al.* [1] tested the characteristics of diesel oil with other vegetable oils, and the results are described in Table 5.2.

According to the findings, plain vegetable oils have viscosities that are typically 13 times higher than those of diesel oils. Vegetable oils are

Table 5.1 Chemical composition of vegetables oils.

Fatty acid composition											
Vegetable Oil	14:0	16:0	18:0	20:0	22:0	24:0	18:1	18:1*	22:1	18:2	18:3
Castor	0	1	3	0	0	0	5	90	0	1	0
Cottonseed	0	28	1	0	0	0	13	0	0	58	0
Crambe	0	2	1	2	1	1	19	0	59	9	7
Linseed	0	5	2	0	0	0	20	0	0	18	55
Peanut	0	11	2	1	2	1	48	0	0	32	1
Raoeseed	0	3	1	0	0	0	64	0	0	22	8
Safflower	0	9	2	0	0	0	12	0	0	78	0
#H.O Safflower	tr.	5	2	tr.	0	0	79	0	0	13	0
Sesame	0	13	4	0	0	0	53	0	0	30	0
Soybean	0	12	3	0	0	0	23	0	0	55	6
Sunflower	0	6	3	0	0	0	17	0	0	74	0

tr. =trace
# High oleic Safflower
* Ricinoleic acid. 12-hvdroxv cis-9-octadecenoic acid

Table 5.2 Fuel properties of vegetable and diesel oils.

Oil	Viscosity cst (38° C)	Cetane number (ASTDM 613)	Gross heating value MJ/Kg	Cloud point °C	Pourf point °C	Flash point °C	Density Kg/ litre
Corn	34.9	37.6	39.5	-1.1	-40	277	0.91
Cottonseed	33.5	41.8	39.47	1.7	-15	234	0.92
Peanut	39.6	41.8	39.78	12.8	-6.7	271	0.9
Rapeseed	37	37.6	39.7	-3.9	-31.7	246	0.91
Soybean	32.6	37.9	39.62	-3.9	-12.2	254	0.91
Sunflower	33.9	37.1	39.58	7.2	-15	274	0.92
Diesel	2.7	47	45.34	-15	-33	52	0.84

significantly hindered in their ability to replace petrol in automobile engines due to their high viscosity [5].

Vegetable oils burn with a relatively consistent Gross Heat of Combustion of diesel fuel [2], which is roughly 90%. Tests have revealed that the overall energy release from vegetable oils is almost identical to diesel fuel due to the additional oxygen present.

The vegetable oils listed span from −3.90°C to 12.80°C. These temperatures are comparatively higher than those of diesel fuel, which varies from −9°C to −220°C [6]. Concerning pour points, vegetable oils exhibit a range of −6.70°C to −150°C, whereas diesel has a pour point range of −23°C to −420°C. It is worth noting exceptions like maize (400°C) and rapeseed (−31.70°C), which possess higher pour points. In extremely cold conditions, the elevated pour and points of certain vegetable oils could pose challenges. In terms of safety [1], vegetable oils have notably higher flash points compared to diesel, which is classified as a generally flammable liquid rather than a highly flammable one, suitable for storage and transport.

5.2.1 Properties of Vegetable Oil Esters

Alcoholic fatty acid esters with low molecular weight have viscosities and volatilities that are substantially more comparable to diesel fuel. They have therefore undergone extensive testing as alternatives to diesel fuel; generally speaking, they show promise; for instance, field trials [3]. An examination of the engine following the test showed that it was in excellent condition [7].

Table 5.2 provides a summary of the major vegetable oils' fuel characteristics. Variations in the chemical structures of the esters have little effect because their heat contents are generally constant and have substantially greater gross temperatures of combustion (Hg) than equivalent vegetable oils [4]. The esters have greater cetane values than diesel fuel. The viscosity of esters is lower than that of ordinary vegetable oils. Thus there should not be any problems flowing through injectors despite the fact that is still high and causing problems like the start of cold weather [8].

In long-term winter, rapeseed oil's methyl ester performed better in engine tests than the reference fuel [5] in terms of wear criterion engine performance and injector coking criteria. Although some oil dilution has been seen (Figure 5.1), it does not significantly affect performance because of rapeseed oil's lubricity and viscosity–temperature characteristics. Because of this, the diesel engine in a conventional car needs lubricating oil changes more frequently—every 6,000 miles as opposed to the usual 9,000.

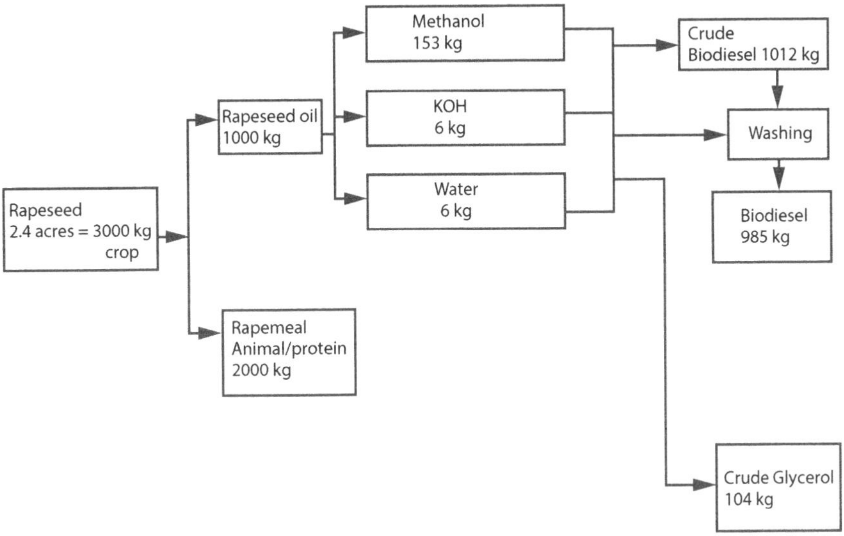

Figure 5.1 Flow chart for the production of biodiesel from rapeseed oil.

5.3 Charcoal Chemistry

5.3.1 Charcoal

In the industrialized world, charcoal is used for a variety of important purposes. Barbecue fuel is frequently made out of charcoal lumps. In the electronics sector, charcoal is used to make pure silica for computer chips [4]. In metallurgical procedures, it is used to reduce metals such as copper, zinc, and lead. The most major market is for activated charcoal, which is used after treatment to boost its adsorptive activity. To make activated charcoal, charcoal is burned to temperatures ranging from 750°C to 9,000°C in the presence of steam or carbon dioxide. As a result, any remaining volatiles are driven out, and finely porous charcoal is produced that. Compared to 1–2 m^2/g for regular charcoal, this is in the range of 200–2,000 m^2/kg. To make one ton of activated charcoal, approximately 3 tons of regular, high-quality charcoal are needed. One of activated charcoal's applications is to remove organic compounds from gases or liquids. It is employed in the industry for odor elimination and solvent vapor recovery. One typical application for activated charcoal is the filtration of water because it may eliminate organic compounds from water and other liquids [5].

5.3.2 Charcoal as Fuel

In less developed countries, cooking using charcoal is a relatively widespread practice. It has the advantage of being a fuel that burns cleanly and needs no preparation. Due to the decrease in volatiles, there are, say, coal. Because charcoal consistently emits very little smoke and has approximately double the energy capacity of wood, it requires less storage space than wood for a comparable amount of energy. It is also great for storing things because it would not rot or be attacked by termites, which is important. In many nations, it serves as the perfect fuel for street vendors. Because wood is typically available for free from the community in rural areas of developing nations, charcoal is primarily used in urban areas.

5.3.3 Industrial Uses

Forever, charcoal is used in industrial settings, industries to big steel smelting firms. Brazil makes extensive use of charcoal as an industrial fuel, but Japan utilized a lot of it up until the early 1950s for domestic and industrial uses. Japan generated over 2 million tons of charcoal in 1950, but by 1968 [6], only 361,000 tons remained.

5.4 The Pyrolysis Phase

Pyrolysis reactions start when the wood reaches a temperature of 2,800°C after all the water has been removed. The cellulose, hemicellulose, and lignin components of wood, as well as other components, are further broken down to produce pyrolysis products. Gases or vapors make up the majority of volatiles, commonly known as pyrolysis byproducts. The pyrolysis phase is underway when volatiles are present because they cause the white smoke to be released to turn dark.

Because once the reaction gets going, less heat is required to maintain it than is required to force the water out during the middle stage, the pyrolysis process is assumed to be strongly exothermic [7].

5.5 Cooling Phase

The temperature starts to steadily drop at the conclusion of the pyrolysis reaction, and the release eventually stops. In case if any air gets inside and lights the charcoal again, the kiln or retort is kept tightly shut to prevent the production of unnecessary combustion products. The retort vessel is only opened once the temperature of the charcoal has dropped below the ignition temperature. Seasoning is the process of letting the charcoal sit in the air for an additional 24 h. Thus, the volatiles that are still present get away and small quantities of oxygen and moisture are also taken in [9].

5.6 Products Obtained from the Carbonization Process

Charcoal production results in the production of solid, liquid, and gaseous byproducts. New techniques for raising yields have also been developed over time.

5.7 SS Charcoal's Qualities

One of charcoal's key properties is how much pure carbon it contains. By taking into account its calorific value, ash content, moisture content, and density, you may decide whether charcoal is acceptable for a specific end application.

5.8 Volatile and Fixed Carbon Content

By subtracting the weight's volatile, ash, and moisture contents of the sample of charcoal, the pure carbon concentration can be determined. It is known as fixed carbon content and is often reported as a percentage. A sample of dry charcoal is heated at 900°C till it reaches a consistent weight to ascertain this. The amount of carbon in the completed product is substantially determined by the greatest temperature reached during carbonization and the time spent at that temperature. The fixed carbon concentration increases significantly up to 500°C as a result. Between 500°C and 1000°C, the amount of fixed carbon concentration is minimal [10].

The weight of the sample of charcoal's moisture, volatile, and ash contents should be subtracted to get the concentration of pure carbon. It is also frequently shown as a percentage or as fixed carbon content. A sample of dry charcoal is heated at 9,000°C till it reaches a consistent weight to ascertain this. The amount of carbon in the final product is significantly influenced by the highest temperature obtained during the carbonization and the time spent at that degree. The fixed carbon concentration increases significantly up to 500°C as a result. Between 500°C and 1000°C, the increase is minimal.

5.8.1 Density of Charcoal

The type of wood used, the manner of manufacturing, and the function for which the charcoal is intended all affect its density. Charcoal typically has a density that is half that of the wood it is made of.

5.8.2 Gross Calorific Value of Charcoal

A specific charcoal's gross calorific value is influenced by the amount of fixed carbon and consequently. The gross calorific values for the majority of charcoal fall between 28 MJ/kg and 33 MJ/kg. The charcoal's water content determines the net calorific value, which is typically around 28 MJ/kg. Ash may also be present in charcoal in amounts ranging from 0.5% to 5%. Ash often comprises oxides clay, silica, magnesium, calcium, and other minerals.

While air moisture provides the only moisture that fresh charcoal receives. The amount of moisture that is gained from the air increases with the volatile content; moisture content might reach 15%.

5.8.3 Pyrolysis Products

Charcoal made from air-dried wood can produce between 60% and 80% of the wood's original mass, depending on the yield. Non-condensable gases and liquid byproducts are also emitted during the pyrolysis process. Non-condensable gases do not condense at room temperature (Table 5.3).

Table 5.3 Non-condensables produced from carbonizing redwood.

Non-condensible gases	% Yield
Carbon dioxide	42.7
Carbon monoxide	34.8
Methane	14.8
Hydrogen	5.4

Non-condensable gases, which account for 10% up to 15% of the wood's air-dry bulk, can be eliminated with the use of a condensing pipe or chimney. The nitrogen present at the start of the reaction, whether in the kiln or the retort tank, combined with the non-condensable gases, mostly methane, carbon dioxide, hydrogen, carbon monoxide, and some oxygen remain after liquid products condense away.

The carbonization temperature once more regulates the various quantities of non-condensable gases that are present in the pyrolysis stage.

The output of carbon dioxide decreased by 10% and the contents of both carbon monoxide and hydrogen increased by 5% when the carbonizing degree of the aforementioned reaction was raised to 600°C [9].

5.8.4 Production of Charcoal from a Specific Wood Species

Using batch metal partial combustion kilns (PCKs), Shah and colleagues' study [10] has illuminated some of the crucial variables related to the manufacture of charcoal. Study subjects included Oak, Meranti, Spruce, and Chestnut trees. As demonstrated in Table 5.4, related species with various moisture levels are investigated. Every test was carried out in the same way.

Table 5.4 The ultimate analysis of named tree species.

		Moisture content	Energy content	Ultimate analysis (wt%)		
	Species	(wt%)	(Kj/kg)	C	H	O
1	Oak	14.7	15.3	52	5.3	42.4
2	Oak	14.7	15.3	51.5	5.5	42.9
3	Meranti	11.5	17.3	53	5.1	41.9
4	Meranti	11.5	12.2	53.3	5.2	42.1
5	Spruce	30.8	12.2	50.9	5.4	43.9
6	Spruce	42.9	9.6	50.2	5.4	44.4
7	Spruce	42.9	9.6	50.5	5.6	44.2
8	Chestnut	12	15.8	15.8	5.2	42.8

5.9 Conclusion

In conclusion, research into the use of bio-ultra carbo-fluids as a diesel fuel substitute offers a viable path toward lowering the emissions of harmful environmental pollutants. In a thorough study using novel formulations of biodegradable by using vegetable oil and charcoal instead of fuel oil and coal, researchers were able to significantly improve the exhaust emissions profile by lowering the levels of carbon monoxide, carbon dioxide, smoke, and hydrocarbons while only slightly reducing brake power output, which ranged from 5.6% to 20.7%. The type and percentage of charcoal were shown to be the key determinants of these formulations' effectiveness, with charcoal having a greater volatile content and making up 30% of the mass producing the best results.

This study underlines the potential of bio-ultra carbo-fluids to replace traditional diesel fuel and provides a solid framework for establishing the best performance criteria for them. The results show a significant reduction in hazardous emissions when compared to conventional diesel fuel while acknowledging the need for more research and potential engine modifications. Although the current economic climate may provide difficulties due to variables like shifting crude oil prices and changing political ideologies, it is important to note that the potential environmental advantages of bio-ultra carbo-fluids should not be underestimated.

Extended durability testing is necessary for a thorough evaluation of the long-term effects of bio-ultra carbo-fluids on engine parts. Overall, this research not only confirms the viability of bio-ultra carbo-fluids as a more environmentally friendly fuel alternative but also indicates the path of future research to improve their use and address related issues. The potential benefits of bio-ultra carbo-fluids in reducing emissions and promoting sustainable practices remain a tempting subject for additional research and development as the global energy environment changes.

References

1. *Reparation and properties of diesel fuels from vegetable oils* Author links open overlay panel A.W. Schwab, M.O. Bagby, B. Freedman, Available: https://doi.org/10.1016/0016-2361(87)90184-0.
2. *Vegetable Oils And Their Derivatives As Fuels For CI Engines: An Overview* A. K. Babu and G. Devaradjane, Available: https://www.jstor.org/stable/44742273.

3. *Durability test on an agricultural tractor engine fuelled with pure biodiesel* (B100) MARCO BIETRESATO DARIO FRISO, Available: 10.3906/tar-1302-51.
4. *Heats of combustion of fatty esters and triglycerides* B. Freedman, M. O. Bagby First published: 01 November 1989, Available: https://doi.org/10.1007/BF02636185.
5. *Results of engine and vehicle testing of semirefined rapeseed oil* Published by the American Society of Agricultural and Biological Engineers, St. Joseph, Michigan www.asabe.org Citation: Transactions of the ASAE. 43(6): 1309-1316. (doi: 10.13031/2013.3027) @2000 Authors: K. P. McDonnell, S. M. Ward, P. B. McNulty, R. Howard-Hildige.
6. *Charcoal Adds to the Good Life*, Written by Sanada Kuniko Photos by Miyake Gaku, available: https://web-japan.org/nipponia/nipponia19/en/topic/index.html.
7. *The pyrolysis of biomass: A general review* C A ZAROR and D L PYLE Department of Chemical Engineering and Chemical Technology, Imperial College, London SW7 2BY, England, available: https://www.ias.ac.in/public/Volumes/sadh/005/04/0269-0285.pdf.
8. *Handbook of Charcoal Making*, The Traditional and Industrial Methods by Walter Emrich.
9. *Thermochemical and biochemical treatment strategies for resource recovery from agri-food industry wastes* Tharaka Rama Krishna C. Doddapaneni, Timo Kikas, in Valorization of Agri-Food Wastes and By-Products, 2021, Available: https://www.sciencedirect.com/topics/agricultural-and-biological-sciences/gasification.
10. *Tree diversity, biomass and carbon storage in sacred groves of Central India* Javid Ahmad Dar, Kothandaraman Subashree, Debojyoti Raha, Ashwani Kumar, Pramod Kumar Khare & Mohammed Latif Khan, Available: https://link.springer.com/article/10.1007/s11356-019-06854-9.

6

Bioinspired Machine Learning: A New Era in Sustainable Energy Systems

Akshaya Kumar Mandal[1*], Pankaj Kumar Deva Sarma[1], Satchidananda Dehuri[2] and Nayanjyoti Mazumdar[1]

[1]*Department of Computer Science, Assam University, A Central University of India, Silchar, Assam, India*
[2]*Department of Computer Science, Fakir Mohan University, Vyasa Vihar, Balasore, Odisha, India*

Abstract

Bioinspired machine learning (Bio-ML) is emerging as a game-changing technique in renewable energy systems. These approaches involve merging computer science and biological analytics to apply machine learning in renewable energy contexts. To propel this technology forward, diverse strategies have been employed to tackle human challenges and expedite progress. These approaches have allowed us to assess the potential of wind and solar energy. Researchers are striving to understand core machine learning concepts and apply them to address difficulties in the power and energy industries, which also involve analytical approaches of computer and biological sciences. Diverse tactics have been used to handle human difficulties and accelerate development with this technology. The entire chapter emphasizes on Bio-ML applications in solar, wind, and other renewable energy sources. Furthermore, it emphasizes the fundamentals of various machine learning technique and their application in solar and wind renewable energy systems. Present trends and future possibilities of Bio-ML technologies are also examined along with highlighting notable achievements in acceptance, modeling, and simulation research in sustainable energy system. The emergence of these technologies have allowed the invention of solutions that limit the hazards of human errors in complex systems, with many innovations in technology based on inspiration drawn from natural systems. These approaches are an interdisciplinary and promising

**Corresponding author*: akshayacs207@gmail.com

O.V. Gnana Swathika, K. Karthikeyan, Milind Shrinivas Dangate and Nicoletta Ravasio (eds.) Biomass and Solar-Powered Sustainable Digital Cities, (57–80) © 2024 Scrivener Publishing LLC

approach that encourages the incorporation of biological activities into machines to improve their intelligence. Bioinspired computing is a branch of computer science which mimics nature to serve a variety of objectives, such as analyzing electric circuits, solving differential equations, designing complicated chemical compounds, and detecting and treating illnesses. Thus, bioinspired computing's potential resides in its ability to handle common issues in a cost-effective and sustainable manner. In conclusion, this chapter delves into Bio-ML applications in solar, wind, and other renewable energy sectors. It highlights various bioinspired approaches, exemplified by solar and wind energy systems, showcasing a pivotal advancement in "Bioinspired Machine Learning: A New Era in Sustainable Energy Systems." It outlines present trends and future potentials of bioinspired methods. In essence, these techniques lower energy usage, stabilize energy load, enhance user comfort, and lower emissions. In addition, the Internet of Energy and bioinspired approaches can enhance distributed and hybrid renewable energy control systems and improve the capabilities of advanced intelligent energy management systems.

Keywords: Bioinspired machine learning (Bio-ML), sustainable energy systems, renewable energy systems, machine learning, bioinspired computing, biological data processing, energy management systems

Abbreviations

The following abbreviations are used in this chapter.

Bio-ML	Bioinspired machine learning
ANN	Artificial neural network
CNN	Convolutional neural network
PV	Photovoltaic
AD	Anno Domini (= in the year of the Lord)
AI	Artificial intelligence
MIT	Massachusetts Institute of Technology
IJCAI	International Joint Conferences on Artificial Intelligence
AAAI	American Association of Artificial Intelligence
ANFIS	Adaptive network–based fuzzy inference systems
GW	Gigawatts
MW	Megawatt
LEC	Levelised Energy Cost
TEC	Thermoelectric cooling
FSI	Floor Space Index
TWh	Terawatt-hour

EMS	Energy management system
IoT	Internet of Things
IoE	Internet of energy
REMS	Residential energy management systems
IRENA	International Renewable Energy Agency
HEMS	Hybrid energy management system
BEMS	Building Energy Management System
ICT	Information and communication Technology
DR	Demand response
DSM	Demand-side management
RTP	Real-time pricing
ToU	Time-of-use
CPP	Critical peak pricing
DE	Differential evolution
RES	Renewable energy sources
EVs	Electric vehicles
HVAC	Heating, venting, and air conditioning
HEM	Home energy management

6.1 Introduction

Bioinspired machine learning (Bio-ML) is emerging as a game-changing technique in renewable energy systems. Many researchers have worked extensively to integrate creative techniques to promote sustainability. Bioinspired approaches have allowed us to assess the potential of wind and solar energy [1]. Researchers are striving to understand core machine learning concepts and apply them to address difficulties in the power and energy industries, which also involve analytical approaches of computer and biological sciences. Diverse tactics have been used to handle human difficulties and accelerate development with this technology. The entire chapter emphasizes on Bio-ML applications in solar, wind, and other renewable energy sources (RESs) as depicted in Figure 6.1. Furthermore, it emphasizes the fundamentals of various deep learning technologies and their application in solar and wind renewable energy systems. Present trends and future possibilities of Bio-ML technologies are also examined along with highlighting notable achievements in acceptance, modeling, and simulation research in sustainable energy system. The emergence of these technologies has allowed the invention of solutions that limit the hazards of human errors in complex systems, with many innovations in technology based on inspiration drawn from natural systems. Bioinspired

Figure 6.1 Renewable energy sources.

computing approaches are an interdisciplinary and promising approach that encourages the incorporation of biological activities into machines to improve their intelligence. In the current world, the strong relationship between science and technology, together with the creation of the defined self, enables us to tackle a variety of difficulties. Bioinspired computing is a branch of computer science that mimics nature to serve a variety of objectives, such as analyzing electric circuits, solving differential equations, designing complicated chemical compounds, and detecting and treating illnesses.

The extensive use of technology has infiltrated many industries throughout the world, including energy systems, health, agriculture, economics, engineering, marine, and others. The increasing energy needs in each sector have resulted in the depletion of fossil fuels, causing many countries to shift their attention to RESs such as wind and solar power. The main goal of Bio-ML is to address human cognitive difficulties, analyze circuits, and regulate expression, train systems, and store information for retrieval. This technology makes use of fuzzy logic, expert systems, artificial neural networks (ANNs), and convolutional neural networks (CNNs) to aid in the design, modeling, control, and fault diagnostics of renewable energy systems [2]. Ant colony optimization, genetic algorithms, and particle swarm optimization, and others are examples of bioinspired techniques that mimic evolution in nature to solve complex optimization problem. Because of the intricacies of traditional technologies, the construction of global technological systems demands rigorous advancement. Explanations must be precise for Bio-ML programs to avoid deviating from correct solutions. Primary research goals of many of the energy fields employ ML research devoted to addressing technical challenges. Artificial neural approaches are a potential innovation that may be used to analyze wind energy sources and track apex power points in photovoltaic (PV) plants. Expert systems have played an important role in the development of fault-tolerant control in current grid systems. Many countries are now investing in research and development toward advanced and efficient smart grids, as earlier versions have proven ineffective in solving specific problems. Thus, bioinspired computing's potential resides in its ability to handle common issues in a cost-effective and sustainable manner. This chapter seeks to provide a complete view on bioinspired technologies with basic nonlinear features, emphasizing the necessity for novel bioinspired applications in sustainable energy systems. The advancement of Bio-ML techniques in the upcoming years will open up numerous opportunities to develop innovative processes tailored to specific applications in renewable energy.

6.2 Historical Backgrounds

The significant contribution of some researchers in the development of bioinspired approaches is gaining momentum, reshaping the world's landscape. The concept of bioinspired computing can be traced back to ancient times when references to machine learning appeared in mythical stories [3]. This belief manifested in various forms of bioinspired computing approaches, aimed at exploring our origins, existence, and creating practical technologies to enhance human life quality. Initially termed as "mythical thinking machines," this idea was exemplified by Heron of Alexandria in 10–70 AD [4], who described machines with simulated intelligence. However, fully achieving intelligent mechanical machines was beyond the technological capabilities of that era. In the 17th century, scholars like Descartes drew analogies between these ideas and animal attributes, laying the early groundwork for bioinspired computing [5, 6]. This viewpoint faced opposition from philosophers and mathematicians but set the stage for future developments.

The first instances of applying mathematical logic to computers emerged in 1642 with the invention of the first mathematical calculator [7, 8]. Over about three decades, progress led to calculators capable of performing complex functions like division and multiplication. The origins of biological computing trace back to 1936 when Alan Turing introduced the abstract concept of a Turing machine. George Boole's creation of Boolean algebra in the late 19th century significantly influenced the structure of modern computers. Samuel Butler's conjecture in the late 19th century, suggesting that machines could evolve following Darwinian principles [9], resonates with present-day research pursuits. In 1942, Isaac Asimov introduced concepts related to responsible robotics and bioinspired computing [10], proposing two laws for robots: avoiding harm to humans and obeying human commands. Another researcher from the University of Connecticut delved into machine ethics [11]. In 1950, John McCarthy organized a pivotal conference that laid the foundation for artificial intelligence (AI) [12, 13], coining the term "artificial intelligence." Marvin Minsky's work introduced the concept of a "society of mind." The Dartmouth Summer Research Project on AI further fostered collaboration among scientists, propelling the field forward. In 1977, Goldstein and Papert presented the Dendral program [14–16]. The 1960s marked the onset of AI's expansion, which surged significantly in the 20th century. MIT and Carnegie Mellon University established AI labs, followed by Rand Corporation, Stanford, and Edinburgh. The first AI conference was held in Edinburgh in 1965. The International Joint Conferences on Artificial Intelligence (IJCAI), starting in 1969, and the establishment of the American Association of Artificial Intelligence (AAAI) in 1980 [17] facilitated AI growth through workshops, journals, and digital libraries.

Bioinspired computing distinguishes itself from traditional AI through its evolutionary approach. This approach begins with basic rules adhered to by simple organisms that gradually evolve within defined parameters. In contrast, conventional AI often involves intelligence programmed from a higher perspective, with human creators instilling intelligence into their creations.

6.3 Applications of Bio-ML

Bioinspired computation involves employing principles from biological systems to tackle complex real-world problems using computational intelligence techniques. This approach includes ant colony algorithm, bee colony algorithm, firefly algorithm, cuckoo search, and particle swarm optimization [18]. Due to their impressive problem-solving capabilities, these techniques have garnered considerable attention and have been successfully applied across various domains including computer science, mechanical engineering, molecular biology, and biomedical engineering. Bioinspired machine learning (Bio-ML) is a form of machine learning approach rooted in biological phenomena. It finds unique applications in solving intricate computational problems, delivering quicker and more accurate predictions compared to manual methods. Bio-ML finds use in areas like fuzzy logic, ANNs, adaptive network–based fuzzy inference systems (ANFIS),

CNNs, and data mining. Applications include mapping multiple inputs to a single output, pattern recognition, association, and generating meaningful patterns [18, 19].

The reach of Bio-ML spans diverse fields such as engineering, mathematics, computer science, meteorology, medicine, neurology, economics, and psychology. It is significantly employed in sound and speech recognition. An example dates back to 1960 with the development of the adaptive linear combiner [20]. Many neural networks rely on algorithms, mathematical models, and nonlinear operations, which incorporate noise maps. Bio-ML also plays a role in market trends, weather forecasting, load prediction, mineral exploration, sensor data prediction, medical analysis, electromyography, explosives detection, and military target identification.

Photovoltaic (PV) electricity production benefits from Bio-ML, aiding real-time design and forecasting of solar radiation and steam generators. Applications extend to designing PV modules, solar collectors, sun tracking systems, and solar air conditioning [21, 22]. In molecular biology, Bio-ML facilitates scientific data analysis, hierarchical pattern recognition, and knowledge acquisition. Cancer patient nutrition planning, particularly during chemotherapy, is an important focus. Bio-ML aids in drug and dosage parameter resolution, with designs like the quadratic phenotypic optimization platform reducing drug toxicity while optimizing outputs. Consequently, the creation and evaluation of new bioinspired algorithms become vital. Current biological heuristic algorithms emphasize simple individuals collaborating to solve complex problems. Emerging bioinspired algorithms focus more on parallelism and intersectionality. Parallelism improves optimization while intersectionality integrates various metaheuristic algorithms, such as immune genetic algorithms and genetic particle swarm optimization, and others. With big data and cloud computing advancements, bioinspired computing's intelligent emergence mechanism gains prominence in areas like human brain modeling, evolvable software, and smart energy management systems (EMSs). Bio-ML's application in renewable energy, supported by bioinspired computation, is illustrated in Figure 6.2.

6.3.1 Wind Renewable Energy Systems

Wind energy is an excellent source of renewable energy. Wind farms are becoming more common in global, and it contributes more electricity to the National Grid. Wind energy is harnessed using turbines that turn generators to produce electricity. Global fossil fuel consumption is rising, creating a higher demand for electricity. Effective power generation methods

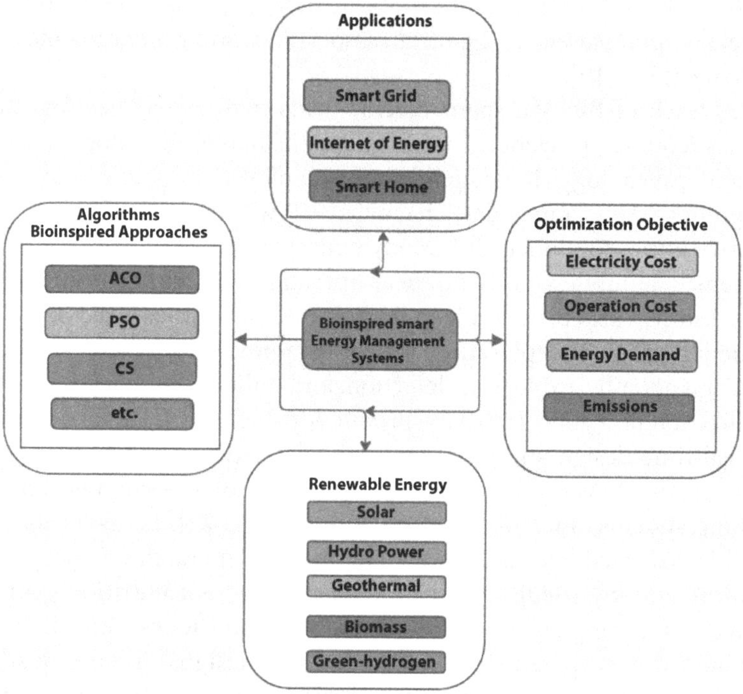

Figure 6.2 Bioinspired smart energy management systems.

depend on the world economy. Wind offers an alternative renewable energy source. In places like Mexico, wind energy availability follows certain patterns [23]. Countries like Mexico are turning to renewable sources due to higher fossil fuel costs and the need to reduce pollution [24]. Researchers are working on better strategies for wind energy systems [31]. However, wind energy is not as reliable as solar due to lower wind speeds. Wind farms work well offshore and at high altitudes. In 2014, Norway got 98% of its energy from renewables, whereas Brazil, New Zealand, Venezuela, and Colombia reached 73.4%, 79%, 62.8%, and 70%, respectively. Generating wind energy traditionally involves windmills turning turbines, converting wind energy into electricity. The first 12-KW wind turbine was set up in Ohio in 1887–1888. Wind energy installations in the European Union grew by 9.96% annually from the previous year, going from 48 GW in 2006 to 141 GW in 2015 [25]. China led wind power installations from 2010 to 2015. Germany led in 2006 and 2007, whereas the United States took over in 2008 and 2009. To make wind turbines more efficient, looking at past wind data at turbine sites can help. While predicting wind speed accurately is tough, using historical wind speed data can provide insights. Using data analytics and AI with bioinspired computing approaches to predicting wind speed are possible.

6.3.2 Solar Renewable Energy Systems

Sunlight is an abundant and free source of energy for our planet. The amount of solar energy entering the Earth's atmosphere in 1 h is greater than the entire annual energy requirement of the planet. It seems to be the ideal source of renewable energy, but the availability of solar energy depends on factors such as time of day, season, and location. In today's world, climate change poses a significant threat. Recognizing these challenges, governments and energy solution providers are seeking sustainable modes of renewable energy. This has turned renewable energy into an alternative to the current reliance on fossil fuels [26]. The uncertainty of climate and dwindling nonrenewable energy sources has spurred research and development in sustainable energy, impacting nations and its businesses [27]. In remote areas, solar energy applications can be expensive due to the limited availability of devices like pyranometers for accurate solar radiation measurement [28]. Growing energy consumption over decades has heightened concerns about resource depletion in the future [29]. The production and depletion of various fossil fuels have significant environmental consequences [30]. IRENA's report states that global solar PV generation surged by a record 179 TWh in 2021, marking a 22% increase from the previous year. Worldwide solar capacity reached 849 GW in 2021, contributing 3.6% of global energy generation. China boasts an impressive solar capacity of 430 GW (as of April 2023), making it the largest solar energy producer. The USA follows with 141.8 GW, a remarkable leap from its modest 0.34 GW in 2008. Japan, with a cumulative capacity of 84.9 GW, secures the third position, whereas Germany leads European countries with a solar capacity of 69.1 GW as of 2021 [25]. India, having an installed solar capacity of over 68 GW, ranks as the fifth largest solar country. The country's vast solar potential, with most Indian states receiving more than 300 days of sunshine annually, is promising [31]. India faces challenges in energy provision and rapid economic growth, necessitating improved energy infrastructure. Governments and energy solution providers have different approaches to leverage data analytics, AI, and bioinspired computing to address these energy challenges.

6.3.3 Hydro Renewable Energy Systems

Hydropower stands out as a well-established renewable energy resource. To govern the water flow that drives turbines to generate electricity, dams or other forms of barriers must be constructed. This energy source is more reliable than solar or wind power because of its ability to store energy during times of high demand. While hydro energy can be commercially viable in specific cases, including for domestic or off-grid use, it depends on its type. Hydro energy is cost-effective and abundant because it harnesses the kinetic energy of flowing water. It can be used as mechanical strength and electrical energy. In a micro-hydropower system, the principle is based on utilizing differences in water levels and flow rates, like rivers, waterfalls, or irrigation canals. The kinetic energy of water's motion turns turbines and generates mechanical energy, which is subsequently converted into electrical energy by a generator. Around 60% of global renewable electricity comes from hydropower, contributing about 16% of total electricity generation.

The International Energy Agency's 2021 report highlights hydropower's potential for flexible electricity supply, with a global installed capacity of 1,330 GW and record generation of 4,370 TWh in 2020. China, Brazil, the USA, Canada, and India are major hydropower producers. Hydropower plays a crucial role in achieving high levels of renewable energy in India, as depicted in Figure 6.3 [32]. In the construction process described by Younas *et al.* [33], micro-hydropower plants start with dams regulating water volume and flow rate. Concrete dams with sluice gates and filters are common. Waterwheels use water power from waterfalls, rivers, and ditches with low investment costs [33]. Murni *et al.* [34] (2012) suggests that communities can independently build such power plants, even in remote areas without electricity grid systems. Water sources can include irrigation systems, dammed rivers, or waterfalls.

6.3.4 Tidal Renewable Energy Systems

Tidal currents are used as an alternative source of hydroelectric power to orating turbines twice a day. Although tidal currents can vary, its regularity compensates for low tides. Currents are used by tidal power systems to turn turbines, especially in areas near islands or beaches where currents are strong. They look like windmills or tidal gates over rivers. Globally, tides can store up to 3,000 GW of energy. Depending on the location and conversion options, the expected output from the tidal dam ranges from

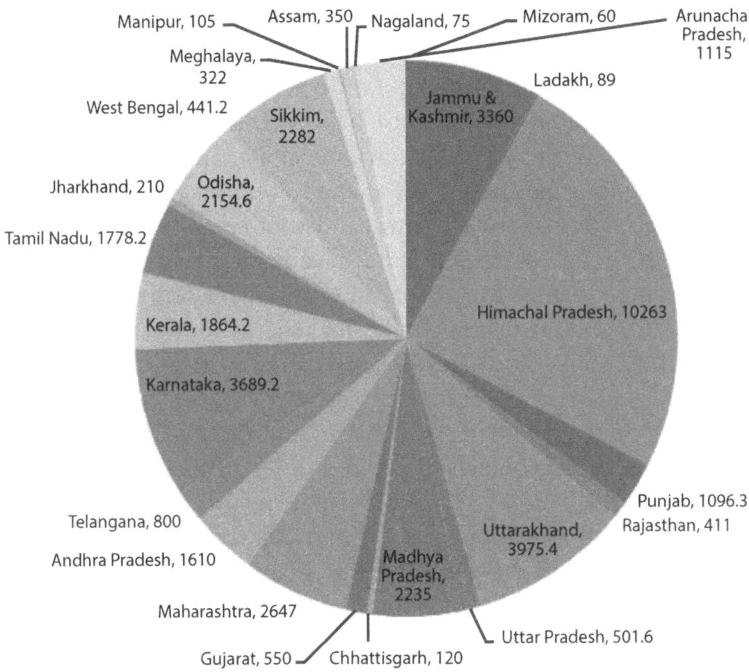

Figure 6.3 Hydroelectric power in India (above 25 MW, as of 28 February 2023) (https://pib.gov.in/PressReleasePage.aspx?PRID=1909276).

120 to 400 GW [35]. Liu and his team [36] were inspired by the curved wings of gliding insects/birds and fish. They offer LEC, TEC, and an integrated model that builds on the best parts of each. Their research suggested that flexible wing structures are pre-set. However, real insects and fish achieve optimal performance through passive deformation of wings/fins with varying stiffness and mass ratios. A completely connected FSI analysis that accounts for wing passive torsion and bending is a future effort. It is reasonable to expect similar outcomes if wing structural properties are predefined. This study could provide crucial guidance for industrial design of similar flapping wing energy devices based on bioinspired approaches.

6.3.5 Geothermal Renewable Energy Systems

The warmth within the earth originates primarily from the decomposition of radioactive substances like uranium, thorium, and potassium and serves as a reservoir of geothermal power. Temperatures beneath the continental crust can range from 200°C to 1,000°C, with heat transferred to the surface primarily by conduction. Some areas have heat closer to the surface due to factors like molten magma, thin crust, or heated groundwater. Such regions are often near tectonic plate boundaries or mountains. Geothermal systems can be divided into high-temperature fields (>180°C) associated with volcanic activity and low-temperature fields (<180°C). Geothermal energy can provide 8.3% of the world's electricity supply and 17% of the world's population. Geothermal energy can provide all the electricity needed in 39 countries, especially in Africa, Central and South America, and the Pacific region [37]. Although solar and wind have gained more attention lately, geothermal energy remains a reliable and sustainable source across diverse societies. Over 78 countries use geothermal energy for electricity, industry, agriculture, and more. However, the growth and installed capacity of geothermal energy plants have been surpassed by wind and solar due to high initial costs, long payback times, construction delays, space constraints, and varying social acceptance. Countries differ in their geothermal use, as discussed by Islam *et al.* [38]. Geothermal energy is also well-suited for building air conditioning applications, particularly in places like Iceland, where a significant portion of energy comes from this source.

6.3.6 Biomass Renewable Energy Systems

Biomass energy generation entails the more environmentally friendly and efficient conversion of plant materials into useful energy. Despite the fact that organic materials are still burned, this novel approach emphasizes cleaner processes. Residential, commercial, and agricultural waste may be converted into solid, liquid, or petrochemical fuels, providing a cost-effective and environmentally conscientious method of creating power for the benefit of both the economy and the environment. In 2020, the global electricity generation from biomass reached 685 TWh. Solid-biomass contributed 69% of all biopower, whereas municipal and industrial waste made up 17%. Asia accounted for 39% of biopower, producing 255 TWh in 2019, followed by Europe at 35% [85]. Colmenares-Quintero *et al.* [39] presented their findings, linking fuel and electrical efficiency with greenhouse gas emissions. Their proposed PSO optimization algorithm identifies the best operational points for each objective, establishing the connection between CO_2 emissions and power plant efficiency. Countries like Colombia, rich in biomass resources, can utilize biofuels to enhance their environmental stance and reduce reliance on fossil fuels.

6.3.7 Green-Hydrogen Renewable Energy Systems

Green hydrogen is produced through the process of electrolyzing water using RESs. It is a cleaner form of hydrogen generated by renewable energy or low-carbon power. Compared to gray hydrogen, derived from fossil fuels without carbon capture, green hydrogen has much lower carbon emissions. Its applications include decarbonizing hard-to-electrify sectors like cement and iron production, producing green ammonia for synthetic fertilizer, and serving as long-duration grid and seasonal energy storage. As of 2021, green hydrogen only made up a small fraction (less than 0.04%) of total hydrogen production due to its higher cost compared to fossil fuel–derived hydrogen. However, governments and international organizations are implementing policies to encourage green hydrogen commercialization, and its production is expected to rise significantly in the near future [40]. Research into green hydrogen using Bio-ML techniques for long-duration grid and seasonal energy storage will likely lead to innovative processes tailored for specific applications.

6.4 Advantages of Using Renewable Energy

Renewable energy offers numerous advantages:

- Lower Electricity Bills: Renewable energy systems can lead to reduced reliance on the conventional grid, resulting in decreased energy expenses.
- Environmental Benefits: Renewable energy sources such as solar, wind, and water provide electricity without producing harmful pollutants or greenhouse gases, resulting in cleaner air and a healthier environment.
- Energy Independence: Harnessing renewable sources reduces dependence on imported fossil fuels, enhancing energy security.
- Job Creation: The renewable energy sector creates jobs in manufacturing, installation, maintenance, and research, fostering economic growth.
- Sustainable Resource: Unlike finite fossil fuels, renewable sources like sunlight and wind are practically inexhaustible, ensuring a long-lasting energy supply.

- Technological Advancement: Investment in renewable energy drives innovation, encouraging the development of advanced technologies.
- Reduced Climate Impact: By reducing carbon emissions, renewable energy helps mitigate climate change and its associated effects.
- Diverse Applications: Renewable energy can power various sectors, from electricity generation to transportation and heating.
- Rural Development: Renewable projects, such as wind farms or solar installations, can stimulate economic activity and development in rural areas.
- Energy Access: In remote or off-grid locations, renewable sources can provide reliable electricity, improving energy access.
- Sell Surplus Electricity to the Grid: If the energy production exceeds our needs and we feed the excess back into the national grid, we may receive an additional payment under the feed-in tariff system.
- Embrace Electric Vehicles: Did you know that our electric vehicle (EV) tariffs are both economical and environmentally friendly?
- Global Leadership: The use of renewable energy demonstrates environmental responsibility and global leadership.

In conclusion, utilizing renewable energy brings a multitude of benefits, ranging from cost savings to environmental preservation and technological progress.

6.5 Bioinspired-Based Smart Energy Management System

The global desire to use energy efficiently and avoid environmental hazards has increased interest in EMSs in recent years [41–45]. The term "smart" signifies the intelligence of EMS in making informed decisions for energy load scheduling and consumption reduction within a set timeframe. EMS employs computer-assisted tools to monitor, optimize, and manage energy distribution, generation, and usage among users. Its main goal is to strike a balance between supply and demand, considering factors like renewable energy resources, energy costs, and consumption patterns, while working within operational constraints and uncertainties [42]. In addition, the benefits of using the Internet of Things (IoT) and machine learning to improve EMS efficiency are growing [42, 44]. Residential EMSs (REMSs) and structural energy control systems (SECSs) are the two most popular types of EMS [36, 41, 45]. Although both HEMS and BEMS share the common aim of minimizing energy consumption, their design and implementation vary, and it presents challenges to the efficient deployment of EMS.

The growing electricity demand and diminishing primary energy sources have led to an emphasis on renewable energy. Addressing this challenge is the concept of a smart grid, which combines advanced Information and Communication Technology (ICT) with distributed and RESs. This combination helps offset the effects of EVs and power plants with high peak loads. Demand response (DR) and demand-side management (DSM) play a central role in EMS smart grid architecture [42]. DSM entails a series of decisions that influence consumer energy usage patterns through planning, execution, and monitoring. DSM allocates available energy through DR schemes to reduce emissions and peak loads and allow users to choose the energy source. DR plans to implement dynamic pricing systems such as real-time pricing (RTP), time-of-use (ToU), inclining blocking rate (IBR), and critical peak pricing (CPP) [43, 46]. The EMS controller uses DR data and pricing rates to optimize and plan energy consumption, taking into account both hard (essential) and soft (desired) constraints. The former ensures feasible solutions, such as adhering to appliance operation schedules, whereas the latter enhances desirability, like starting appliances as early as possible within their allowable usage timeframe [45]. Figure 6.2 shows the bioinspired intelligent EMS, which shows energy sources, optimization targets, methods, and applications. It employs both conventional and renewable sources for energy generation, with considerations like peak-to-average ratio, energy demand, pricing, and user comfort impacting energy selection. EMS controllers manage these factors by optimizing them based on input information. Incorporating bioinspired algorithms into the smart EMS enhances its performance by enabling flexible, global search optimization. These algorithms can quickly explore potential areas to determine the best energy management solutions for buildings, homes, Internet of Energy (IoE), and the power grids. Bioinspired approaches outperform exact methods for optimization, effectively seeking global optima in feasible regions [47–49]. The subsequent sections delve into the concepts, structures, optimization goals, and bioinspired strategies within the smart EMS framework. For example, a smart HEMS acts as an emergency response platform to monitor and schedule home appliances in real time, taking into account customer needs to save money, improve energy efficiency, and maintain user comfort [41, 43, 50].

6.6 Opportunities and Challenges of Future Research

This part addresses the constraints observed in prior research, which could offer potential avenues for further exploration in applying bioinspired methods to smart energy management.

6.6.1 Hybridizing Bioinspired Approaches

It is noticed that a significant growth has taken place in research using bioinspired methods for addressing sustainability and smart energy management challenges. However, no single approach fits all types of issues perfectly. The formulation of the problem must match the algorithm's characteristics. Additionally, real-world problems often involve multiple conflicting objectives that require careful trade-offs. The complexity increases when both binary and continuous variables are involved. Exploring bioinspired algorithms with parallel and multi-objective processing seems promising. Hybridizing these algorithms could also offer valuable research directions. Numerous novel bioinspired approaches have recently emerged. For instance, the Social Spider Prey Algorithm considers vibrations caused by trapped prey on spider webs to identify feasible solutions in multi-objective optimization. The Whale Optimization Algorithm [54], inspired by Humpback whale behaviors, optimizes the number of distributed generators. Ant-lion colony optimization [51, 52] based on theirs food searching behavior aids in identifying optimal distributed generation sizes and microgrid energy scheduling. The Earliglow Algorithm, combining Jaya and strawberry algorithms, proves successful in HEMS. The Dragonfly Algorithm, mimicking dragonfly hunting and migration, optimizes energy cost, PAR, and delay in HEMS. Various other novel algorithms like Grasshopper Optimization Algorithm, Wind-Driven Optimization, Bacterial Foraging Algorithm, Glowworm Swarm Optimization, Flower Pollination Algorithm, Kestrel-based Search Algorithm, and Artificial Fish Swarm Algorithm address different energy management issues [51–62]. Hybridizing these algorithms can reduce complexity and improve solutions. Examples include combining GA and PSO, Bird Swarm with Cuckoo

Search, Bat Algorithm with Flower Pollination Algorithm, and enhanced DE with Harmony Search Algorithm [18, 51]. However, when selecting hybrid algorithms, factors like problem type, optimization approach (local or global), efficiency, and reliability should be carefully considered. These hybrid approaches hold potential to enhance convergence and computational efficiency in solving energy optimization and scheduling problems within EMS.

6.6.2 Energy Management Systems (EMSs)

Bioinspired algorithms might encounter challenges when dealing with load optimization in dynamic and diverse environments. Centralized application of these algorithms could lead to higher computational costs as the environment becomes more complex. To address this, decentralized frameworks could be employed. For instance, each energy consumer could act as an agent within a multi-agent system, collaborating to achieve optimal energy usage. This approach can introduce flexibility to coordinated management. Leveraging big data from IoT devices and smart meters, integrating bioinspired methods with machine learning techniques can enhance load prediction, automate context exploration, and enable artificial context perception. This transition could lead to self-adapting and self-configuring EMS [63–66]. While renewable energy sources (RESs) offer promising and sustainable energy options, they also bring irregular and unpredictable characteristics. This can lead to an imbalance in the system if only one RES is relied upon. Hybrid renewable energy systems (HRESs) were created as a result. HRES can operate independently or in conjunction with the primary network [67]. Extensive real-time research, cost-effective AER methods, and the appropriate ESS solutions must be implemented to improve reliability and performance. In a large HRES, for instance, it is crucial to address issues with scalability, effective energy resource deployment, and cost-optimization of capital and operating expenditures. Lastly, EVs play a dual role as consumers and generators, making their inclusion in EMS essential [68].

6.6.3 Internet of Energy (IoE)

Recent developments in computers and networking give strong reasons to create new techniques for difficult tasks. Integrating traditional bioinspired approaches with these new paradigms can dramatically enhance computing efficiency. For example, the IoE is an Internet-like system that allows for the two-way flow of energy and data [69–72]. The Whole Internet integrates smart grid and IoT features. The IoT involves the exchange of services with smart devices, whereas the smart grid establishes a two-way link between the network and the EMS, allowing monitoring and control of manufacturing facilities [41, 50]. Numerous foundational concepts and aspects of IoE have been explored in existing research. The transition from conventional energy networks with extra capacity to a more stable system that generates electricity to meet peak demand has been addressed [73]. A microgrid interconnection framework using energy routers for electrical transmissions has been proposed [74]. A distributed computing intelligence paradigm has been proposed for privacy-preserving energy management, where each energy device acts as an agent to perform local calculations without revealing sensitive information [70]. Focusing on cooperation with smart devices, a decentralized HVAC control scheme and a home energy management (HEM) system have been established [75]. Research also suggests that dedicated IoE communication infrastructure can enable peer-to-peer (P2P) communication between microgrids [71]. The difficulties and drawbacks of applying IoT in sustainable energy systems have been discussed [37]. Additionally, block chain technology has been suggested for energy trading and transfers without intermediaries [72]. Buildings, EVs, distributed energy systems, homes and businesses, and the entire IoE are expected to become more widespread. This will require complex monitoring and management of distributed energy generation and storage. Distributed energy trading, the growing number of charging stations for EVs, the management of energy optimization, and the exchange of data and energy through smart grids are just some of the new challenges. Exploring bioinspired approaches to address these challenges in the IoE is therefore an exciting area of research.

6.7 Conclusion

Bio-ML, an emerging technology in renewable energy systems, has garnered attention as researchers strive to integrate innovative approaches for sustainability. Various energy sources like wind, solar, hydro, tidal, and geothermal have been assessed using bioinspired methods. These approaches involve merging computer science and biological analytics to apply machine learning in renewable energy contexts. To propel this technology forward, diverse strategies have been employed to tackle human challenges and expedite progress. This chapter delves into Bio-ML applications in solar, wind, and other renewable energy sectors. It highlights various bioinspired approaches, exemplified by solar and wind energy systems, showcasing a pivotal advancement in "Bioinspired Machine Learning: A New Era in Sustainable Energy Systems." This chapter outlines present trends and future potentials of bioinspired methods. In essence, these techniques lower energy usage, stabilize energy load, enhance user comfort, and lower emissions. In addition, the IoE and bioinspired approaches can enhance distributed and hybrid renewable energy control systems and improve the capabilities of advanced intelligent EMSs. Further development of conventional biological data processing will open up new possibilities for the development of specific mechanisms in the future.

Acknowledgments

We gratefully acknowledge the anonymous referees for specific suggestions that aided in improving the presentation of this chapter.

Credit Authorship Contribution Statement

Akshaya Kumar Mandal (https://orcid.org/0000-0003-0955-4096): Conceptualization, data correction, resources, and writing—review and editing—original draft.
Pankaj Kumar Deva Sarma: Supervision, and review and editing.
Satchidananda Dehuri: Supervision, and review and editing.
Nayanjyoti Mazumdar: Review and editing.

Funding

This research has received no specific funding.

Competing Interest

The researchers indicate that they have no known financial conflicts of interest or close personal ties that may seem to have impacted the research described in this paper.

References

1. B. K. Bose, "Artificial intelligence techniques in smart grid and renewable energy systems—some example applications." *Proc. IEEE* 105, no. 11 (2017): 2262-2273.
2. N. J. Nilsson, "*Principles of Artificial Intelligence*", Springer-Verlag, New York (1980).
3. M. J. Apter, "Machines Who Think: A Personal Inquiry into the History and Prospects of Artificial Intelligence by Pamela McCorduck." *Leonardo* 15, no. 3 (1982): 242-242.
4. https://monoskop.org/File:McCorduck_Pamela_Machines_Who_Think_2nd_ed.pdf
5. Heron of Alexandria. *Encyclopedia Britannica.* Encyclopedia Britannica, 1-24-017. https://www.britannica.com/biography/Heron-of-Alexandria
6. J. Cottingham, "A Brute to the Brutes?': Descartes' Treatment of Animals." *Philosophy* 53, no. 206 (1978): 551-559.
7. K. N. Laland, W. Hoppitt, "Do animals have culture?." *Evol. Anthropol.: News Issues Rev.* 12, no. 3 (2003): 150-159.
8. D. Karwatka, "Blaise Pascal and the first calculator." *Tech Directions* 64, no. 4 (2004): 10.
9. D. D. S. Price, "A history of calculating machines." *IEEE Micro* 4, no. 1 (1984): 22-52.
10. B. Samuel, "The Book of the Machines: Chapters in Erewhon". *The Floating Press* (1873). https://oheg.org/hyr.pdf
11. S. L. Anderson, "Asimov's "three laws of robotics" and machine metaethics." *Ai & Society* 22 (2008): 477-493.
12. M. Minsky, "*Semantic Information Processing*. MIT Press, Cambridge, MA." (1968).
13. M. Minsky, "Steps toward artificial intelligence." *Proc. IRE* 49, no. 1 (1961): 8-30.
14. A. Newell, H. A. Simon, "*Human problem solving*". Vol. 104, no. 9. Englewood Cliffs, NJ: Prentice-hall, 1972.
15. I. Goldstein, S. Papert, "Artificial intelligence, language, and the study of knowledge." *Cognit. Sci.* 1, no. 1 (1977): 84-123.
16. R. K. Lindsay, E. A. Feigenbaum, B. G. Buchanan, J. Lederberg, "*Applications of artificial intelligence for chemical inference: The DENDRAL project.*" (1980).
17. American Association for Artificial Intelligence (AAAI). (2005). Menlo Park, CA. www.aaai.org/aitopics/history
18. A. K. Mandal, P. K. D. Sarma, S. Dehuri, "A Study of Bio-inspired Computing in Bioinformatics: A State-of-the-art Literature Survey." *The Open Bioinformatics Journal* 16, no. 1 (2023). http://dx.doi.org/10.2174/18750362-v16-e230517-2022-17
19. S. A. Kalogirou "Applications of artificial neural-networks for energy systems." *Appl. Energy* 67, no. 1-2 (2000): 17-35.

20. S. A. Kalogirou, "Artificial neural networks in renewable energy systems applications: a review." *Renew. Sustain. Energy Rev.*, 5, no. 4 (2001): 373-401.
21. A. Mellit, S. A. Kalogirou, S. Shaari, H. Salhi, A. H. Arab, "Methodology for predicting sequences of mean monthly clearness index and daily solar radiation data in remote areas: Application for sizing a stand-alone PV system." *Renew. Energy* 33, no. 7 (2008): 1570-1590.
22. A. Mellit, A. M. Pavan, "A 24-h forecast of solar irradiance using artificial neural network: Application for performance prediction of a grid-connected PV plant at Trieste, Italy." *Sol. Energy,* 84, no. 5 (2010): 807-821.
23. Q. Hernandez-Escobedo, F. Manzano-Agugliaro, J. A. Gazquez-Parra, A. Zapata-Sierra, "Is the wind a periodical phenomenon? The case of Mexico." *Renew. Sustain. Energy Rev.*, 15, no. 1 (2011): 721-728.
24. Q. Hernández-Escobedo, F. Manzano-Agugliaro, A. Zapata-Sierra, "The wind power of Mexico." *Renew. Sustain. Energy Rev.*, 14, no. 9 (2010): 2830-2840.
25. https://www.worldbioenergy.org/uploads/221223%20WBA%20GBS%20 2022.pdf
26. Saurenergy.com. https://www.saurenergy.com/solar-energy-articles/the-role-of-ai-andml-in-solar-energy.
27. A. J. C Trappey, P. P. J. Chen, C. V. Trappey, L. Ma, "A machine learning approach for solar power technology review and patent evolution analysis." *Appl. Sci.*, 9, no. 7 (2019): 1478.
28. S. M. Al-Alawi, H. A. Al-Hinai, "An ANN-based approach for predicting global radiation in locations with no direct measurement instrumentation." *Renew. Energy,* 14, no. 1-4 (1998): 199-204.
29. N. K. Sharma, P. K. Tiwari, Y. R. Sood, "Solar energy in India: Strategies, policies, perspectives and future potential." *Renew. Sustain. Energy Rev.*, 16, no. 1 (2012): 933-941.
30. Z. Ramedani, M. Omid, A. Keyhani, S. Shamshirband, B. Khoshnevisan, "Potential of radial basis function based support vector regression for global solar radiation prediction." *Renew. Sustain. Energy Rev.*, 39 (2014): 1005-1011.
31. A. Kumar, K. Kumar, N. Kaushik, S. Sharma, S. Mishra, "Renewable energy in India: current status and future potentials." *Renew. Sustain. Energy Rev.*, 14, no. 8 (2010): 2434-2442.
32. https://www.hydropower.org/iha/discover-facts-about-hydropower.
33. M. Younas, R. Kamal, M. S Khalid, A. Qamar, A. Qamar, "Economic planning for remote community microgrid containing solar PV, biomass gasifier and microhydro." In *2018 Clemson University Power Systems Conference (PSC)*, pp. 1-7. IEEE, 2018.
34. S. Murni, J. Whale, T. Urmee, J. Davis, D. Harries, "The role of micro hydro power systems in remote rural electrification: a case study in the Bawan Valley, Borneo." *Procedia Eng.*, 49 (2012): 189-196.
35. https://www.britannica.com/science/tidal-power.

36. W. Liu, Q. Xiao, F. Cheng, "A bio-inspired study on tidal energy extraction with flexible flapping wings." *Bioinspir. Biomim.*, 8, no. 3 (2013): 036011.
37. https://www.greenfacts.org/en/geothermal-energy/index.htm.
38. M. T. Islam, M. N. Nabi, M. A. Arefin, K. Mostakim, F. Rashid, N. M. S. Hassan, S. M. A. Rahman, S. McIntosh, B. J. Mullins, S. M. Muyeen, "Trends and prospects of geothermal energy as an alternative source of power: A comprehensive review." *Heliyon*, 8, no. 12 (2022).
39. R. F. Colmenares-Quintero, G. D. Góez-Sánchez, J. C. Colmenares-Quintero, L. F. Latorre-Noguera, D. Kasperczyk, "Application of a simulation tool based on a bio-inspired algorithm for optimisation of distributed power generation systems." *Cogent Eng.* 8, no. 1 (2021): 1909791.
40. R. Cho, "Why We Need Green Hydrogen." *Columbia Climate School: New York, NY, USA* (2021).
41. M. Beaudin, H. Zareipour, "Home energy management systems: A review of modelling and complexity." *Renew. Sustain. Energy Rev.*, 45 (2015): 318-335.
42. S. K. Rathor, D. Saxena, "Energy management system for smart grid: An overview and key issues." *Int. J. Energy Res.*, 44, no. 6 (2020): 4067-4109.
43. B. Zhou, W. Li, K. W. Chan, Y. Cao, Y. Kuang, X. Liu, X. Wang, "Smart home energy management systems: Concept, configurations, and scheduling strategies." *Renew. Sustain. Energy Rev.*, 61 (2016): 30-40.
44. D. Minoli, K. Sohraby, B. Occhiogrosso, "IoT considerations, requirements, and architectures for smart buildings—Energy optimization and next-generation building management systems." *IEEE Internet Things J.* 4, no. 1 (2017): 269-283.
45. S. N. Makhadmeh, A. T. Khader, M. A. Al-Betar, S. Naim, A. K. Abasi, Z. A. A. Alyasseri, "Optimization methods for power scheduling problems in smart home: Survey." *Renew. Sustain. Energy Rev.*, 115 (2019): 109362.
46. Y. Chen, P. Xu, J. Gu, F Schmidt, W. Li, "Measures to improve energy demand flexibility in buildings for demand response (DR): A review." *Energy Build.*, 177 (2018): 125-139.
47. J. Del Ser, E. Osaba, D. Molina, X. S Yang, S. Salcedo-Sanz, D. Camacho, S. Das, P. N. Suganthan, C. A. Coello Coello, F, Herrera, "Bio-inspired computation: Where we stand and what's next." *Swarm Evol. Comput.*, 48 (2019): 220-250.
48. X. S. Yang, "Nature-inspired optimization algorithms: Challenges and open problems." *J. Comput. Sci.*, 46 (2020): 101104.
49. K. Hussain, M. N. Mohd Salleh, S. Cheng, Y. Shi, and Y. Shi, "Metaheuristic research: a comprehensive survey." *Artif. Intell. Rev.*, 52 (2019): 2191-2233.
50. D. Xenias, C. J. Axon, L. Whitmarsh, P. M. Connor, N. Balta-Ozkan, A. Spence. "UK smart grid development: An expert assessment of the benefits, pitfalls and functions." *Renew. Energy*, 81 (2015): 89-102.
51. A. K. Mandal, S. Dehuri, "A survey on ant colony optimization for solving some of the selected np-hard problem." In *Biologically Inspired Techniques in Many-Criteria Decision Making: International Conference on Biologically*

Inspired Techniques in Many-Criteria Decision Making (BITMDM-2019), pp. 85-100. Springer International Publishing, 2020.
52. S. Mirjalili, "The ant lion optimizer." *Adv. Eng. Softw.*, 83 (2015): 80-98.
53. V. C. V. Reddy, "Ant Lion optimization algorithm for optimal sizing of renewable energy resources for loss reduction in distribution systems." *J. Electr. Syst. Inf. Technol.*, 5.3 (2018): 663-680.
54. P. D. P. Reddy, V. C. V. Reddy, T. G. Manohar, "Whale optimization algorithm for optimal sizing of renewable resources for loss reduction in distribution systems." *Renewables: Wind, Water, Solar*, 4 (2017): 1-13.
55. A. K. Mandal, P. K. D. Sarma, S. Dehuri, "Machine Learning Approaches and Particle Swarm Optimization Based Clustering for the Human Monkeypox Viruses: A Study." *International Conference on Innovations in Intelligent Computing and Communications*. Cham: Springer International Publishing, 2022.
56. A. K. Mandal, P. K. D. Sarma, "Novel Applications of Ant Colony Optimization with the Traveling Salesman Problem in DNA Sequence Optimization." *2022 IEEE 2nd International Symposium on Sustainable Energy, Signal Processing and Cyber Security (iSSSC)*. IEEE, 2022.
57. K. Roy, K. K. Mandal, A. C. Mandal, "Ant-Lion Optimizer algorithm and recurrent neural network for energy management of micro grid connected system." *Energy*, 167 (2019): 402-416.
58. O. Samuel, S. Javaid, N. Javaid, S. H. Ahmed, M. K. Afzal, F. Ishmanov, "An efficient power scheduling in smart homes using Jaya based optimization with time-of-use and critical peak pricing schemes." *Energies*, 11.11 (2018): 3155.
59. I. Hussain, M. Ullah, I. Ullah, A. Bibi, M. Naeem, M. Singh, D. Singh, "Optimizing energy consumption in the home energy management system via a bio-inspired dragonfly algorithm and the genetic algorithm." *Electronics*, 9.3 (2020): 406.
60. I. Ullah, Z. Khitab, M. N. Khan, S. Hussain, "An efficient energy management in office using bio-inspired energy optimization algorithms." *Processes*, 7.3 (2019): 142.
61. M. Awais, N. Javaid, K. Aurangzeb, S. I. Haider, Z. A. Khan, D. Mahmood, "Towards effective and efficient energy management of single home and a smart community exploiting heuristic optimization algorithms with critical peak and real-time pricing tariffs in smart grids." *Energies*, 11.11 (2018): 3125.
62. S. J. B. Christopher, M. C. Mabel, "A bio-inspired approach for probabilistic energy management of micro-grid incorporating uncertainty in statistical cost estimation." *Energy*, 203 (2020): 117810.
63. S. Rahim, N. Javaid, A. Ahmad, S. A. Khan, Z. A. Khan, N. Alrajeh, U. Qasim, "Exploiting heuristic algorithms to efficiently utilize energy management controllers with renewable energy sources." *Energy Build.*, 129 (2016): 452-470.

64. K. H. N. Bui, I. E. Agbehadji, R. Millham, D. Camacho, J. J. Jung, "Distributed artificial bee colony approach for connected appliances in smart home energy management system." *Expert Syst.*, 37.6 (2020): e12521.
65. S. Ali, D. H. Kim, "Simulation and energy management in smart environment using ensemble of GA and PSO." *Wirel. Pers. Commun.*, 114 (2020): 49-67.
66. Z. A. Khan, A. Khalid, N. Javaid, A. Haseeb, T. Saba, M. Shafiq, "Exploiting nature-inspired-based artificial intelligence techniques for coordinated day-ahead scheduling to efficiently manage energy in smart grid." *IEEE Access*, 7 (2019): 140102-140125.
67. Z. A Khan, A. Zafar, S. Javaid, S. Aslam, M. H. Rahim, N. Javaid, "Hybrid meta-heuristic optimization based home energy management system in smart grid." *J. Ambient Intell. Hum. Comput.*, 10 (2019): 4837-4853.
68. M. Naz, Z. Iqbal, N. Javaid, Z. A. Khan, W. Abdul, A. Almogren, A. Alamri, "Efficient power scheduling in smart homes using hybrid grey wolf differential evolution optimization technique with real time and critical peak pricing schemes." *Energies*, 11.2 (2018): 384.
69. K. Wang, J. Yu, Y. Yu, Y. Qian, D. Zeng, S. Guo, Y. Xiang, J. Wu, "A survey on energy internet: Architecture, approach, and emerging technologies." *IEEE Syst. J.* 12.3 (2017): 2403-2416.
70. W. Zhong, K. Xie, Y. Liu, C. Yang, S. Xie, Y. Zhang, "Admm empowered distributed computational intelligence for Internet of energy." *IEEE Comput. Intell. Mag.*, 14.4 (2019): 42-51.
71. S. Marzal, R. González-Medina, R. Salas-Puente, G. Garcerá, E. Figueres, "An embedded internet of energy communication platform for the future smart microgrids management." *IEEE Internet Things J.* 6.4 (2019): 7241-7252.
72. A. Miglani, N. Kumar, V. Chamola, S. Zeadally, "Blockchain for Internet of Energy management: Review, solutions, and challenges." *Comput. Commun.*, 151 (2020): 395-418.
73. S. Kalyanaraman, "Back to the future: lessons for internet of energy networks." *IEEE Internet Comput.*, 20.1 (2016): 60-65.
74. Y. Liu, Y. Fang, J. Li, "Interconnecting microgrids via the energy router with smart energy management." *Energies*, 10.9 (2017): 1297.
75. S. Wang, J. Xing, Z. Jiang, Y. Dai, "A decentralized swarm intelligence algorithm for global optimization of HVAC system." *IEEE Access*, 7 (2019): 64744-64757.

7

The Impact of the Internet of Things in the Smart City From the Point of View of Energy Consumption Optimization

Sanjeevikumar Padmanaban[1], Mostafa Azimi Nasab[1], Mohsen Hatami[2], Omid Halimi Milani[3], Mohammad Ali Dashtaki[2], Morteza Azimi Nasab[1*] and Mohammad Zand[1]

[1]*Department of Electrical Engineering, IT and Cybernetic, University of South-Eastern Norway, Porsgrunn, Norway*
[2]*Department of Electrical and Computer Engineering, Binghamton University-SUNY, Binghamton, NY, USA*
[3]*Department of Electrical and Computer Engineering, University of Illinois at Chicago, Chicago, IL, USA*

Abstract

The Internet of Things (IoT) is a new concept in the world of information and communication technology, and, in recent years, it has gained a lot of attention in research communities and users in smart cities. IoT refers to a network in which every physical object is identified by an identity and forms a network with other objects. IoT is the future of the internet, and it will simplify many complex services in the future. The new applications of IoT provide innovative actions in smart cities worldwide, and the explosive growth of smart cities and IoT applications poses many scientific challenges that require research efforts from industry and academia, especially for the development of smart cities based on IoT. Protocols, architectures, and new services are among the basic needs to respond to these challenges. With the increase in energy tariffs, the issue of energy consumption and its related costs in cities has taken on new dimensions. Because the energy of sensor nodes is usually provided by batteries, the most important issue that needs to be examined in these networks is to reduce the energy consumption of nodes so that the network's lifetime increases to an acceptable level.

**Corresponding author*: morteza1368morteza@gmail.com

O.V. Gnana Swathika, K. Karthikeyan, Milind Shrinivas Dangate and Nicoletta Ravasio (eds.) Biomass and Solar-Powered Sustainable Digital Cities, (81–122) © 2024 Scrivener Publishing LLC

Keywords: Internet of things, smart city, framework, smart energy

7.1 Introduction

The emergence of the Internet of Things (IoT) has brought significant differences in human life and made it easier. Nowadays, IoT has found its way into every field. With significant advances in communication technology devices such as radio wave identification, field communication devices, and located sensor nodes, IoT has become very prominent. With the emergence of communication devices and services, IoT is becoming the next-generation revolution for current mobile and fixed telephone network infrastructures. Recent predictions show that IoT will be an essential part of the future internet [1].

IoT draws future scenarios in which physical objects and devices around us, such as various sensors, RFID tags, location equipment, and IoT-enabled mobile devices, come together, and these objects and devices provide conditions for connecting, collaborating, and communicating in social, personal, and environmental content to achieve common goals [1].

Integration of IoT technologies in smart cities helps citizens to gain knowledge about their environment and participate in obtaining new data with mobile devices. In addition, data become accessible and are available as searchable data resources through common mechanisms, leading to uniform access to data providers in smart cities [1].

7.1.1 Infrastructure of Information Technology for Energy Data Collection

Many technologies are needed to collect energy data, and many of these technologies are currently in use. At the lowest level of the framework, there are smart sensors and meters that are connected via wireless or wired networks. Energy meters available on the market can have several parameters (energy consumption, power factor, and minimum/maximum peak voltage) and, therefore, provide a high level of flexibility in monitoring and analyzing energy consumption. The first layer consists of all IoT devices that can produce real-time or batch data. The generated data can be structured, semi-structured, or unstructured. In most cases, a small amount of data is created using various sensors. This layer includes temperature sensors, humidity sensors, pressure sensors, motion sensors, mobile device sensors, and other IoT devices that generate data. Data are repeatedly generated, and the amount of collected data is accumulated in thousands of databases. With the help of various tools and IoT-based technologies, this network brings many benefits and advantages to the value chain members of a country's or even the world's energy network. Some of these energy-efficient technologies include the following [2, 3]:

1. Smart meters: A smart meter is an electronic device that records electricity, water, and gas consumption to optimize energy consumption in real time and with a focus on real-time consumption. Smart measuring devices equipped with advanced microprocessors (smart measuring devices and extensive monitoring systems) are the most popular measuring devices in the energy and service sector, and this popularity will increase in the coming years. New smart measuring devices in homes and factories provide information that helps to understand energy consumption and locations that can be used to save energy. Smart measuring devices provide a wide range of useful tools and services for consumers, suppliers, network operators, regulators, and more, ultimately creating a smarter energy world. These devices perform accurate measurements and data transmission of electricity, gas, water, or heat consumption and provide two-way information gateways and communication infrastructure between meters and related systems.

Factors affecting the selection of measuring devices: accuracy, ease of deployment, communication protocols, cost, availability, and granularity.

2. Smart counters: Smart networks that work with IoT use real-time data to manage energy distribution. Along with smart meters, they provide conditions for energy services and tariffs that can significantly reduce business energy costs and increase their credibility and stability. The smart energy network has created smart networks by receiving and analyzing information, which will have the power to predict energy levels at a specific time. Smart metering and the possibility of continuous data transmission to them lead to the realization of accurate analysis of energy consumption patterns. In this regard, AMI systems are very useful. This integrated system includes hardware, software, and telecommunications networks that provide two-way communication to read,

configure, monitor, and remotely control meters, collect, manage, and process collected data.

The AMI system consists of three subsystems, and each subsystem needs to communicate with other subsystems in addition to taking its specific features and requirements into account:

- Smart measuring system: Electricity, water, and gas meters are each measuring systems in energy networks.
- Telecommunications system: After measuring energy consumption, data are transferred to the next stage, which is the central system. The selection of meter communication modules is done according to the design of the communication network and its features.
- Central system: Finally, the data center and software packages work to analyze and manage the system. This vast amount of information is useful only if it is analyzed and provides meaningful information for better system management [4].

In this study, three types of smart electricity, water, and gas meters are examined to optimize energy consumption.

Smart electricity meters are devices that measure the electrical parameters of consumption and record electricity consumption in hourly or shorter intervals. The recorded information is sent to a central system for monitoring and billing. Smart meters can also analyze the collected data and prepare a report. They have the ability to detect possible power outages and perform necessary services, making them useful for residential, commercial, and industrial customers. Smart meters eliminate the need for sending personnel to read the meters, reducing vehicle traffic on the streets and carbon dioxide emissions, and playing a significant role in environmental protection. Smart electricity meters consist of committed modules to perform advanced energy measurement and provide additional information to electrical installations through bidirectional communication. They are equipped with internal circuit breakers, allowing the electricity company to remotely establish or disconnect the power supply to customers, resulting in ease of controlling customers' electricity and economic benefits. The necessary equipment for implementing these systems includes local centers, servers, network routers and switches, firewalls, printers, modems, software, GPS, collectors, repeaters, antennas, meters, communication modules, display, and smart meter coding modules [5].

Smart water meters are devices that repeatedly store and transmit water consumption data, in addition to recording the volume of consumption, date, and time. Smart water meters continuously read information remotely and are part of the AMI system. AMI systems use smart water meters to measure, collect, and analyze water consumption data and then send this information to customers via the internet based on a request or a fixed schedule. The system includes hardware, software, customer water consumption portals, controllers, and smart meter information management systems and business support systems. AMI technology extends AMR technologies by allowing bidirectional communication of meters, sending information and commands to end-users for the following purposes: real-time consumption and pricing information, leak detection and abnormal consumption detection, water consumption efficiency messaging, measuring changes in water consumption, and remote service disconnection.

Smart gas meters use IoT technology. They consist of a smart gas meter, data collector, and central control network. The smart gas sensor is connected to the gateway after installation and sends information to the data center via communication technology. A smart gas system has the following functions: remote reading, bidirectional communication, tariff application, customer billing methods, remote valve activation or deactivation, and communication with other smart devices. The four main components for sending gas meter information to the gas company are as follows: the gas meter as a measuring and information collection device; the collector/concentrator that receives or sends information to the meters; communication methods such as power line communication (PLC), digital subscriber line (DSL), global system for mobile (GSM), or general packet radio service (GPRS); and the information management system for analyzing and managing the sent information. Various communication and communication modules are used in gas meters, such as Mbus, Zigbee, and other radio technologies.

7.1.2 Meter Reading Information Collection Service

This service is focused on collecting and aggregating data resulting from operations. Using tags installed on the meters, their numbers are automatically sent without the need for presence. The real advantage of this system is storing consumption behavior, which can identify consumer behavior throughout the year.

7.1.3 Smart Water Sensor

Monitoring water consumption and providing useful information to consumers are done through smart water sensors. Smart water sensors determine the quality, temperature, pressure, and water consumption. These devices usually communicate with a water and electricity company, which, using software to analyze the data, converts it into a comprehensible format for the consumer [6]. Users can then understand how their consumption is calculated compared to the city average, previous months, etc. Another important function is identifying potential water leaks, which identifies defective pipes and devices that are leaking and prevents water waste.

7.1.4 Temperature and Humidity Sensor

The range of techniques used to measure temperature is extensive and can be based on phenomena such as thermoelectricity, fluorescence, and spectral properties. Most sensors for measuring air temperature are based on electrical and mechanical operating principles. Mechanical devices are generally inexpensive and reliable but are usually used for direct control and virtualization purposes, the most common example being the internal thermostat. In contrast, electrical sensors have the ability to convert measured values into a digitally encoded signal for direct communication for measurement and control purposes. A temperature sensor is used to record and control the temperature of the environment or object being measured. Temperature sensors are made in various types with the ability to measure high temperatures and high accuracy. Temperature measurement methods are divided into two categories: contact and non-contact. In non-contact temperature measurement, infrared light is used to measure the temperature without the sensor coming into contact with the substance being measured. Temperature sensors produce an electrical signal based on the amount of heat sensed by their sensor, which is proportional to the amount of heat [7].

Humidity sensors are electronic components that react to humidity and measure and report the relative humidity of the air.

7.1.5 Daylight Sensor

Daylight sensors continuously evaluate the level of brightness and compare it to the ambient light to adjust the maximum level of ambient light. In addition to standardizing light for daily activities, these sensors will be very useful in reducing energy consumption. By placing these sensors in different spaces and collecting information about the level of light intensity in different parts of the smart environment and sending it to the central controller, it is possible to provide the necessary level of brightness in different spaces with central control system processing this information. This lighting management method will minimize energy consumption. Using innovative smart solutions, modifying light sources and designing lighting can reduce energy consumption in the lighting sector. Smart equipment modification not only reduces consumption significantly but also helps identify high-consumption equipment and, if necessary and economically justified, replace them with low-consumption equipment [7].

Lighting control related to light can be used to turn lights on or off when using daylight. A daylight-responsive system consists of a light-sensitive element such as a light-sensitive sensor, a lighting controller, and an electronic light meter battery. Various types of light sensors include PIR sensors, ultrasonic sensors, photodiodes, and photosensors. Wireless sensors are more expensive than wired sensors. Regarding space sensors, ultrasonic sensors are more expensive than RIP sensors, and, for lighting sensors, light-sensitive sensors are more expensive than photodiodes [8].

7.1.6 Mobile Device Sensors

A mobile phone is a portable device with an operating system that can run applications such as business, organizational resource planning, financial applications, etc. Most portable devices are equipped with Wi-Fi, Bluetooth, NFC technology, and GPS global positioning system, which allow them to connect to the internet and other devices. Smartphones and personal digital assistants are suitable for users who want to use some computer features and are on the move at the same time.

7.2 Network Technologies and Communications

A network is a collection of nodes and connections between them that communicate securely, accurately, and quickly with each other using specific rules and regulations (protocols) and share resources such as information, software, and hardware with each other [9].

Networks are divided into the following four categories based on their geographical scale:

- WAN: a wide network used to exchange information over long distances, such as an entire country or continent.
- MAN: urban networks that connect multiple networks located in different areas of a large or small city.

- LAN: a high-speed local network used to communicate and exchange data in a small geographical area.
- PAN: a personal network with a short range that provides communication between the devices around a person.

Wide-ranging protocols are used in IoT communications. Internet of Things devices must communicate with each other quickly and with many capabilities to achieve efficiency and exchange information in a unified manner. Using IoT protocols, information exchange and delivery of applications and services are possible from anywhere. Many protocols and standards have been developed to support the IoT, and Table 7.1 shows a summary of the most popular protocols. In the following, protocols that are more effective in optimizing energy consumption are introduced [10].

7.2.1 Low-Power Wi-Fi (Halow) Protocol

Halow technology has been introduced for IoT applications and products compatible with IEEE 802.11 Ah technology, which operates at frequencies lower than 1 GHz and provides wider range and much lower energy consumption for Wi-Fi devices. Halow technology includes a wide range of energy optimization solutions, such as use in smart homes, smart cars, smart cities, wearable devices, as well as industries and agriculture. In this technology, thousands of battery-operated sensors can be connected to a Wi-Fi node. Also, due to the low-frequency radio wave propagation capabilities, it can easily pass through walls and obstacles, and its working range is about twice that of current Wi-Fi, and it can easily connect to cloud systems, making it the most effective protocol used in IoT [11].

7.2.2 Zigbee Protocol

This protocol is very low-cost and low-power and has key capabilities to support mesh network communications using low-power digital transmitters and receivers based on the 802 IEEE standard for wireless personal networks with low data rates. Zigbee enables fast communication between other devices and is an ideal choice for energy-efficient smart devices [12]. Its low power consumption allows for longer operation with smaller batteries. Because Zigbee nodes can be put to sleep, power consumption is reduced on average, resulting in longer battery life. This protocol has five layers: the physical layer, the media access control layer, the transport layer, the network layer, and the application layer. Zigbee was first introduced in 2004, and since then, numerous updates have been made to it.

Table 7.1 Common protocols in the Internet of Things.

Application protocol		DDS	CoAP	AMQP	MQTT	MQTT-SN	XMPP	HTTP REST
Service discovery		mDNS			DNS-SD			
Infrastructure protocol	Routing protocol	PRL						
	Network layer	6LoWPAN			IPv4/IPv6			
	Link layer	IEEE 802.15.4						
	Physical/device layer	LTE-A		EPCglobal	IEEE 802.15.4		Z-Wave	
Influential protocols		IEEE 1888.3, IPsec			IEEE 1905.1			

7.2.3 RPL Protocol

RPL protocol is a routing protocol for energy-limited and noisy networks (such as networks composed of interconnected IoT devices). This protocol supports the minimum routing requirements by building a stable topology on stable paths and is also inherently an optimized internet routing protocol for sensor networks based on 6IPV. A directed acyclic graph represents the core of this protocol, which shows the routing diagram. In this graph, each node only knows its parent and does not keep any information about its children. This protocol is suitable for high-loss and low-power networks and leads to maintaining message confidentiality and integrity, improving performance, ensuring package delivery in the order they were sent, supporting various link-layer protocols, and responding to wireless network limitations [13].

7.2.4 802.15.4 IEEE Protocol

The 802.15.4 IEEE protocol is a set of LR-PAN personal wireless network standards. This standard is based on high-level communication protocol specifications such as Zigbee. This standard provides data rates of 40 kb/s to 250 kb/s and operates at a frequency of 915/868 MHz for low data rates and a frequency of 4/2 MHz for high data rates. This protocol is defined by IEEE in the data link layer of the Open Systems Interconnection (OSI) model and is one of the personal wireless network protocols used for low-speed and low-power in personal wireless networks. This protocol is a suitable option for wireless sensor networks due to its capabilities and advantages, such as energy consumption optimization, high throughput rate, and low power. It is used for applications that have a large number of nodes and their batteries need to last for several years. This protocol has the ability to run on most different platforms and can handle a very large number of nodes. Additionally, this protocol has high security and provides encryption and authentication services [14]. Figure 7.1 shows the 802.15.4 IEEE packet structure, which is essential for understanding the communication protocols used in IoT devices.

Answer: The given text is a technical description of low-power Wi-Fi (Halow), Zigbee, RPL, and 802.15.4 IEEE protocols. These protocols are used in IoT applications and products. Halow technology provides wider range and much lower energy consumption for Wi-Fi devices. Zigbee protocol is a low-cost and low-power protocol that enables fast communication between other devices and is an ideal choice for energy-efficient smart devices. RPL protocol is a routing protocol for energy-limited and noisy networks. 802.15.4 IEEE protocol provides data rates of 40 kb/s to 250 kb/s

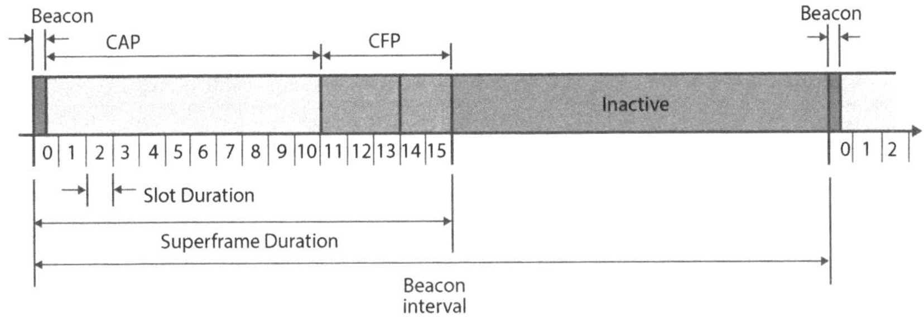

Figure 7.1 802.15.4 IEEE packet structure.

and is one of the personal wireless network protocols used for low speed and low power in personal wireless networks [15].

7.2.5 6LoWPAN Protocol

6LoWPAN protocol is used by IoT applications, which is a low-power personal wireless network based on 6IPv technology. This protocol uses header compression and encapsulation mechanisms. It is designed specifically for home and building automation due to its low cost and low energy consumption, making it suitable for IoT applications that have a large number of low-cost devices. This technology can use several different communication platforms. Currently, packet formats and necessary mechanisms for packet transmission through the 802.15.4 IEEE standard have been defined on low-power personal wireless networks. Because 6IPv supports 1280-byte packets and 802.15.4 IEEE sends 127-byte packets, the information is segmented. Many companies have started to support this open standard protocol for IoT, which is cost-effective, has low energy consumption, and is expandable by everyone due to its openness [16].

7.2.6 MQTT Protocol

MQTT is a publish/subscribe protocol used for machine-to-machine (M2M) communication and IoT with low bandwidth and high latency networks. It is a standard M2M protocol for IoT devices that communicates with sensors. This protocol provides an easy way to connect devices to limited computational resources and lower bandwidth networks, while ensuring reliable message delivery. The architecture of this protocol includes a central server that communicates with one or more devices in a star topology, and the basis of communication is based on messages sent to a node (device), and the central server is responsible for managing the network and transmitting messages.

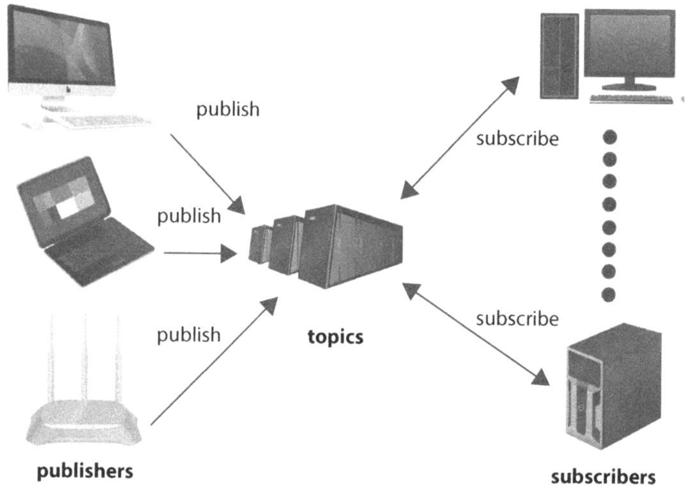

Figure 7.2 MQTT architecture.

MQTT design principles minimize the bandwidth and resources required by devices and are suitable for M2M communications, IoT devices, and mobile applications that have low battery consumption and low bandwidth. MQTT can be implemented on various platforms to connect IoT devices to the internet. It can also act as a messaging protocol between sensors and servers. The components of MQTT are Publisher, Subscriber, and Broker. A specific device is considered a message receiver or subscriber for a specific message in the eyes of the Broker whenever the sender sends that particular message. The sender or Publisher is considered the message producer and sends the message to its own created receiver through the Broker. Security is also provided by the Broker as it checks the identity of the message receiver and sender [17]. Figure 7.2 illustrates the architecture of a smart water sensor system, highlighting how these sensors monitor water consumption and provide useful information to consumers.

CoAP protocol is an application layer protocol for IoT applications and a messaging protocol. Due to the limited resources and processing capabilities of IoT devices, HTTP protocol cannot be used in IoT applications. Generally, the main goal of CoAP protocol is to enable devices with limited resources to have RESTful interactions. REST is an easy way to transfer data between the receiver and the provider of services over HTTP. In addition, the CoAP protocol changes some of the main operations and basics of HTTP to meet the needs of IoT. For example, it has low energy consumption and can work in the presence of unstable communication links. The main function of the CoAP protocol is shown in Figure 7.3 [18].

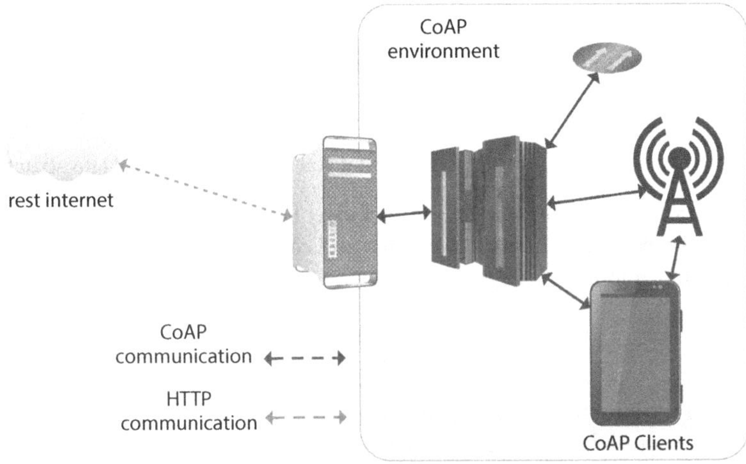

Figure 7.3 CoAP protocol performance.

7.2.7 Management and Storage Systems for Data

IoT data management includes the process of collecting, processing, storing, and analyzing data. The first and most important goal is the accurate and reliable collection and cleansing of data from various sources and managing heterogeneous data structures. Then, high-level information must be extracted, summarized, and concluded so that raw data can be used to support decisions. Data are the main element of the information technology and the IoT industry, and, therefore, the most important task of managers in any organization is to preserve, maintain, and manage information that is constantly increasing in quantity and diversity [19].

Database banks designed and implemented for working with the IoT must have special conditions and features. Providing cloud computing-based platforms in the field of the IoT paves the way for the entry of new databases and the use of their many achievements and services. Data generated by IoT devices are time-series and event-based data, and a database is needed that can respond to data nature in dimensions such as volume, speed, and diversity. Relational databases in software with heavy data have low performance and are not suitable for deploying the IoT and are only usable when data are small or there are not many requests for reading and writing in the database. Most relational databases can grow to a reasonable size and this operation is well done on one server, but, when the work extends to more than one server, special complexities must be addressed to make these databases shared between servers. Given the conditions of

Table 7.2 Comparison of relational and non-relational database parameters.

Non-relational databases	Relational databases	Model
NOSQL	SQL	Data storage
non-relational	relational (tabular)	Data structure
Data storage in different forms such as documents, columns, graphs and -key pairs	In the form of a table with rows and columns	flexibility
value	Suitable for structured data	You
Suitable for semi-structured or unstructured data	You are limited to communication	security
You are flexible	static	types of
dynamic	Has authentication mechanisms	samples
Due to the presence of unstructured data and the lack of encryption, it affects security	a type	Model

data in the IoT, new database models should be used. Table 7.2 compares important parameters of relational and non-relational databases [20].

NoSQL databases are a new generation of databases that have distinct features such as secure storage, fast retrieval, updating, analysis, and processing of high-volume data simultaneously. Such databases can quickly grow and handle high volumes of requests. Data in these systems have the following characteristics [21]:

- Data are not in a relational (tabular) format.
- The software system is open source.
- The database is horizontally scalable.
- Easy distribution and support for large volumes of data.
- Suitable for today's cloud-based data storage.

MongoDB is a flexible and open-source database known as a NoSQL database that stores data in BSON format, which is a binary JSON format. It uses documents with a JSON pattern and stores data in BSON format.

This database provides efficient and high-performance use of RAM for its cache and query optimization. It is suitable for the following:

- Event logging, content and document management systems, high-volume issues, and mobile applications. MongoDB has benefits that can address data-related challenges, including:
- Storing large volumes of unstructured and highly flexible data.
- Maximizing the use of cloud-based computing and storage.
- Providing a cost-effective storage solution.
- Enabling faster development and deployment.
- Efficient and cost-effective database architecture scaling.
- Creating various data with MongoDB, including data analysis, location-based operations, real-time reporting, sensor data, and connected IoT devices, and empowering content management systems.

Hadoop is a tool for implementing NoSQL thinking and non-relational databases and an open-source software framework. Hadoop acts as a launcher for specific types of distributed databases, allowing data to be distributed and spread across thousands of servers. The main Hadoop framework consists of the following modules:

- Hadoop Common: This includes libraries and utilities needed by other Hadoop modules.
- Hadoop Distributed File System (HDFS): a distributed file system that stores data on cluster machines and creates a wide bandwidth.
- Hadoop YARN: a resource management platform that manages computational resources in clusters.
- Hadoop Map/Reduce: a programming model for processing data on a large scale.

7.2.8 Cloud Computing

Cloud computing is a computational paradigm in which all computational resources such as memory, storage, and processing capabilities are provided by a third-party provider over the internet. Unlike traditional computing methods where users have a complete set of resources on a personal computer to perform any computational task, in cloud computing, all resources are accessed through a web-based program. Cloud computing provides companies, organizations, individuals, and researchers the ability to scale resources up or down as needed. To achieve the potential of cloud computing and measure its widespread use, a combined framework of the IoT and cloud computing seems more reliable [22]. Cloud computing services are classified into three categories:

- SaaS: Provision of application programs that run in the cloud and are internet-based, providing a way to deliver software applications over the internet instead of installing and maintaining software physically on-site. Providing applications as a service has many advantages over traditional software packages, including ease of use, lower costs, improved customer service, and minimal maintenance.
- PaaS: When the cloud provides a development environment for users to develop software, the provided environment is called a platform. With PaaS, users build and run their own application programs within the environment provided by the service provider. In the cloud service PaaS, infrastructure and application programs are managed for users, and support is also provided. Some of the features that can be provided with the PaaS service include operating systems, storage, database management systems, support, server software, and tools for design and development.
- IaaS: Provision of network, computational, and processing resources, allowing the consumer to control the operating systems and storage. In this model, the service provided by the cloud includes processing resources such as hardware, servers, network bandwidth, or other infrastructure services such as processors, storage space, and other resources as needed.

By leveraging cloud computing, it is possible to process large volumes of data. Because the data collected by the IoT nodes is of high volume, cloud computing is used as an infrastructure to maintain and process this data. Internet of Things devices have limited processing resources, which do not allow them to perform complex and heavy processing. Usually, the collected data are sent to more powerful nodes, where processing and aggregation of information is possible. However, without a suitable infrastructure, scalability is a challenge, which is why the cloud provides unlimited processing capabilities [23]. Figure 7.4 depicts the data flow in a typical MQTT-based IoT network, demonstrating the interaction between publishers, subscribers, and the broker.

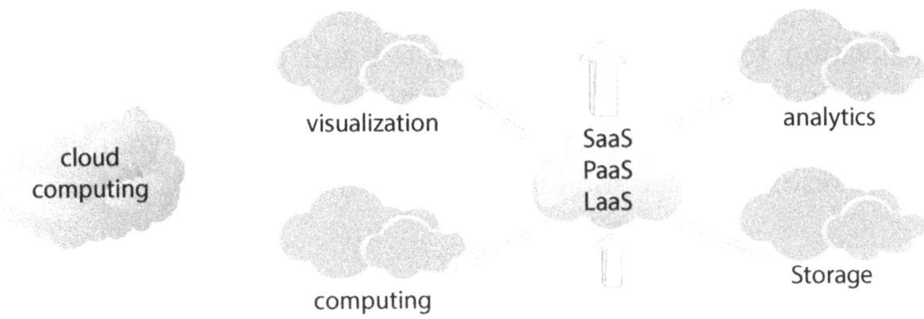

Figure 7.4 Cloud processing.

7.3 Data Centers and Server

Data centers and server rooms are a collection of servers, services, communication infrastructure, and electronic equipment used to maintain and support network services. Data centers can provide web server hosting for the general public by having high-speed internet lines, powerful and multiple servers, advanced power supply equipment, and air conditioning. Large companies also rent a server in a data center and remotely control it with special tools. With the rapid growth of the internet, the number of data centers is increasing day by day. Energy costs are a key challenge for information technology departments and one of the major concerns for data center owners. The deployment of the IoT produces large amounts of data that must be processed and analyzed in real time. Data center managers need capacity management systems to actively meet business priorities related to IoT. Although data storage is the main concern, data centers must also provide space for storing this volume of data. In addition to storage space, companies must focus on their ability to access and use the data generated in the IoT space with less cost and more efficiency. The following are ways to reduce costs and optimize energy consumption in a data center [24].

7.3.1 Virtualization of Servers

Server virtualization is a solution for installing and running multiple independent operating systems on a single hardware server by combining different servers on a physical host. A virtual server runs programs like a real server, so instead of having many servers with low efficiency, virtualization combines processing power on fewer servers that have higher efficiency. Physical servers generate a lot of heat and require expensive cooling systems in addition to the electrical power needed to keep them running. Virtualization reduces the heat generated and eliminates the need for expensive and power-hungry cooling systems, resulting in energy savings. Additionally, server virtualization can be very cost-effective in reducing hardware resource usage and increasing productivity [25].

7.3.2 Data Cleansing

Data cleansing is essential for interpreting and analyzing IoT data to ensure the reliability of IoT systems. Its goal is to reduce data uncertainty, and, given the vast amount of information, the uneven resources of errors and processing requirements in real time are challenging. This concept was first introduced in data warehouses, where its purpose was to clean incorrect, incomplete, and repetitive data, solving data quality issues in databases and system files. The reason for such problems and the production of incomplete data can be signal interference between sensors, environmental complexity, and incorrect calibration. Also, invalid data produced due to deviation from the natural range can be identified and their quality significantly increased by covering them with more than one sensor before they reach the end-user by cleaning them to a large extent. Data cleansing technology in the IoT has been widely studied. The data cleansing framework has different organizational structures, data cleansing approaches in layers, and different processing perspectives. A flexible structure for cleaning sensor data consists of five stages: placement, shaping, integration, arbitration, and virtualization. In the first stage, data are filtered for quality issues. The shaping and integration stages are designed to process the data. The shaping stage uses time windows in multiple quantities to correct incorrect or external values. The integration stage is also used to read a number of sensors related to a unit type. In the arbitration stage, logical errors are removed, and it is used to resolve disputes between surrounding spatial groups. Finally, in the last stage, data are combined from multiple sources [26].

7.3.3 Data Compression

In this mechanism, information is encoded in such a way that the number of bits required to represent the original message is reduced. This is energy-efficient because the transmission time is shorter compared to smaller packet sizes. However, existing compression algorithms for sensors are not usable due to the limitations of these resources. Therefore, special techniques must be developed to adapt to the computational capabilities and power of wireless devices. Data compression is used to reduce transmission traffic, reduce the amount of stored data, and organize and improve data. Information that is rarely accessed is a better candidate for compression. Compression also helps to reduce the consumption of valuable resources such as hard disk space or bandwidth. The goal of data compression is to reduce their capacity, and, therefore, less data lead to less energy consumption. During data usage, they must be returned to their original state, and this process may be detrimental to some applications. All compression systems require encryption and decryption algorithms [27].

7.3.4 Reduction of Data Volume by Removing Redundant Information

More than half of the information may be in the form of duplicate and unnecessary copies. There are software programs that can reduce the amount of stored information by finding and removing unnecessary copies. Storing less data requires fewer hardware resources, which, in turn, consume less energy. This method is a data compression technique for eliminating redundant data and can also be used for transferring data over the network to reduce the number of bytes that need to be sent. Therefore, this method reduces the required storage space for storing data and helps optimize consumption [28].

7.3.5 Energy Efficiency in IoT Applications

In many IoT applications, energy efficiency is a key design pattern. The use of devices that have the ability to control energy consumption through their capabilities is highly desirable.

7.3.6 Energy-Efficient Servers

New servers use energy-efficient technologies for power management, better voltage regulators, processors that consume less power, and cooling fans that consume more energy. These optimizations reduce overall power consumption and heat production, making cooling operations easier and more efficient. These devices reduce overall power consumption by minimizing the power consumption of idle processors and using energy-saving technologies that reduce the speed and voltage of the processor, which also reduces heat production and allows for easier and cooler cooling operations [1, 4].

7.3.7 Ventilation and Cooling of Servers and Data Centers

Server rooms or data centers are areas that produce a lot of heat due to the placement of servers and a large volume of electronic components. This heat increases with the number of racks placed in the server room, and, therefore, cooling the server rooms with different capacities and technologies is determined based on their cooling load. Evaluating airflow can help identify ways to improve the efficiency of cooling systems in data centers by better utilizing these devices. Airflow evaluation should include examining factors such as the amount of airflow in relation to the hottest server racks, the bandwidth of the lower floor, pressure in subfloor storage potentials, the size of the return apartment or ceiling height, and the size of the pressure tiles. Cabinets with air conditioning keep the equipment cool and save energy. Because most of the power used by equipment is lost through heat, they can actually put pressure on the cooling capacity of the data center. The components placed in a data center can easily produce enough heat to heat a house at the end of winter. Therefore, because, when network components are heated, they are exposed to damage and reduced lifespan, cooling should be used. The damage caused by heat is not always immediately apparent as a catastrophic crisis. Heat damage symptoms include damage to nodes and hardware errors that can occur during a weekly or monthly cycle, leading to severe failure. It is important to ensure that single cabinets provide adequate ventilation for network equipment. The temperature inside the cabinet is influenced by various factors such as door holes, cabinet size, and types of equipment placed inside the cabinet. Most direct methods of cooling network equipment aim to provide the necessary amount of air to cool the devices regardless of how it moves up and down in the cabinet. A specific amount of air is considered to cool the devices within their ideal temperature range [29].

7.3.8 Applications and User Interfaces

In the IoT, various tools collaborate to create new applications and services and achieve common goals together. The challenges of development are to build a large and intelligent world, which is truly digital and virtual, toward creating smart environments that have brought about intelligent energy and many other areas. In providing services related to things, services should be provided in this area and an interface for the user to use the IoT is provided. The application layer is the highest layer of smart city architecture, which mediates between citizens and data management layer. The performance of the application layer has a direct impact on the outlook and satisfaction of users with the smart city operations and is directly involved in interaction with citizens. The key services of the application layer include community development, network distribution, intelligent transportation, and weather forecasting. The application layer improves the performance of the city by using various programs that use stored and processed data. Different components that make up a smart city include a smart community, smart energy, smart transportation, smart network, and health. The application layer provides the services requested by the customer. For example, the application layer can measure humidity and temperature for customers who request this information. The importance of this layer for IoT is that it has the ability to provide high-quality smart services and meet customer needs. This layer covers the needs of vertical markets such as smart homes, smart buildings, intelligent transportation, industrial automation, and healthcare [30].

Expectations for IoT applications in various industries are very high. IoT applications are very diverse and different, and services that lead to interaction with the real world are IoT applications and are not limited to a specific industry. The ability of IoT applications to meet the high-quality smart services and meet customer needs requires compliance with requirements such as reliability, capability, reasonable cost, security and safety, ease of use, standardization, and integration. Also, at this stage, issues such as security and access levels will be very important. From a business perspective, the IoT represents an extraordinary opportunity for various organizations, including IoT service and application providers, IoT platform and integrators, telecommunications operators, and software vendors. In this section, to achieve the goals of the project regarding IoT applications in smart cities that lead to optimizing energy consumption, the following items have been considered: identifying IoT applications in smart cities, identifying IoT applications with a focus on energy consumption management (water, electricity, and gas), and focusing on the following applications: smart homes and home automation, smart street lighting, smart roads, smart environment, smart offices, intelligent transportation, smart networks, and smart waste collection [31].

7.3.9 Smart Homes and Home Automation

Smart homes and home automation rely on the use of information and communication technologies (ICTs) to control homes. A key element of smart homes is the use of intelligent scheduling algorithms that enable residents to make the best choices regarding electricity consumption to reduce energy consumption. Another term commonly used in this context is home automation. Home automation is managed without human intervention, and we can control objects in the home through devices. IoT involves transferring data over a network without human-to-human or human-to-machine communication. Home automation depends entirely on sensors. Smart homes make it possible for tasks to be performed automatically with much less activity, and in addition to convenience, smart homes also increase energy efficiency. Optimal use of energy resources is made in these buildings. The proposed system for smart homes consists of the following sections [31]:

- Central server/cloud: a monitoring and coordination system with a local database that communicates remotely with smart devices.
- Identification sensors including gas, temperature, sound, motion, and light sensors: Sensors transmit environmental status to the system.
- Smart software and mobile application: sending requests to the server to access and adjust smart home devices.
- Communication technologies for data transfer to the cloud: The communication protocol is a standard and acceptable route for all system components to exchange information.
- Smart bulbs
- Doors and windows
- Smart TV and refrigerator
- Smart sockets: By using sockets that have the ability to send and receive information through communication lines, we can turn on and off connected consumers or be informed of their status.
- Smart thermostat: a device for controlling home heating, and this device allows the user to adjust the temperature via the internet (with the help of smartphones).
- Central air conditioning
- Security cameras

These systems are self-controlled and include the following:

- Heating, cooling, and air conditioning systems
- Fire extinguishing system
- Security system
- Energy management
- Lighting system

- Audio and video systems
- Irrigation system

7.3.10 Smart Street Lighting System

Currently, a large amount of electrical energy is consumed by street lamps, and their number is also very high. By implementing a smart lighting system, it is possible to prevent energy waste. The components required to create this system include the following [32]:

- LED lights with adjustable power
- Light sensors
- Motion sensors
- Controllers
- Cloud servers
- Communication devices such as Zigbee
- Vehicles and pedestrians [32]

7.3.11 Communication in the System

In this system, a communication technology must be used to communicate with servers, sensors, and smart lamps. Figure 7.5 presents the performance metrics of the CoAP protocol, showcasing its efficiency in enabling RESTful interactions for devices with limited resources.

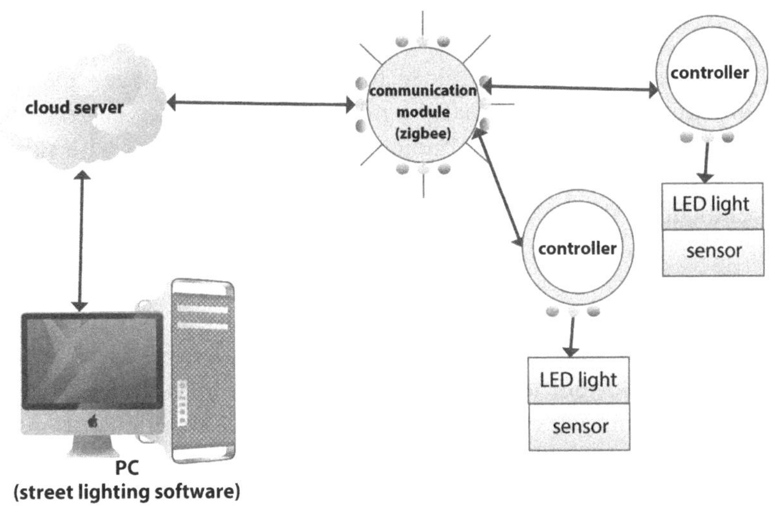

Figure 7.5 The structure of the smart street lighting system.

7.3.12 System Operation

The operation of this system is as follows [33]:

1. When the sensors detect vehicles or pedestrians, they inform the server and send the information for analysis and processing.
2. The server decides whether to turn the lights on or off.
3. The server sends a message to the smart street lamps to turn them on.

7.3.13 Light Control

When the motion sensors do not detect any movement, the system reduces the ambient light. Conversely, if the presence of individuals is detected, the ambient light increases.

7.3.14 Remote Monitoring and Management

Some companies have proposed software solutions to remotely monitor and manage street lights wirelessly [14, 17].

7.3.15 Smart Roads

Smart roads refer to roads that automatically collect garbage and have a high level of safety. They can even move themselves, reducing the costs associated with controlling and maintaining roads. In addition, solar energy can be stored and converted into electricity for home use, which highlights the importance of this technology. Instead of large street lighting on highways and roads that extend for hundreds of kilometers, smart roads can use glowing sign technology at night that charges with sunlight during the day and returns light to the environment at night. Sensors are also embedded throughout the path of these roads that are sensitive to vehicle movement. When a vehicle reaches a specific area of the road, the sensors detect its movement and produce light only in that area. This is more suitable for roads that may be less frequently used and have high electricity consumption costs. Another feature of these roads is the ability to charge electric vehicles by moving on magnetic lines embedded in the roads, which eliminates the need for electric vehicle charging stations and saves time during travel [33].

7.3.16 Smart Parking Ecosystems

Smart parking ecosystems consist of two streams: information and traffic. Drivers receive information about the availability of parking, are directed to the desired parking lots, and then park. When a vehicle enters or exits a parking lot, the parking inventory information changes and is communicated to other drivers who are searching for a parking space. The flow of information from the moment it is identified by sensors until it is displayed on drivers' devices is called the information flow [34].

7.3.17 Sensor Parking and Management Tools for City Parking

Sensors are installed in street parking to identify vehicles. Then, the sensors form a network and send the latest information to storage devices. Additionally, a smartphone application needs to be developed to allow drivers to check parking inventory information. The factors that need to be considered in the deployment of this technology include the robustness of the sensor devices, stability and timely operation of the sensor networks, quality and agility of urban services, user-centric information dissemination and allocation, uncertainty of driver behavior, and travel tracking [34].

7.3.18 Algorithm for Smart Parking System

An algorithm for the operation of the smart parking system based on IoT is designed such that, when a user is searching for a parking space, they must first enter the system. After successful entry, a request message is sent to search for a parking space. Then, the system sends a response message that includes information about the address of the parking lot and the route to get there. The parking location is selected based on a function [$f(\alpha, \beta)$] calculated based on the current location of the vehicle and the location of the parking space. When the user enters the parking location, their access rights must be checked for entry, which is verified by RFID technology or scanning the car card. If the information is correct, then the user is allowed to enter the parking lot. If the current parking location is full, then the system sends a proposed message that includes the location of a new parking lot and its address and route to get there with minimal cost [35].

7.4 Benefits of IoT in Smart Parking Systems

IoT provides solutions to help drivers find parking spaces and prevent road congestion, which not only provides a direct advantage to users but also helps reduce traffic density and greenhouse gas emissions, benefiting the entire community. In smart parking systems, sensors are embedded in the parking lot area, and data related to the availability of that parking space are sent to the server, and drivers can be informed of empty parking spaces by installing a mobile application. Additionally, by adding RFID tags to each of the existing parking spaces, it is possible to use validation devices that calculate parking time and provide parking bills based on real-time information, which can satisfy drivers and ultimately lead to a reduction in carbon emissions, fuel savings, and reduced time to find suitable parking spaces, control traffic flow in the area, and optimize parking management [11, 30].

7.4.1 IoT Sensors for Smart City Monitoring

IoT technology enables monitoring and planning at the city level, including monitoring sensors for the following:

- Noise
- Pollution control
- Traffic monitoring

7.4.2 Useful Reports from IoT Sensors

IoT sensors provide useful reports for the following:

- Monitoring noise pollution in traffic-heavy cities
- Detecting water pipe leaks
- Monitoring temperature and humidity and alerting for extreme weather conditions
- Real-time monitoring and reporting for CO_2 levels
- Early earthquake detection
- Hazardous air and gas quality [36].

7.4.3 Smart Offices

Smart offices are physical environments where energy management and security are benefits that the IoT provides through its own equipment.

Compared to other types of buildings, office buildings consume more energy. The amount of energy consumption in office buildings varies depending on the location and size of the building, lighting systems, air conditioning systems, and the number and types of equipment used in them. Smart control of equipment will significantly reduce energy consumption and provide comfort for users. In smart offices, IoT focuses on the following to reduce energy consumption: automation of offices, lighting, and HVAC control, access control, intrusion detection, fire, and accident detection. Smart control of office buildings includes control of lighting, cooling and heating systems, audio systems, electric curtains, smart control of electrical and electronic devices, smart control of parking and vehicle entry and exit systems, smart security systems control, fire alarm and extinguishing system control, and smart gas leak detection. Smart lighting control involves measuring the ambient light level with a light sensor. These sensors detect the movement of people in rooms, hallways, and staircases and turn on the lights. If the sensors detect that multiple people are present in one place, they increase the light level. In addition, cloud computing allows users to remotely control lighting and monitor energy consumption through smartphones. Smart control of heating and cooling systems allows continuous monitoring of system conditions and performance using smart sensors connected to the internet. When information is collected by these sensors, it is immediately shared with the building manager, project engineers, and other relevant individuals. New systems have advanced operating systems such as Linux and advanced node processing, which provide the ability to monitor and control equipment using real-time data with a suitable user interface for designers. Smart thermostats can learn the desired schedule and adjust accordingly on a daily basis or use a motion detector for automatic air conditioning and temperature settings. Smart thermostats can also work with Wi-Fi and can be controlled and adjusted via smartphones or tablets. In this structure, a digital connection is established between wireless sensors and the HVAC controller. When people leave the house, the wireless control technology can turn off the system, and when people enter, it can increase the heat and prevent unnecessary heating or cooling. Smart control of security systems includes a wide range of security sensors and controllers that are installed in different spaces to ensure security. Different types of sensors such as infrared sensors can detect the presence of a person or intruder even in the dark and report it [37]. Smart control of electric curtains allows easy control of curtains through various smart control systems. In the smart system, electric curtains can be opened and closed automatically at different times or under certain conditions, such as the difference in light intensity inside and outside the building or the intensity of the environment's heat. Smart sensors can measure the light level inside and outside the building and, by activating energy-saving scenarios, open the curtains during the day so that there is no need to turn on the lights. Smart curtains can also be opened and closed according to a specific schedule.

7.4.4 Smart Transportation

With the increase in urban population, it is necessary to apply more sustainable solutions for people's movement toward their destination to overcome traffic and air pollution. Therefore, a significant coverage of electrical and electronic infrastructures is required for transportation technology. Environmental control, monitoring of service infrastructures in trains, subways, stations, airports, and at street levels are supported. Currently, transportation problems such as environmental pollution, depletion of energy resources, problems with monitoring of intercity transportation, and the need for time and cost savings in daily life have become one of the serious challenges in today's cities (metropolises). Therefore, the use of smart transportation systems will have many benefits such as reducing fuel and energy consumption, reducing effective travel time, reducing environmental pollution. Smart transportation systems use different technologies. Control systems for traffic lights, automatic vehicle number identification systems, speedometer cameras, traffic status control, etc., are among the most commonly used technologies in this field [38].

Smart transportation systems use various technologies. The following are essential components for the operation of these systems:

- Data collection: Data are generated by sensors installed in the city, mobile devices, vehicles, video cameras, RFID scanners, and GPS data.
- Data analysis and processing: Processing and storage technologies such as relational databases, Big Data technologies for processing generated data such as Spark, MapReduce, and Hadoop.
- Data and information transfer: Through communication protocols, location-based applications.

In public smart transportation, accurate information about public transportation scheduling, accidents, routing, vehicle status, and travel time is transferred and analyzed based on the equipment and components embedded in the vehicles. Smart transportation includes various communications and sensor systems in vehicles that enable real-time processing of transportation data. These data include the number of vehicles, their average speed, traffic intensity, violations, and accidents. In smart transportation systems, it is possible to monitor the status of drivers and vehicles such as buses, taxis, emergency vehicles, and fire trucks and reduce delays and stops, and maximize safety on the streets with control systems. Traffic movement sensors control traffic density in the city and report it to the system so that drivers can choose their route intelligently and prevent wasting time and energy.

7.4.5 Some of the Solutions for Smart City Traffic Control

- Receiving traffic signals using IoT technology and notifying other drivers to prevent them from entering the traffic route.
- Receiving information about sudden weather changes and accidents on the route and transferring it to drivers.
- Registering vehicle information and driver performance using fleet management systems and trackers to improve driving behavior and reduce traffic caused by route changes, sudden braking, and acceleration.
- Announcing the presence of pedestrians on the routes and implementing punitive solutions for pedestrians who move on vehicle routes or cross the street without paying attention to traffic lights.
- Using shared vehicles instead of personal vehicles [39].

A smart grid is an electrical power network that can withstand potential failures by using modern computers and communication technologies. The smart grid is used for generating, distributing, managing, and consuming energy. Smart grids are composed of technologies such as transmission lines, controllers, power lines, computers, and equipment that work together to provide two-way communication between electricity producers and consumers. The use of IoT in smart grids allows energy producers to control and manage resources efficiently, especially with the increase in population. IoT in smart grids connects millions or billions of buildings to the power distribution network through meters, which are used for collecting, analyzing, controlling, monitoring, and managing energy consumption. IoT enables energy distributors to improve their services to meet customer needs, reduce errors, increase efficiency, and improve service quality [40].

Smart grids consist of three main processes:

- Power generation: a process in which energy is used from a source to produce electrical energy.
- Transmission: the transfer of electrical energy from the producer to the distribution centers near cities or industrial centers.
- Distribution and consumption: the delivery of electrical energy to consumers [10].

7.4.6 Smart Grids

A smart grid is an electrical power network that can withstand potential failures by using modern computers and communication technologies. The smart grid is used for generating, distributing, managing, and consuming energy. Smart grids are composed of technologies such as transmission lines, controllers, power lines, computers, and equipment that work together to provide two-way communication between electricity producers and consumers.

7.4.7 IoT in Smart Grids

The use of IoT in smart grids allows energy producers to control and manage resources efficiently, especially with the increase in population. IoT in smart grids connects millions or billions of buildings to the power distribution network through meters, which are used for collecting, analyzing, controlling, monitoring, and managing energy consumption. IoT enables energy distributors to improve their services to meet customer needs, reduce errors, increase efficiency, and improve service quality [41].

7.4.8 Processes in Smart Grids

Smart grids consist of three main processes:

- Power generation: a process in which energy is used from a source to produce electrical energy.
- Transmission: the transfer of electrical energy from the producer to the distribution centers near cities or industrial centers.
- Distribution and consumption: the delivery of electrical energy to consumers [1]. Table 7.3 provides a detailed comparison of the standard status of four key technologies—GSM, VPAN, PLC, and Zigbee—highlighting their differences in profitability, data rate, and adherence to standards.

Table 7.3 Comparison of smart energy grid technologies.

Technology	Advantage	Data rate	Standard
GSM	Low energy consumption for indoor equipment	2.5G: 144 kbps	2.5G HSCSD, GPRS
WPAN	Very low energy consumption, low installation cost	IEEE 802.15.4: 256 kbps	IEEE 802.15.4
PLC	Lower maintenance and operational costs, the possibility of physical disconnection based on other networks	1–10 Mbps	BB-PLC: TIA-(1113) HomePlug1.0
Zigbee	Compatible with 6IPv based networks	250 kbps	Zigbee

7.4.9 Technologies in Smart Grids

Technologies such as integrated communications and communication technologies, advanced sensors and measurement devices, distributed power control systems, and smart grid stations are required for a smart grid. These technologies are currently being used [21, 42].

7.4.10 Smart Energy Management Systems

On the consumer side, smart energy management systems consist of two systems:

- A smart measuring device that measures the consumption of any type of energy (electricity, water, and gas).
- An energy consumption information display device that displays various parameters of this energy consumption for the user, including the amount of consumption, cost, consumption at different times, and actions to better manage energy consumption at different times.

These devices are usually connected to the energy provider network (gas, electricity, or water) through a communication network and can send weekly, daily, and even hourly energy consumption information to the energy provider [43].

7.4.11 Smart Waste

Smart waste collection is a key solution to the increasing amount of waste produced in cities. The process of waste collection, transportation, sorting, recycling, and disposal requires extensive management and costs. The deeper penetration of ICT solutions in this area leads to significant cost savings and economic and ecological benefits. For example, the use of smart waste bins with load-level sensors can optimize the waste collection truck route, reduce waste collection costs, and improve recycling quality. IoT connects end devices such as smart waste bins to the waste management system, processes the data in the software control center, and determines the desired level of waste collection truck. Smart waste bins equipped with ultrasonic fill-level sensors optimize waste management and reduce environmental pollution. The sensors report the level of waste in each bin, allowing truck drivers to choose which bins to collect, reducing fuel consumption and service hours. The sensors in the waste bins report data on capacity, pollution level, etc. The online routing system also provides traffic status information, which is very important for effective routing. The main components of a cloud-based DSS system are shown in Figure 7.6 [43].

In recent years, smartphones have become very popular. Smartphones usually have multiple networks such as G3, WiMAX, Wi-Fi, and Bluetooth. They also have various sensors, including GPS systems. Smartphones are suitable for monitoring, controlling, and managing remote energy control systems from any location at any time. After proper authentication and access authorization, authorized users can modify their energy-saving policies online and communicate with servers related to their office and residential buildings. IoT requires the use of smartphones for part of its chain. Mobile applications are used to control devices or monitor information sent from objects. Security is one of the most important issues in IoT architecture. Due to the fact that IoT sensors often collect very private information and transfer the real world to the virtual world in a way, IoT can bring new useful solutions, but it can also bring new vulnerabilities to businesses, and organizations need to consider its security implications before moving toward IoT. Security means the confidentiality, integrity, and availability of a service network. IoT security is in the field of protecting devices and networks connected to the IoT ecosystem. Solutions to support user access to services must be considered, and various threats such as component

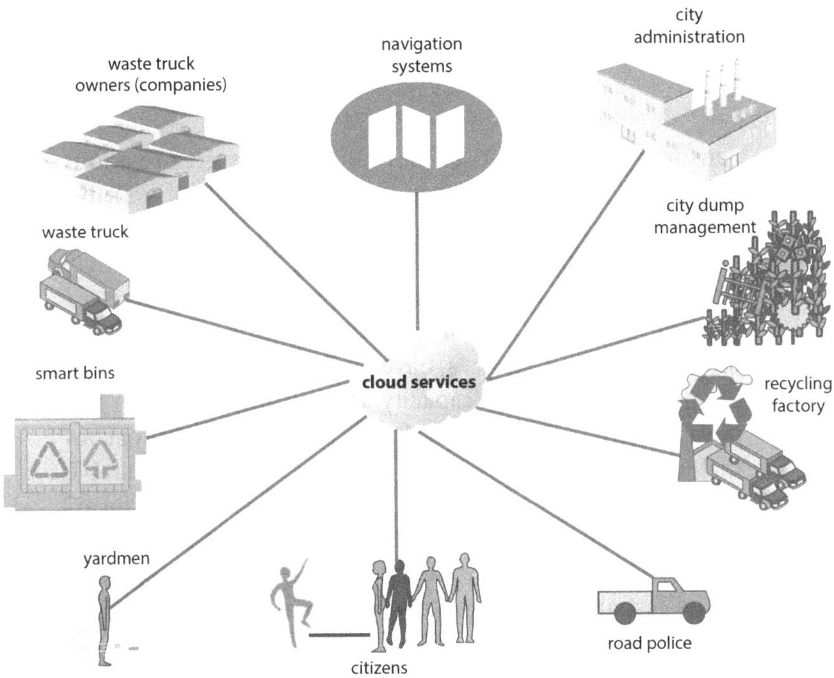

Figure 7.6 Waste collection management system.

and infrastructure failures and cyber-attacks must be addressed. All IoT applications require access to the internet, and all devices require an IP address for identification. Therefore, security solutions for internet-connected embedded devices are essential and important at all levels of IoT architecture. In fact, all of these issues must be considered and corrected in the initial stages of system design. In the following, we will examine the factors related to each of the discussed layers in the proposed IoT framework [10, 15, 44].

7.5 Layer of Communication Equipment and Sensors

In the divided layers, security is most important in the sensor layer because it collects all user information. The main equipment in the sensor layer includes sensors, RFID and Zigbee systems, and other devices that are placed in public places. If appropriate security measures are not taken, then data can be easily stolen, and it can be monitored. Therefore, security solutions for IoT devices are essential to protect against network traffic manipulation, data theft, and network control. At this level, factors that need attention and necessary measures to strengthen them are introduced [45].

1. Sensor security: Smart sensors can be the target of various attackers. Hackers may forge smart data to attack systems. Opponents can obtain smart meter data to extract user behavior patterns and sensitive information. Attackers may also hack smart sensors to extract information about users' physical presence in their homes, their habits, and much more. Many IoT devices have limited memory, processing power, and storage capacity and must operate with less power, such as when using batteries. Security approaches that strongly emphasize data encryption are not suitable for such devices because they are not capable of performing complex encryption and decryption operations quickly enough to securely transfer data. Therefore, it is necessary to use fast and lightweight encryption algorithms for these sensors and devices.
2. Storage security: Storage security is a set of parameters and settings that make storage resources available to authorized users and trusted networks and block them for other individuals. These parameters can be applied to hardware, programming, communication protocols, and organizational policies.
3. Hardware testing: A strong testing framework is required to ensure the security of IoT hardware. This measurement includes testing the range, capacity, and latency of IoT devices. In addition, IoT device manufacturers must test all components and modules used in their devices extensively to ensure their proper performance with IoT applications.
4. Secure network: Securing the IoT network can be done by connecting IoT devices to the network using security features such as antivirus, anti-malware, firewalls, and intrusion prevention systems.
5. Secure communication: After IoT devices are secured, the next challenge is to ensure that the connection between the network and cloud services and applications is secure. The best way is to use transport layer security (TLS) encryption, which is compatible with these conditions.

6. Protocol security: In some communication protocols, there are solutions that can provide integration, credibility, and reliability for communications. Examples include TLS/SSL or IPsec, which are designed to encrypt communication at the network layer and transport layer, respectively.

7.5.1 Information Management Layer

Data management is a crucial task in smart cities as it makes the production, transfer, and extraction of data more efficient and easier than in a regular city. However, the instability of cybersecurity poses a challenge to data-centric smart cities. When services and vital infrastructures are digitally connected and data are dependent on the smart network, the risk of serious and dangerous cyber-attacks increases. Therefore, if a smart city decides to use private cloud services, that organization must have higher standards than the normal private sector. As privacy protection and security are essential for the continuity of a smart city, organizations managing the data of a smart city must be regulated and directed toward achieving these goals. A delicate balance must be struck between the organization's responsibility for security gaps and the economic justification for the organization's risk. Cloud security includes a wide range of issues, including confidentiality, data integration, and access control. Cloud security provides different levels of control over the network infrastructure to ensure the continuity and protection of assets such as websites and web applications. Cloud security provides unique security considerations. In the cloud environment, all interfaces, such as API and web-based interfaces, must be reviewed for security vulnerabilities. Factors such as strong passwords, account lockout mechanisms, two-factor authentication, encryption of transportation, vulnerability testing, password expiration dates, and enabling HTTPS instead of HTTP are all highly significant in web-based interfaces. Data encryption is used to protect users' security and prevent information breaches. Data are encrypted between IoT edge devices and final systems using standard encryption algorithms to ensure data integrity and prevent data corruption by hackers. However, the wide range of IoT devices and hardware may not have the ability to process standard encryption algorithms and may limit protocols. Therefore, all IoT encryption must be equipped with complete encryption key processes. Data encryption is usually performed using keys created through PKC (public key encryption keys). Instead of using PKC, local elements (static) such as PSK or certificates are used. This is because public keys may be very slow, and time, calculations, bandwidth, and memory resources are important to us. Local PSKs can provide a reasonable level of encryption by allowing the user to select a password [46].

7.5.2 Application Layer

This layer requires security standards that are appropriate to the needs of the software, which makes the important and complex task of ensuring security. Standardization organizations such as IEEE and IETF work to secure the IoT by designing essential communication technologies. These technologies are essential for creating a reliable and efficient IoT. The IoT has an unprecedented ability to scale and be flexible. One of the main goals is to ensure that suitable identity verification mechanisms are available and prevent any type of attack that jeopardizes the access and integration of services and data. One of the primary requirements for IoT security is to ensure that data are available to authorized users at all times, which will be further examined below [47].

Access control levels: Data and related information in the IoT must be verified, and the process in which data senders and information receivers provide information to identify each other to establish trust in information exchange. The simplest and most basic solution for identity verification is to include a username and password, and if necessary, the process of controlling and securing identity can be made more complex by adding identification elements.

Another common technique for identity verification and identification is the use of biometric factors. Various types of biometric factors, including fingerprints, facial scans, iris scans, palm scans, sound patterns, keystroke patterns, and signature movements, exist.

In the world of data security, there is a modification called AAA, which stands for Authorization, Authentication, and Accounting. The most important part of these operations is authentication, which is the process by which processes prove the real or legal identity of their users. Authentication is the ability to check and verify the identity of any communication device or object. For example, every energy provider needs to verify each smart meter for payment from the relevant user. In this process, after providing identification elements by the applicant, the system compares the username and password with the database specific to identification codes and issues acceptance or rejection of access to resources. In the next concept, authorization is a process by which users and/or applicants are given access to resources, the ability to use resources, or resources located on the data network, and this process provides network managers with the ability to determine the type of access to each of the network resources for individual access or groups of them. Finally, the last process, Accounting, produces a report on the performance of users or hardware whose identities have been verified in the previous two operations. This process determines what share of the system resources and services the operation or user will have and what price they have paid for them [48].

Backup and data recovery: Backup is the process of creating a copy of data that is used to retrieve the original data after damage, loss of information, security attacks and hacker or viral attacks, loss of data, errors in application programs, or server downtime. For organizations with large and important data volumes, it is necessary to have a separate computer system to store backup files so that access to information and its retrieval is possible at any time in the event of an incident for the main server.

API security methods: An API is an application programming interface that implements communication between different software and serves as an interface between a library and programs that request services from that library. APIs provide this capability [49].

7.6 Conclusion

The IoT is a new concept in the world of ICT that has gained a lot of attention in research communities and users in smart cities. IoT refers to a network in which every physical object is identified by an identity and forms a network with other objects. With the increase in energy tariffs, the issue of energy consumption and its related costs in cities has taken on new dimensions. The most important issue that needs to be examined in these networks is to reduce the energy consumption of nodes so that the network's lifetime increases to an acceptable level. Smart meters, smart counters, and smart gas meters are some of the energy-efficient technologies that can be used to optimize energy consumption. One of the primary requirements for IoT security is to ensure that data are available to authorized users at all times. Access control levels must be verified, and the process in which data senders and information receivers provide information to identify each other to establish trust in information exchange. The simplest and most basic solution for identity verification is to include a username and password, and, if necessary, the process of controlling and securing identity can be made more complex by adding identification elements. Another common technique for identity verification and identification is the use of biometric factors. The most important part of IoT security is authentication, which is the process by which processes prove the real or legal identity of their users. In this process, after providing identification elements by the applicant, the system compares the username and password with the database specific to identification codes and issues acceptance or rejection of access to resources. In the next concept, authorization is a process by which users and/or applicants are given access.

References

1. M. Zand, M. A. Nasab, A. Hatami, M. Kargar and H. R. Chamorro, "Using Adaptive Fuzzy Logic for Intelligent Energy Management in Hybrid Vehicles," *2020 28th ICEE*, pp. 1-7, doi: 10.1109/ICEE50131.2020.9260941.IEEE Index.
2. Ahmadi-Nezamabad, H., Zand, M., Alizadeh, A., Vosoogh, M., & Nojavan, S. (2019). Multi-objective optimization based robust scheduling of electric vehicles aggregator. *Sustainable Cities Soc.*, *47*, 101494.
3. Ding, Xuefeng, Qihong Gan, and Mir Pasha Shaker. "Optimal management of parking lots as a big data for electric vehicles using internet of things and Long–Short term Memory." *Energy*, 268 (2023): 126613.
4. Milani, O. H., Nguyen, T., Parekh, A., Cetin, A. E., & Prasad, B. (2023). 0537 Incident Hypertension Prediction in Obstructive Sleep Apnea using Machine Learning. *Sleep*, *46*(Supplement_1), A236-A237.
5. Zand, M., Nasab, M. A., Sanjeevikumar, P., Maroti, P. K., & Holm-Nielsen, J. B. (2020). Energy management strategy for solid-state transformer-based solar charging station for electric vehicles in smart grids. *IET Renewable Power Gener.*, *14*(18), 3843-3852.
6. Piechowiak, M., Zwierzykowski, P., & Musznicki, B. (2023). LoRaWAN Metering Infrastructure Planning in Smart Cities. *Applied Sciences*, *13*(14), 8431.
7. Ghasemi, M., Akbari, E., Zand, M., Hadipour, M., Ghavidel, S., & Li, L. (2019). An efficient modified HPSO-TVAC-based dynamic economic dispatch of generating units. *Electr. Power Compon. Syst.*, *47*(19-20), 1826-1840.
8. Nasri, S., Nowdeh, S. A., Davoudkhani, I. F., Moghaddam, M. J. H., Kalam, A., Shahrokhi, S., & Zand, M. (2021). Maximum Power point tracking of Photovoltaic Renewable Energy System using a New method based on turbulent flow of water-based optimization (TFWO) under Partial shading conditions. In *Fundamentals and Innovations in Solar Energy* (pp. 285-310). Springer, Singapore.
9. Rohani, A., Joorabian, M., Abasi, M., & Zand, M. (2019). Three-phase amplitude adaptive notch filter control design of DSTATCOM under unbalanced/distorted utility voltage conditions. *J. Intell. Fuzzy Syst.*, *37*(1), 847-865.
10. Ehsanifar, M., Dekamini, F., Spulbar, C., Birau, R., Khazaei, M., & Bărbăcioru, I. C. (2023). A Sustainable Pattern of Waste Management and Energy Efficiency in Smart Homes Using the Internet of Things (IoT). *Sustainability*, *15*(6), 5081.

11. Zand, M., Nasab, M. A., Neghabi, O., Khalili, M., & Goli, A. (2019, December). Fault locating transmission lines with thyristor-controlled series capacitors By fuzzy logic method. In *2020 14th International Conference on Protection and Automation of Power Systems (IPAPS)* (pp. 62-70). IEEE.
12. Z. Zand, M. Hayati and G. Karimi, "Short-Channel Effects Improvement of Carbon Nanotube Field Effect Transistors," *2020 28th Iranian Conference on Electrical Engineering (ICEE), Tabriz, Iran*, 2020, pp. 1-6, doi: 10.1109/ICEE50131.2020.9260850.
13. Tightiz, L., Nasab, M. A., Yang, H., & Addeh, A. (2020). An intelligent system based on optimized ANFIS and association rules for power transformer fault diagnosis. *ISA Trans.*, *103*, 63-74.
14. Zand, M., Neghabi, O., Nasab, M. A., Eskandari, M., & Abedini, M. (2020, December). A hybrid scheme for fault locating in transmission lines compensated by the TCSC. In *2020 15th International Conference on Protection and Automation of Power Systems (IPAPS)* (pp. 130-135). IEEE.
15. Zand, M., Nasab, M. A., Khoobani, M., Jahangiri, A., Hosseinian, S. H., & Kimiai, A. H. (2021, February). Robust speed control for induction motor drives using STSM control. In *2021 12th Power Electronics, Drive Systems, and Technologies Conference (PEDSTC)* (pp. 1-6). IEEE.
16. Sanjeevikumar, P., Zand, M., Nasab, M. A., Hanif, M. A., & Bhaskar, M. S. (2021, May). Spider community optimization algorithm to determine UPFC optimal size and location for improve dynamic stability. In *2021 IEEE 12th Energy Conversion Congress & Exposition-Asia (ECCE-Asia)* (pp. 2318-2323). IEEE.
17. Azimi Nasab, M., Zand, M., Eskandari, M., Sanjeevikumar, P., & Siano, P. (2021). Optimal planning of electrical appliance of residential units in a smart home network using cloud services. *Smart Cities*, *4*(3), 1173-1195.
18. Khattak, S. B. A., Nasralla, M. M., Farman, H., & Choudhury, N. (2023). Performance Evaluation of an IEEE 802.15. 4-Based Thread Network for Efficient Internet of Things Communications in Smart Cities. *Applied Sciences*, *13*(13), 7745.
19. Nasab, M. A., Zand, M., Padmanaban, S., Bhaskar, M. S., & Guerrero, J. M. (2022). An efficient, robust optimization model for the unit commitment considering renewable uncertainty and pumped-storage hydropower. *Comput. Electr. Eng.*, *100*, 107846.
20. Azimi Nasab, M., Zand, M., Padmanaban, S., & Khan, B. (2021). Simultaneous long-term planning of flexible electric vehicle photovoltaic charging stations in terms of load response and technical and economic indicators. *World Electric Vehicle Journal*, *12*(4), 190.
21. Zand, M., Nasab, M. A., Padmanaban, S., & Khoobani, M. (2022). Big data for SMART Sensor and intelligent Electronic devices–Building application. In *Smart buildings digitalization* (pp. 11-28). CRC Press.

22. Padmanaban, S., Khalili, M., Nasab, M. A., Zand, M., Shamim, A. G., & Khan, B. (2022). Determination of Power Transformers Health Index Using Parameters Affecting the Transformer's Life. *IETE J. Res.*, 1-22.
23. Sanjeevikumar, P., Samavat, T., Nasab, M. A., Zand, M., & Khoobani, M. (2022). Machine learning-based hybrid demand-side controller for renewable energy management. In *Sustainable Developments by Artificial Intelligence and Machine Learning for Renewable Energies* (pp. 291-307). Elsevier.
24. Khalili, M., Ali Dashtaki, M., Nasab, M. A., Reza Hanif, H., Padmanaban, S., & Khan, B. (2022). Optimal instantaneous prediction of voltage instability due to transient faults in power networks taking into account the dynamic effect of generators. *Cogent Eng.*, *9*(1), 2072568.
25. Asha, A., Arunachalam, R., Poonguzhali, I., Urooj, S., & Alelyani, S. (2023). Optimized RNN-based performance prediction of IoT and WSN-oriented smart city application using improved honey badger algorithm. *Measurement, 210*, 112505.
26. Asadi, A. H. K., Jahangiri, A., Zand, M., Eskandari, M., Nasab, M. A., & Meyar-Naimi, H. (2022, January). Optimal Design of High Density HTS-SMES Step-Shaped Cross-Sectional Solenoid to mechanical stress reduction. In *2022 International Conference on Protection and Automation of Power Systems (IPAPS)* (Vol. 16, pp. 1-6). IEEE.
27. Nasab, M. A., Zand, M., Hatami, A., Nikoukar, F., Padmanaban, S., & Kimiai, A. H. (2022, January). A Hybrid Scheme for Fault Locating for Transmission Lines with TCSC. In *2022 International Conference on Protection and Automation of Power Systems (IPAPS)* (Vol. 16, pp. 1-10). IEEE.
28. Dashtaki, A. A., Khaki, M., Zand, M., Nasab, M. A., Sanjeevikumar, P., Samavat, T., ... & Khan, B. (2022). A Day Ahead Electrical Appliance Planning of Residential Units in a Smart Home Network Using ITS-BF Algorithm. *Int. Trans. Electr. Energy Syst., 2022.*
29. Rastgoo, Samaneh, Zahra Mahdavi, Morteza Azimi Nasab, Mohammad Zand, and Sanjeevikumar Padmanaban. 2022. "Using an Intelligent Control Method for Electric Vehicle Charging in Microgrids" *World Electric Vehicle Journal,* 13, no. 12: 222. https://doi.org/10.3390/wevj13120222.
30. Srinivasulu, M., G. Shivamurthy, and B. Venkataramana. "Quality of service aware energy efficient multipath routing protocol for internet of things using hybrid optimization algorithm." *Multimedia Tools Appl.,* 82.17 (2023): 26829-26858.
31. Seresht, R. M., Miri, M., Zand, M., Azimi Nasab, M., Sanjeevikumar, P., & Khan, B. (2023). Frequency control scheme of an AC Islanded microgrid based on modified new self-organizing hierarchical PSO with jumping time-varying acceleration coefficients. *Cogent Eng.*, *10*(1), 2157982.
32. Padmanaban, S., Samavat, T., Nasab, M. A., Nasab, M. A., Zand, M., & Nikokar, F. (2023). Electric Vehicles and IoT in Smart Cities. *Artificial Intelligence-based Smart Power Systems*, 273-290.

33. Liu, C., & Ke, L. (2023). Cloud assisted Internet of things intelligent transportation system and the traffic control system in the smart city. *J. Control Decis.*, *10*(2), 174-187.
34. Padmanaban, S., Nasab, M. A., Shiri, M. E., Javadi, H. H. S., Nasab, M. A., Zand, M., & Samavat, T. (2023). The Role of Internet of Things in Smart Homes. *Artificial Intelligence-based Smart Power Systems*, 259-271.
35. Ahmadi, M., Rastgoo, S.,Mahdavi, Z., Nasab, M.A., Zand, M., Sanjeevikumar, P.,Khan, B.: Optimal allocation of EVs parking lots and DG in micro grid using two-stage GA-PSO. *J. Eng.*, 2023, e12237 (2023). https://doi.org/10.1049/tje2.12237.
36. Zand, M., Nasab, M. A., Padmanaban, S., Maroti, P. K., & Muyeen, S. M. (2023). Sensitivity analysis index to determine the optimal location of multi-objective UPFC for improvement of power quality parameters. *Energy Rep.*, *10*, 431-438.
37. Sagar, A., Kashyap, A., Nasab, M. A., Padmanaban, S., Bertoluzzo, M., Kumar, A., & Blaabjerg, F. (2023). A Comprehensive Review of the Recent Development of Wireless Power Transfer technologies for Electric Vehicle Charging Systems. *IEEE Access*.
38. Nasab, M. A., Zand, M., Dashtaki, A. A., Nasab, M. A., Padmanaban, S., & Blaabjerg, F. (2023). Uncertainty compensation with coordinated control of EVs and DER systems in smart grids. *Solar Energy*, *263*, 111920.
39. Nikoufard, M., & Hatami, M. (2016). Ultra-short novel transverse magnetic to transverse electric polarization rotator in hybrid integration of InGaAsP/silicon-on-insulator technologies. *Indian J. Phys.*, *90*, 211-217.
40. Gava, Matheus Araujo, *et al.* "Optimizing Resources and Increasing the Coverage of Internet-of-Things (IoT) Networks: An Approach Based on LoRaWAN." *Sensors*, 23.3 (2023): 1239.
41. Padmanaban, S., Samavat, T., Nasab, M. A., Nasab, M. A., Zand, M., & Nikokar, F. (2023). Electric vehicles and IoT in smart cities. *Artificial Intelligence-based Smart Power Systems*, 273-290.
42. Samavat, T., Nazari, M., Ghalehnoie, M., Nasab, M. A., Zand, M., Sanjeevikumar, P., & Khan, B. (2023). A Comparative Analysis of the Mamdani and Sugeno Fuzzy Inference Systems for MPPT of an Islanded PV System. *Int. J. Energy Res., 2023.*
43. Padmanaban, S., Nasab, M. A., Shiri, M. E., Javadi, H. H. S., Nasab, M. A., Zand, M., & Samavat, T. (2023). The role of internet of things in smart homes. *Artificial Intelligence-based Smart Power Systems*, 259-271.
44. Zand, M., Nasab, M. A., Hanif, H. R., Ghahri, M. R., Padmanaban, S., & Khan, B. (2023). A hybrid scheme for optimal performance of photovoltaic system converters with multilayer structure. *The J. Eng.*, *2023*(1), e12218.
45. Padmanaban, S., Nasab, M. A., Samavat, T., Nasab, M. A., & Zand, M. Securing Smart Power Grids Against Cyber-Attacks. In *IoT and Analytics in Renewable Energy Systems (Volume 1)* (pp. 17-36). CRC Press.

46. Padmanaban, S., Nasab, M. A., Samavat, T., Zand, M., Nasab, M. A., & Hashemi, E. Cyber Security in Smart Energy Networks. In *IoT and Analytics in Renewable Energy Systems (Volume 2)* (pp. 309-325). CRC Press.
47. Padmanaban, S., Nasab, M. A., Samavat, T., Zand, M., & Nasab, M. A. Artificial Intelligence Techniques for Smart Power Systems. In *IoT and Analytics in Renewable Energy Systems (Volume 1)* (pp. 107-123). CRC Press.
48. Nikoufard, M., & Hatami, M. (2018). Analysis of ultra-compact TE to TM polarization rotator in InGaAsP and SOI technologies. *Optik-International Journal for Light and Electron Optics, 153*, 9-15.
49. Ghazal, Taher M., *et al.* "Internet of Things Connected Wireless Sensor Networks for Smart Cities." *The Effect of Information Technology on Business and Marketing Intelligence Systems.* Cham: Springer International Publishing, 2023. 1953-1968.

8

Bio-Based Fuel Production, Environmental Impact, and Economic Assessment

Kavinkumar Ravikumar[2], Nicoletta Ravasio[1*] and Milind Shrinivas Dangate[2†]

[1]National Research Council-Institute of Chemical Sciences and Technology (CNR-SCITEC) "G. Natta", Milano, Italy
[2]Chemistry Division, School of Advanced Sciences, Vellore Institute of Technology, Chennai, India

Abstract

The development of sustainable and environmentally acceptable alternative energy sources has been driven by growing environmental concerns over the effects of using fossil fuels for transportation. Biomass-derived liquid hydrocarbon fuels offer a convenient replacement for petroleum-derived transportation fuels, but their carbon and energy content pose challenges. The US is expected to have 1.3 billion tons of harvestable biomass by 2030, but this amount is insufficient for the transportation industry alone. Traditional bioenergy technologies, such as cellulose fermentations to ethanol, fail to use lignin, which accounts for around 40% of biomass energy, and lose one-third of the biomass carbon to CO_2. This necessitates the development of biomass conversion technologies that capture the intrinsic biomass energy in the liquid fuels while retaining the majority of biomass carbon. One approach is to quickly pyrolyze all biomass, including lignin, to produce mostly liquid bio-oil. However, this unstable bio-oil is unsuitable as a fuel or stable intermediate due to its low energy content. Electrocatalytic hydrogenation (ECH) can be used to create renewable power from solar and wind farms, allowing for hydrogenation at a nearby small-scale facility. This approach offers a crucial benefit in bioenergy systems where transporting low bulk density biomass may result in high transportation costs. This paper discusses the creation of a bioenergy system [pyrolysis-ECH (Py-ECH)] that combines quick pyrolysis with ECH in decentralized depots, followed by hydro processing at a central refinery. Estimates of the system's mass, carbon, and energy fluxes were made, and the fuel yields for

**Corresponding author*: nicoletta.ravasio@scitec.cnr.it
†*Corresponding author*: milind.shrinivas@vit.ac.in

O.V. Gnana Swathika, K. Karthikeyan, Milind Shrinivas Dangate and Nicoletta Ravasio (eds.) Biomass and Solar-Powered Sustainable Digital Cities, (123–146) © 2024 Scrivener Publishing LLC

the Py-ECH systems were superior to cellulosic ethanol in terms of energy, mass, and carbon.

Keywords: Biomass, pyrolysis, cellulosic fermentations to ethanol (CE), electrocatalytic hydrogenation (ECH), energy

8.1 Motivation

As a result of climate change, nowadays, it is commonly acknowledged that severe reducing the greenhouse gas (GHG) and CO_2 emissions and others from manmade activities will be required to prevent significant economic and social upheaval. Because automobiles must have their energy supply with them, reducing CO_2 emissions from transport is far more difficult than from fixed uses. Therefore, when liquid hydrocarbons reign, specific and volumetric energy density are strongly sought. Gasoline transports about 47 times more energy per unit mass [44.4 MJ kg^{-1} vs. 0.95 MJ kg^{-1} on a higher heating value (HHV) basis], and when containers and insulation are taken into account, about five times that of liquid hydrogen. This is in comparison which are shown in Figure 8.1 to the most technologically advanced lithium battery currently available on the market [1]. A renewable liquid hydrocarbon fuel is greatly desired due to these features as well as compatibility with current infrastructure and vehicles. For the year 2015, the estimated annual energy consumption for transportation in the United States is 29 EJ. This number is unlikely to change much until 2030, with a growth rate of 0.1% annually for transportation usage of energy for the years 2012–2040. A specific energy of 48 MJ kg^{-1} (HHV) is equivalent to almost 0.6×10^9 tons of petroleum.

In this regard, it is important to keep in mind that the carbon in biomass can only be used in a small portion of the conventional biofuel manufacturing processes. By taking into account the generation of cellulosic ethanol (CE), this carbon loss is demonstrated. Although biological and thermochemical methods of converting biomass to ethanol have made significant strides lately, their inherent stoichiometry still loses at least one-third of the carbon in the form of CO_2 while concentrating the energy in the feedstock into a smaller portion of the initial mass. A portion of the biomass used in most of these biofuel techniques is also burned, producing extra CO_2 waste.

However, the yearly amount of carbon provided by biomass is presently less than that used in fuels. Therefore, carbon must be acknowledged as the primary limiting element rather than the energy level of the

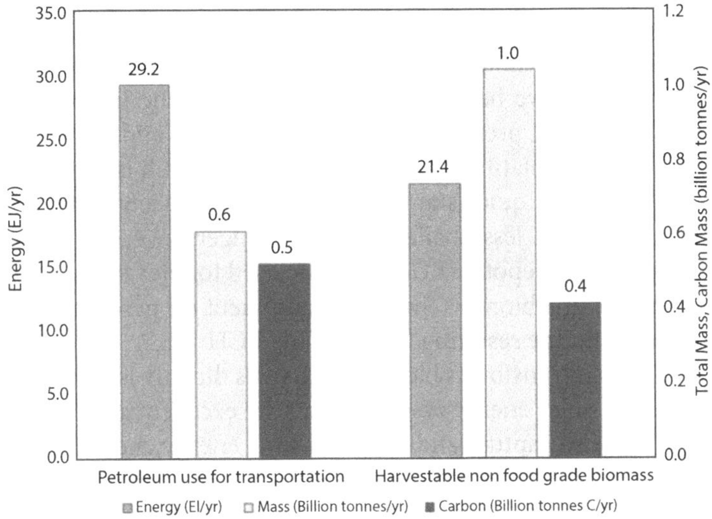

Figure 8.1 Comparison of the predicted harvestable biomass and the yearly energy balance for the US transportation industry. Based on the biomass that will be accessible by 2030 (U.S. Billion Ton Study 2016), projections are generated. Petroleum has a specific energy of 48 MJ kg^{-1}, whereas dry biomass has a specific energy of 20.6 MJ kg^{-1}. Exajoules (EJ), which are 1,018 joules (J), are used to measure energy. A "billion tons" of mass is equal to 10^{12} kg.

feedstock. Thus, the goal of fuel conversion from biomass should be to retain the most carbon. This increases the appeal of a fuel method that mixes biomass and renewable power. Given the scarcity of biomass carbon, it is crucial to develop biomass-to-biofuel conversion techniques that increase energy content while reducing carbon loss. Pyrolysis–electrocatalytic hydrogenation (Py-ECH) is a case in point of this method. Fast pyrolysis is a thermal technique that transforms biomass into bio-oil by rapidly heating the material without oxygen [1, 2]. The feedstock has an impact on the complicated composition of bio-oil. Organic compounds and roughly 20 weight percent of water are among the several lower carbon number organic molecules that it often includes. Because of these reactive oxygenation moieties, bio-oil is susceptible to polymerization, which raises viscosity, even when stored at room temperature. Its low energy content is equivalent to that of the initial biomass. Deoxygenation and stability of bio-oil can be accomplished by a process known as hydrodeoxygenation (HDO), which is utilized for petroleum upgrading. Such a procedure may benefit from the current petroleum infrastructure [2, 3]. However, it would be expensive and energy inefficient to transport biomass to a far-off central location for pyrolysis and upgrading. Furthermore, such transportation is troublesome due to the previously noted volatility of untreated bio-oil. To date, two popular methods—dilution in the organic solvent and store at low-temperature—have been suggested to lengthen the life of bio-oil in storage. Small-scale H_2 production has very high costs, and the necessity to handle explosive or volatile gases raises security issues. It is recommended that ECH be used as a quick stabilizing procedure adjacent to the biomass pyrolysis sites. It is a less challenging and dangerous option. The term "biomass upgrading depots" (BUDs) will be used to refer to these regional depots for processing biomass that have equipment for pyrolyzing the biomass and maintain the resulting bio-oil using ECH [4].

Any device that transforms biomass into fuels directly is basically a carbon capture and solar energy system. Plants are excellent at fixing carbon, but they struggle to capture and store the sun's energy chemically. This is because there is no evolutionary pressure for plants to retain more energy than is required for their development and reproduction. The majority of plants in nature only use around 1% to 2% of the incident solar energy for photosynthetic purposes, losing the remaining 98%–99% through optimum wavelength mismatch, respiration, reflection, and transmission [12]. Compared to plants, human technologies are far more effective at capturing solar energy, but they do it by using electricity, which is difficult to store. The majority of the lower-end photovoltaic cells that are sold commercially have an energy efficiency of 12%–15% and are manufactured of single crystalline silicon [6]. The replacement of fossil fuels with renewable fuels, especially liquid fuels, is a national priority. By increasing the energy content of biomass with renewable power, carbon is used more efficiently, resulting in higher yields of fuels made from biomass. The method employs traditional HDO at quite extreme pressures and temperatures, which consumes H_2 gas from water electrolysis. The direct electrocatalytic reduction system (at the BUD) described in this paper, on the other hand, functions at almost ambient temperatures without the utilization of high pressures or H_2 gas.

8.2 Methods

A quick and easy approach for stabilizing bio-oil made from the rapid pyrolysis of biomass feedstocks is the ECH scheme, which is covered in the sections above. After stabilization, the bio-oil can be sent to larger petroleum refinery complexes for further upgrading. As a result, the conversion of biomass consists of a single phase called ECH reduction. It is essential in this case to adopt a complete strategy and investigate the impacts of including ECH in the entire biomass conversion technique. This study uses ECH to upgrade the light while simulating a complete biomass conversion [5]. The raw feedstock is ground and dried in the procedure, which is now known as the Py-ECH process, and a fuel that resembles gasoline is produced at the end. Biomass rapid pyrolysis (BFP), electrochemical hydrolysis (ECH), and hydro processing are the main techniques used.

8.2.1 Cellulosic Ethanol (CE) by Fermentation

The procedure outlined in National Renewable Energy Laboratory (NREL)'s 2011 report is followed by the CE strategy taken into consideration in this study [11]. In addition to the primary end product, ethanol, which is produced, power is also produced by incinerating the non-fermentable materials. In actuality, the power produced meets a significant portion of the process's energy needs. NREL used Aspen software to complete the whole design for the CE procedure. As a result, for the purposes of our study, all information needed for the analysis was taken from this paper. The non-fermentable components are incinerated, yielding power as a by-product in addition to the primary end product, which is ethanol.

8.3 Results

Sankey diagrams, which are shown in Figure 8.2, are used to illustrate the energy assessments for the two renewable fuel generating systems. The Py-ECH (as shown in Figure 8.4) system was balanced on a part-by-part basis, and the CE system was balanced on a system-wide basis. According to the study, the quantity of fuel energy generated by the two systems differs noticeably, with Py-ECH (as shown in Figure 8.5) producing around twice as much fuel in the form of energy as CE with the same quantity of input biomass. Energy is lost throughout each step of CE synthesis; however, fermentation causes the greatest reduction in the initial fuel stream. Final ethanol contains 44% of the energy (HHV) from the starting biomass.

As a consequence, the Py-ECH system uses around 89% of the biomass energy to make fuel products from hydrocarbon, losing less energy during the synthesis of bio-oil and getting it back during the ECH and hydro processing. These variations in fuel energy production are due to variations in usage of electricity, loss of heat, and surplus electricity generation. The Py-ECH approach is less efficient than the CE procedure in terms of energy output. To offset the positive effects of the extra electricity produced during

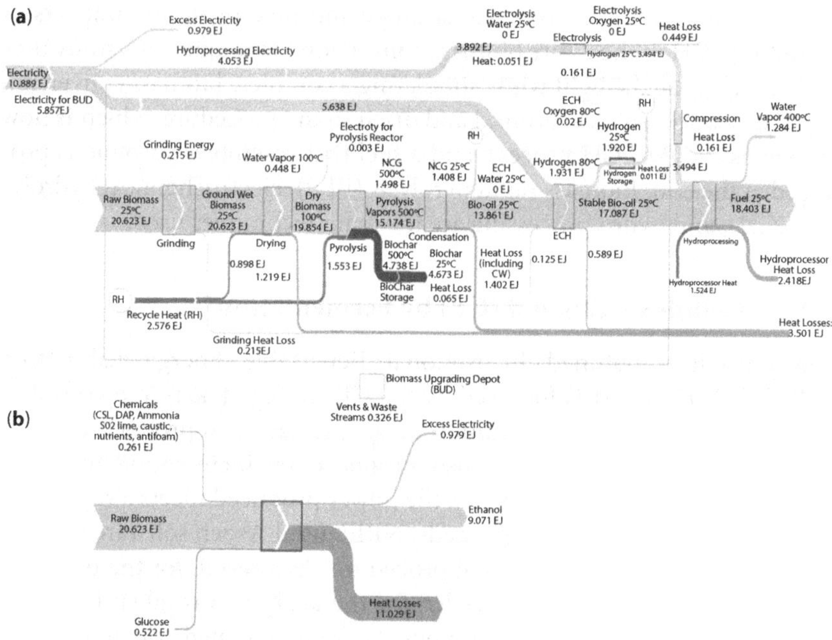

Figure 8.2 Sankey diagrams for energy flow. (a) Energy analysis of Py-ECH with the yield of 89%; (b) energy analysis of CE with the yield of 44%.

the CE process, a 0.979-EJ electrical power output was introduced to the Py-ECH process (as can be seen in Figure 8.2a).

While the CE system consumes zero electricity, the Py-ECH system needs 10.9 EJ. The conversion of renewable energy into chemical energy, which is stored in the fuel generated, improves the Py-ECH system. About half of the losses from the CE process are caused by net heat losses. Both the CE and Py-ECH processes generate enough heat on their own to function without the need for external heat sources. Reheating stable bio-oil in the central refinery using the heat from the hot water vapor stream eliminates the requirement for external fuel burning. The yield of fuel energy can be raised by introducing further process modifications. The process's power requirements can be reduced by using the combined combustion energy of the NCG and H_2 streams. To sequester carbon, biochar can be spread on the ground. The CE method is anticipated to be more sensitive to the kind of biomass, especially in regard to lignin content.

The air need for burning NCG and H_2 for heat integration as well as any subsequent combustion by-products have been excluded from the analysis shown in the Sankey diagram for the Py-ECH process. Because the C-C and C-H bonds in hydrocarbon fuels serve as the fuel's energy carriers, carbon must be present in the finished fuel [7]. Another study found that 98.4% of the sugar that was initially ingested may be recovered via acid hydrolysis. Additionally, char and non-condensable gases (NCGs) are two additional forms of carbon loss in Py-ECH. However, it is possible to heat up the released NCG (approximately 12%).

Due to electrolytic hydrogen's high specific energy and low molecular weight, both of these processes require adding it to the pyrolysis bio-oil. These characteristics make electrolytic hydrogen good for upgrading because of its low molecular weight. Stabilized bio-oil contains much more energy than raw bio-oil while maintaining an approximately same mass. On other words, compared to the raw bio-oil, the stabilized bio-oil has more energy as it leaves the BUD. In reality, as shown in Figure 8.3, ECH completes around 71% of the overall upgrade before leaving the depot. Additionally, the fuel's whole energy content would have come from renewable sources if the power was generated from them. Although electrocatalysis is less effective at deoxygenating bio-oil than conventional hydro processing, electrocatalysis following pyrolysis at local BUDs may significantly increase the energy and financial value of bio-oil under benign conditions.

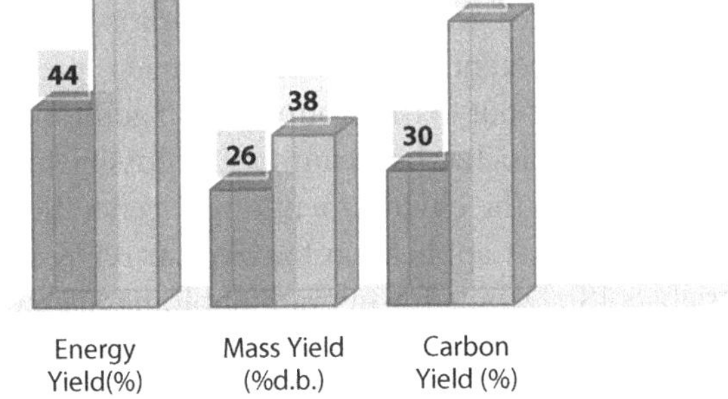

Figure 8.3 Comparison of the yields of energy, mass, and carbon of Py-ECH and CE. The use of electrical power considerably increases the energy production.

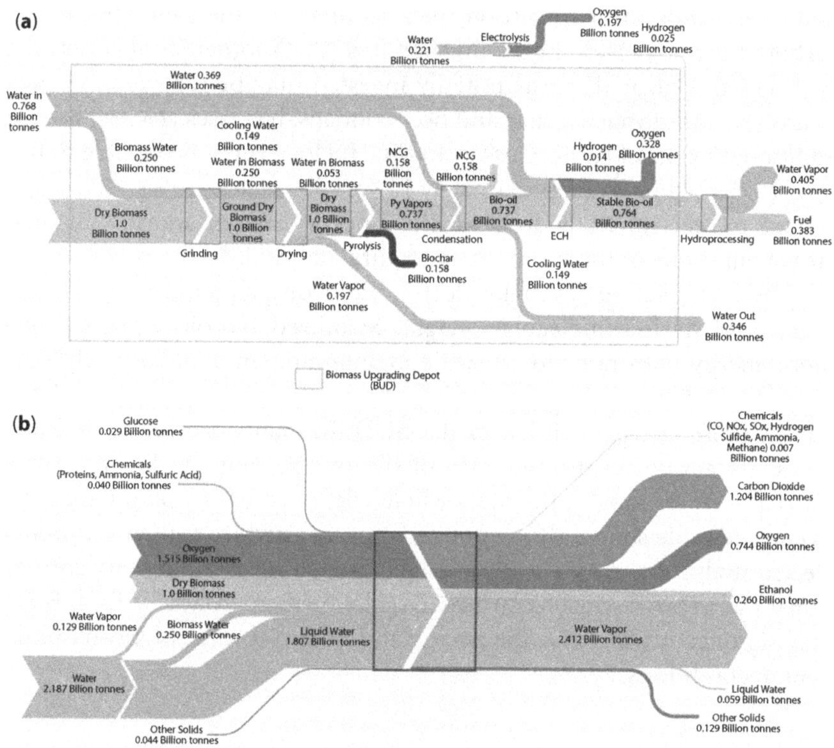

Figure 8.4 Sankey diagrams for mass flow. (a) Mass analysis of Py-ECH showing the yield of 38%. (b) Mass analysis of CE with the yield of 26%.

8.4 Discussion and Outlook

A thorough life-cycle evaluation is required to determine the environmental advantages of the Py-ECH process with and without the application of biochar to the soil, even though it generates less carbon dioxide than the CE process. It is notable that the potential for climate change is greatly diminished when co-product biochar is put to the ground [8]. In addition to counteracting the greenhouse effect, biochar has a number of possible uses, such as the removal of pollutants from the environment or the generation of stationary electricity. However, the best application for biochar relies on a number of site-specific variables [8]. Depots for upgrading biomass are essential components of the Py-ECH idea. A huge refinery complex's economies of scale need be balanced with depots' logistical advantage in order for this technology to be used commercially. Therefore, technoeconomic and life-cycle evaluations should be carried out, along with system-wide optimization [9].

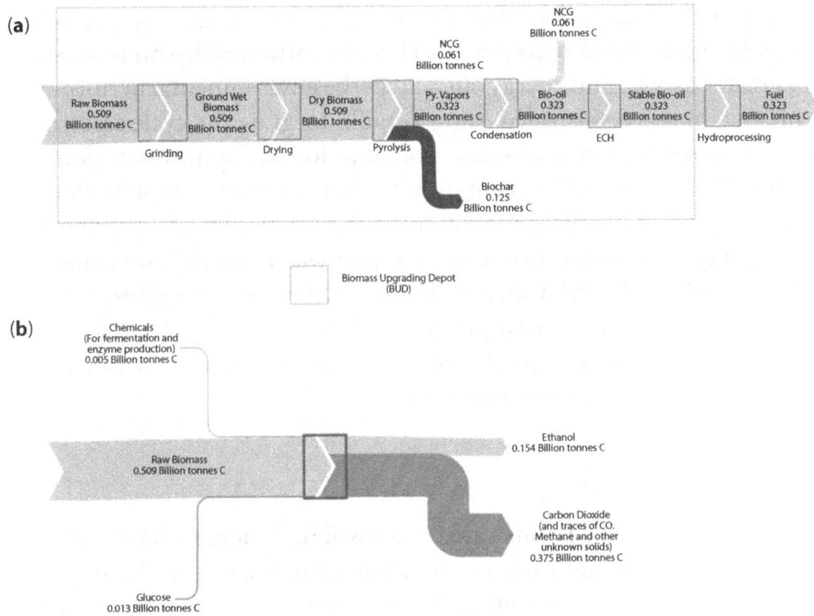

Figure 8.5 Sankey diagrams for carbon flow. (a) Carbon analysis of Py-ECH with the yield of 63%; (b) cellulosic ethanol (CE) carbon analysis with the yield of 30%.

8.5 Conclusions

The distributed Py-ECH approach has the capacity to extend the life of bio-oil as a sustainable alternative fuel. This process is more efficient and less expensive than CE because it retains twice as much carbon and consumes less water. As the amount of changeable renewable energy sources rises, the distributed conversion units flexible electrical load may become increasingly important for system management. Future energy systems devoid of carbon dioxide and fossil fuels may be built on the synergies of biomass, liquid hydrocarbon fuel, and renewable power.

8.6 Life Cycle Assessment

8.6.1 Abstract

It is essential to quantify the environmental effects of bioenergy systems if fossil-based systems are to be replaced. The most accurate approach for calculating these consequences is life cycle evaluations. In comparison to conventional cellulosic biorefineries, Py-ECH bioenergy systems, which employ localized rapid pyrolysis, ECH, and centralized hydro processing, offer greater carbon and energy efficiency. The environmental performance of this system was determined using a cradle-to-grave life cycle assessment (LCA) in three impact categories: potential for eutrophication, potential for global warming, and potential for water shortages. Results demonstrated that, when compared to CE biorefineries, the Py-ECH system had much reduced eutrophication potential and water scarcity footprint. The study underlines the relevance of finding renewable energy sources for electricity in order to achieve positive global warming potentials as well as the significance of the yearly carbon sequestration rate in calculating system-wide global warming potential.

8.6.2 Introduction

It has frequently been demonstrated that fossil fuel energy, such as the variety of liquid fuels made from crude oil or natural gas, significantly contributes to air and water pollution. Because the pace of filling is far slower than the rate of usage, such fuels are not renewable. Finding less-polluting, sustainable alternative energy generating techniques is necessary.

The current work advances by using LCA to compare the environmental implications after having previously established a roadmap for the Py-ECH system's thermodynamic and financial favorability in contrast to the CE system. This experiment was done to preserve identical assumptions for the two systems under discussion, despite the fact that several CE LCAs have previously been carried out using various biomass feedstocks. The present comparative environmental assessment of the Py-ECH and CE systems is a crucial step toward the commercial application of the Py-ECH system as a workable alternative for fossil fuel systems. It should be noted that biomass pyrolysis, which is improved using hydrogen gas from diverse sources, has also undergone life cycle analyses. These analyses have demonstrated the superiority of pyrolysis upgrading systems over fossil fuel systems. A road to carbon-negative cellulosic biofuels was also revealed by a recent life cycle study of a depot-based bioenergy system [10]. However, the Py-ECH system combines a decentralized system approach with moderate ECH to upgrade biomass pyrolysis.

8.6.3 Methodology

The Py-ECH and the CE processes underwent a comprehensive "cradle-to-grave" LCA for three environmental effect categories: the potential for global warming, the potential for eutrophication, and the imprint of water shortages. In order to compare the carbon footprint of the two systems, the energy return was also calculated. The life cycle inventory was created using spreadsheets, incorporating pertinent variables from recognized models like The Greenhouse Gases, Regulated Emissions, and Energy Use in Transportation (GREET) and Carbon Calculator for Land Use and Land Management Change from Biofuels Production (CCLUB), with process flow information from the work of Lam *et al.* [12].

8.6.3.1 *Life Cycle Impact Categories*

8.6.3.1.1 Eutrophication Potential (EUP)

The production of biomass, a crucial component of all bioenergy systems, is linked to the use of NPK fertilizers. These fertilizers are a significant contributor to eutrophication, which is described as the exponential development of cyanobacteria and algae due to an excess of nutrients in the environment that affects marine ecosystems. Using the TRACI Model, which assumes little volatility in the characterization parameters for the eutrophication potentials for the two systems over the 20-year and 100-year timescales were determined. While the transit factor takes into consideration the emissions' release into various media (such as air or water) before they eventually reach aquatic systems, the potency factor measures the impact of a specific nutrient. The following gives the expression used to calculate the Eutrophication Potential (EUP):

$$EUP_i = e_i . CF_{air}/Pat: r$$

$CF_{air/Pat:r}$ is the TRACI characterization factor for substance and, in air or water medium, e_i is inventory data of substance (in kg).

8.6.3.2 *System Definition*

Depending on where system boundaries are placed, the outcomes of an LCA analysis might differ significantly. For the sake of this analysis, the growing of the maize plant (to create stover) at the front end and the burning of the produced hydrocarbon fuel at the rear end define the limits of the Py-ECH system. Similar limits are established for the CE system. Figures 8.6 and 8.7, which depict the two limits of the two bioenergy systems, are used to visually represent this. The main elements of the entire system are shown in various hues, including the fuel combustion process (black), the transportation network (purple), the supply chain (green), and processing at refineries (dark blue) or depots (light blue).

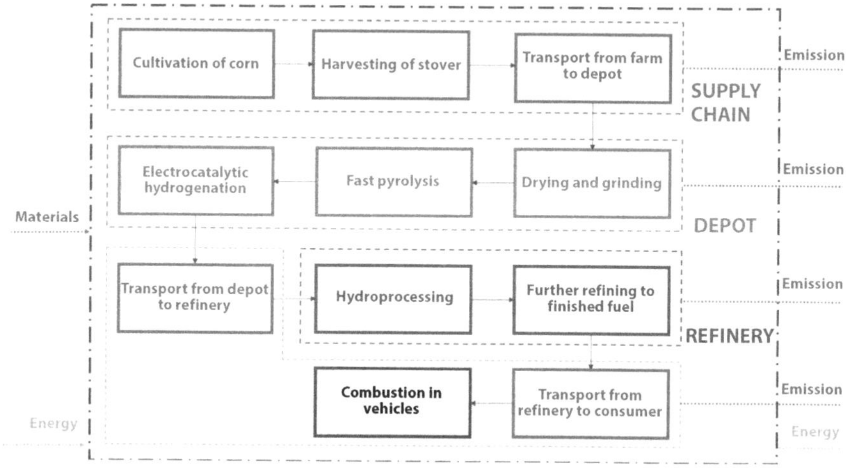

Figure 8.6 System boundaries for Py-ECH system.

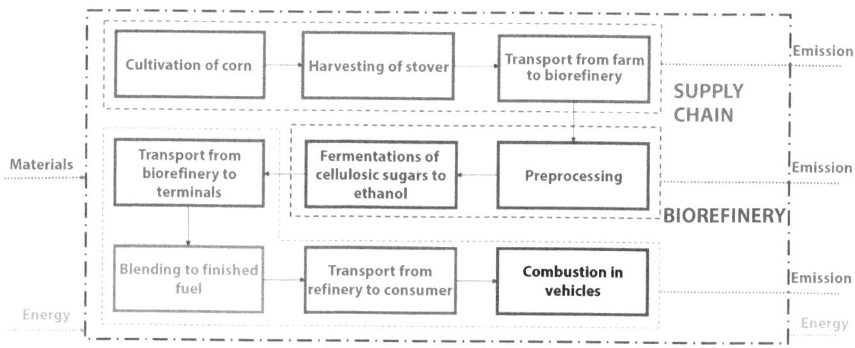

Figure 8.7 System boundaries for CE system.

8.6.3.3 Life Cycle Inventory (LCI)

The study by Lam *et al.* served as the basis for the process flow data for the Py-ECH system [12]. The GREET and CCLUB models from Argonne National Laboratories were used to determine all-important LCI data. NREL, Environmental Protection Agency (EPA), and other reputable data sources were used to extract additional pertinent data that were missing from these models. The modified Weidema technique was used to give data quality indicators to the collected inventory in order to qualify it. Toffel *et al.* changed the names of the completeness and temporal correlation parameters to "representativeness" and "data age," respectively, in 2004. To more accurately depict the dependability of the data, they divided the reliability parameter into characteristics related to the data collecting process and the independence of the data source [13].

8.6.3.3.1 Supply Chain

The supply chain consists of mainly two sub-processes, namely, the cultivation of the corn and the collection of the generated stover.

8.6.3.3.2 Cultivation of Corn

With the premise that corn is cultivated in an ongoing growing process without tillage, the research focuses on the cultivation of maize using the GREET and CCLUB models. Corn-growing terrain is said to have previously been utilized as pasture or cropland. The amount of stover removal has a considerable impact on the final GWP and EUP readings. Only 60% of the harvested corn stover is taken out of the fields, with GREET giving customers the option between 60% and 30% by weight. To enhance harvest production, the option of removing 60% of the stover was chosen. The CCLUB model calculated that carbon arrangement is 0.174 Mg C/ha/year for a 60% removal, assuming a linear relationship on stover removal. GREET accounts for emissions from production and soil application. Data on soil carbon sequestration is dependent on a number of variables, such as crop rotation techniques, assessments of the depth of the soil, rainfall, tillage, stover removal, and soil texture. Alvarez has thoroughly examined the values range and their relationship to these factors. A sensitivity analysis was conducted to determine the effect of the annual carbon sequestration rate on the global warming potential. Using the GREET assumption for the yearly carbon sequestration rate, Figure 8.8 illustrates the carbon flow for the CE and Py-ECH systems.

Once more, the GREET model was used to determine how much water is needed for agriculture. Three types of water may be distinguished: blue water, green water, and gray water, which is used to absorb the contaminants released into freshwater bodies. Blue water includes surface and ground water. Green water is related with precipitation. GREET exclusively counts blue water consumption for crop cultivation, which is the volume of freshwater utilized for anthropogenic activities but not recycled owing to evapotranspiration or deterioration in quality. Due to the fact that it has no impact on blue water use, green water use in agriculture may be overlooked. The fundamental presumption is that crop cultivation had no impact on how much water was consumed by green vegetation on the land. In fact, it is also stated in the literature that when evaluating emissions owing to changes in land use, the impacts of green water consumption are best addressed as

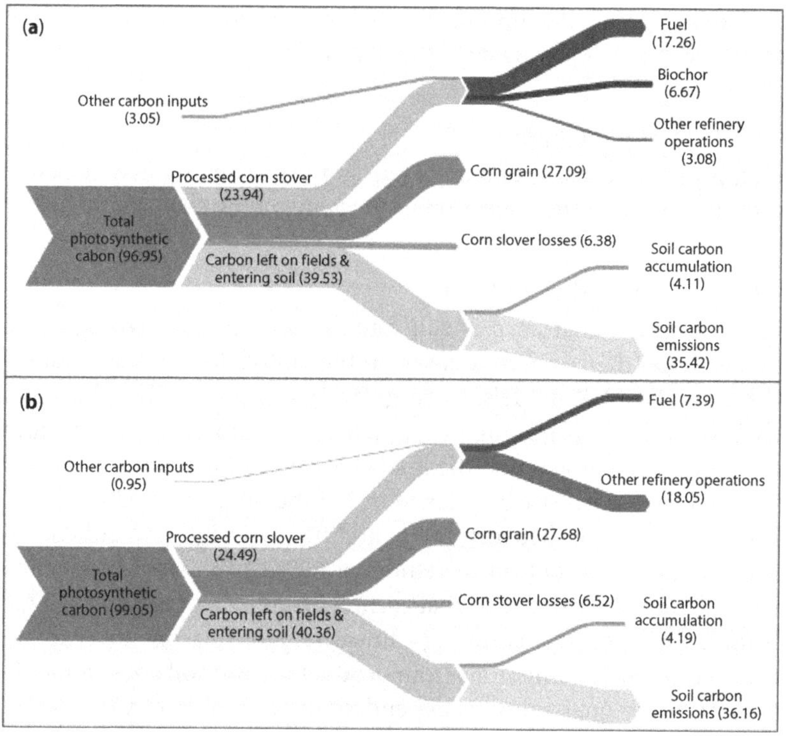

Figure 8.8 System carbon flow for (a) Py-ECH and (b) CE, for GREET assumed value of carbon sequestration.

a contributing element. Gray water has not been taken into account either because its environmental effects in this study overlap with eutrophication and other impact categories. The atmospheric emissions of nitrogen oxides (NOx), which fertilizer manufacture and application contribute to throughout the cultivation stage, were taken from GREET. From Intergovernmental Panel on Climate Change (IPCC) data, ammonia emissions caused by the volatilization of some of the applied fertilizers were calculated.

All three fertilizer nutrients (N, P, and K) were taken into consideration. Information on emissions resulting from fertilizer runoff of N and P into water resources was collected from a report released by NREL in 2005. Given that runoff to groundwater was assumed to be low based on geographic factors, study of N runoff to surface water only took this into account [14]. In accordance with the GREET assumption, the figure for N surface runoff (as nitrates) was set at 24% of the total N added. Only surface water is contaminated by phosphorus as a result of runoff. Due to P's high sorption to soil minerals, there are very little amounts of P in the atmosphere and no impact to ground water contamination [15]. The amount of phosphorus that was applied as fertilizer was found to range from 1% to 14%, with the average amount of P discharge (as phosphates) to surface water bodies being calculated at 7%. According to the paper, it is also expected that potassium does not partition to air or water and has a negligible effect on eutrophication of the two.

8.6.3.3.3 Collection of Stover

The windrowing, baling, and collection processes are the three fundamental steps in the harvesting of stover. In order to prevent accumulating extraneous material with the stover feedstock, a stalk chopper/windrower was used. Since square bales can be carried more readily, square balers are expected for baling. Additionally, square bales offer the benefit of a larger bulk density, which lowers the cost of transportation. Tractors with bale forks, which can handle a single bale at a time, are used for collecting. The total amount of fuel needed for the whole harvest of stover, excluding grain, was calculated from GREET at 3.58 gallons of diesel per acre. All collecting equipment was assumed to run on diesel. From GREET, the emissions brought on by using diesel were also calculated.

Stover collection emissions are computed after taking into account losses of 8.4%, 2%, and 2% of dry stover from storage, shipping, and farm handling, respectively. The breakdown of this lost biomass results in emissions for the CE processes that are estimated to vary from 2.3 to 8.4 g CO_2 e/MJ depending on the temperature and moisture content of the feedstock [16]. The CE method was therefore projected to produce 5.35 g CO_2 e/MJ on average. A value of 2.3 g CO_2 e/MJ for Py-ECH was calculated based on the carbon content of the feedstock biomass for the two processes.

8.6.3.3.4 Processing

The work by Lam *et al.* was used to obtain LCI data for the Py-ECH process's constituent parts [12]. In a similar vein, the NREL paper by Humbird *et al.* on cellulosic sugar fermentations to ethanol served as the source of all the data for the CE system [11]. Using information from the literature, it was determined that the resident carbon percentage of biochar that is buried in the soil ranges from 65 to 100 weight percent. In order to conduct this study, a mean value of 82.5% was used. Additional soil advantages of biochar application include a decrease in the need for fertilizer, a decrease in NOx emissions, and a decrease in the leaching of nutrients from the soil, all of which reduce emissions. These reductions, however, are challenging to predict and were not taken into account in our investigation.

The Py-ECH system needs heat at the central refinery. The combustion of natural gas, which has a heating value of 52.2 MJ/kg, was considered to be the source of this heat. While related NOx emissions from the combustion of natural gas at the Py-ECH refinery were calculated using GREET, carbon dioxide emissions from the combustion of natural gas were measured stoichiometrically from methane combustion. The CE system does not need any outside heat or power since it burns wastewater treatment sludge and biomass lignin to provide enough heat and energy for all plant utilities. In reality, selling extra power to the grid results in a credit. Using a value of 0.31 kg/MWh of the heating value of combusted fuel, the NOx emissions resulting from the combustion of the lignin and sludge at the CE biorefinery and the NCG at the Py-ECH depots were estimated [11].

8.6.3.3.5 Combustion

For the entire combustion of ethanol and all hydrocarbons to CO_2 for the CE and Py-ECH processes, respectively, the emissions for completed fuel combustion were computed. GREET was used to assess the NOx emissions that affect the potential for eutrophication.

8.6.4 Results and Discussion

8.6.4.1 Contribution Analyses

8.6.4.1.1 Global Warming Potential (GWP)

The GWP was computed and contrasted for each system and allocation scenario because the rate of global climate change is increasing [13]. GWP contributors for the two systems for both scenarios are highlighted in Figure 8.9 to demonstrate the contributions of various system parts. These contributors are mentioned in the plot legend. The Py-ECH system is shown for both the scenario where upgrading takes place during ECH and hydroprocessing using fossil fuels and the scenario where upgrading takes place during ECH and hydroprocessing using renewable electricity in this graphic. When using energy generated from renewable sources, the Py-ECH system outperforms the CE system in terms of GWP.

Because of the Py-ECH system's high-power consumption, the significance of clean electricity is highlighted. Given that it generates less energy per unit of biomass, the CE system needs more biomass feedstock to provide the equal amount of fuel energy. The biogenic carbon fixation by the feedstock, however, helps the CE system more. While there are higher emissions created during the production, synthesis, and application of

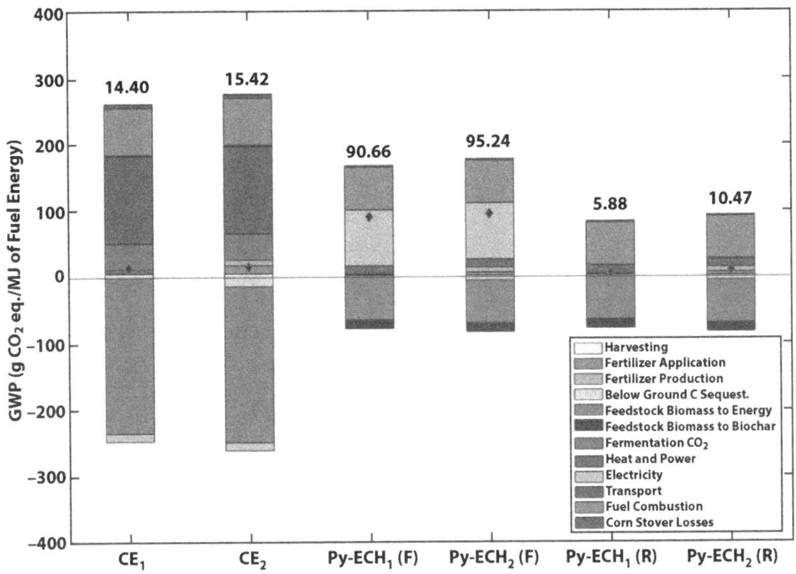

Figure 8.9 Global warming potential (GWP) contribution analysis. Subscript 1 refers to Scenario 1, whereas Subscript 2 refers to Scenario 2. "F" refers to a fossil fuel electrical grid, and "R" refers to renewable power.

fertilizers, these emissions are balanced by an increase in the quantity of feedstock carbon that is fixed. Four further sub-components make up the processing parts of the GWP contributions of the two systems. Burning lignin and wastewater sludge is the first sub-component, and it is this system that produces the most heat and power.

The required power is provided by the grid, and the use of fossil fuels is eliminated by selling surplus power to the grid. The Py-ECH system does not contain the third processing sub-component, carbon dioxide generated by the fermentation of holocellulose sugar. The Py-ECH system generates grid-supplied excess energy as well as biochar that is applied to soil. When applied to the soil, biochar sequesters carbon, although neither method includes a transit phase. For both systems, emissions from burning fuel are essentially equal. As employing mass allocation marginally raises all values, there is minimal difference between the various allocation scenarios inside a single system.

8.6.4.2 *Energy Ratio*

There are a number of alternative energy ratios that may be calculated to assess the system's effectiveness and renewability in terms of renewable

Table 8.1 Description of energy ratios [17].

Energy ratio	Abbreviation	Description
Total Energy Ratio	ERt	Total usable energy output and input
Renewability Factor	RF	Fuel energy output/fossil energy input
Energy Yield	Ey	Fuel energy/feedstock energy
Energy Return on Investment	EROI	Total energy of fuel and co-products/energy
		invested in the process
Fossil Energy Ratio	ERf	Fuel energy + co-product energy output
		Fossil energy input

energy generated compared to fossil energy used (Table 8.1) [17]. For the two distinct processes and the Py-ECH scenario with renewable power, five such energy ratios have been determined and are shown in Table 8.2.

Unlike the Ey, which only analyzes the amount of feedstock energy contained in the primary product, the fuel, the ERt evaluates productivity of the whole thing and takes both the final products and other by-products into account. As a result, the ERt accounts for both the excess electricity in the CE system as well as the excess hydrogen gas in the Py-ECH system.

Biochar is put to the soil to capture carbon; hence, the Py-ECH method does not classify it as an energy co-product.

Table 8.2 Energy ratios for the CE and Py-ECH systems. ERt stands for total energy ratio.

Energy ratio	CE	Py-ECH (FE)	Py-ECH(RE)
ERt	0.44	0.70	0.70
RF	9.66	1.98	8.22
Ey	0.42	0.91	0.91
EROI	10.43	1.85	1.85
ERf	10.40	2.42	10.03

EROI is defined similarly to ERt, with the distinction that it excludes the energy related to the input of the biomass feedstock. It is more focused on the extra energy inputs required for producing the fuel and by-products, which are necessary for the process. However, ERt, Ey, and EROI do not account for the system's renewability and do not differentiate between different energy sources used as inputs. Although there are alternative methods, like Energy Renewability Efficiency, RF is the most straightforward and transparent approach to assess a system's renewability. It is the ratio of all non-renewable energy inputs to the energy of the primary output, in this instance, the fuel. The system's renewability increases with increasing RF. The ratio between all inputs of fossil energy and all outputs of energy (fuel and by-products) is shown by ERf on the other side. According to Table 8.2, the CE system outperforms the Py-ECH system in terms of RF, ERf, and EROI. By burning part of its biomass feed, the CE system will be able to produce some of its own heat and power, significantly reducing the demand for non-renewable inputs and raising RF and ERf. The CE system performs better than the Py-ECH system in terms of EROI because it uses some of the biomass feed as heat and electricity instead of counting it as an input like the Py-ECH system does. Because the Py-ECH depends on grid power, RF, ERf, and EROI are reduced. However, the Py-ECH outperforms the CE system if the energy ratios incorporate biomass feed as an energy input because it is more energy efficient overall and biomass feed energy makes up the majority of all energy inputs. As a result, the ERt and Ey are higher in the Py-ECH. The RF and ERf of the Py-ECH system significantly rise while utilizing renewable power as a result of the significantly reduced input of fossil energy. However, the system's ERt, Ey, and EROI remain unchanged.

Renewability factor, energy yield, energy return on investment, and fossil energy ratio are all abbreviations for the same concept. Whereas RE stands for renewable power, FE stands for fossil electricity. The use of fossil grid power and the central refinery's burning of natural gas for heat are the causes of the Py-ECH system's lower RF and ERf. The RF and ERf of the Py-ECH system significantly rise if the grid energy is more renewable and the central refinery's heat source is made renewable, as shown in Figure 8.10. Because the biomass lignin is burned for both heat and electricity, it is obvious that the CE system would have a high RF of roughly 10 due to the little amount of fossil energy input. Figure 8.10a depicts the refinery's fluctuation in RF and ERf when renewable heat is used, whereas Figure 8.10b depicts the refinery's change in RF and ERf when renewable electricity is used. The Py-ECH system will never be able to equal the RF and ERf of the CE system if the source of either the electricity or the heat is solely made

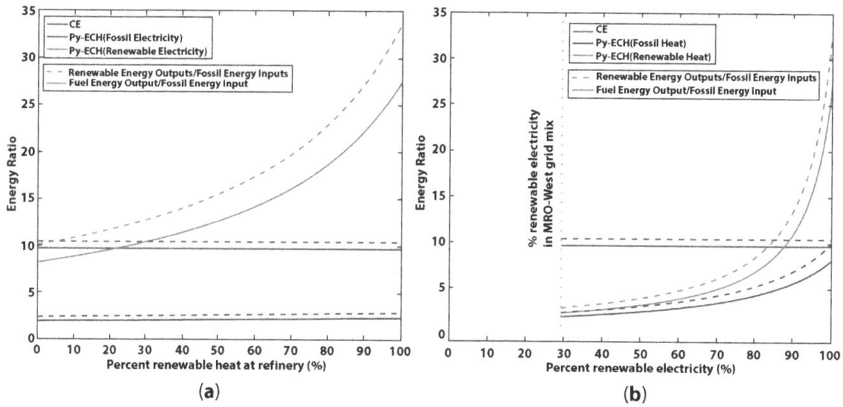

Figure 8.10 Sensitivity of system EROI with (a) % renewable heat at refinery (b) % renewable electricity

up of fossil fuels. Even though it is assumed that the refinery's heat supply is entirely renewable, the Py-ECH system has a more RF and ERf at about 80–90% renewable power.

8.6.5 Conclusion

The Py-ECH system always has a lesser potential for eutrophication and water shortages than the CE system. Even the renewability factor outperforms the CE system when the source of power is roughly 85% renewable, which assesses the quantity of renewable energy created depending on the amount of fossil energy inputs. This is noteworthy because the CE system has a very high renewability factor and generates its own heat and electricity using the lignin of the feedstock biomass. In conclusion, the Py-ECH system is more ecologically friendly than the CE system in all the areas that have been examined in this research due to its power supply, which is more than 87% renewable. This is noteworthy since, over a 20-year time horizon, the GHG emissions of the CE system, as given above, are itself 10%–15% lower than those of gasoline produced from crude oil. As a result, Py-ECH system, which utilize 87% renewable electricity, outperforms the CE system in terms of its environmental performance while continuing to generate fuel with an equivalent energy content to gasoline. Even without an enhancement in the recyclability of electrical sources, the Py-ECH system emits fewer GHGs than gasoline. It must be acknowledged that the Py-ECH technology is not yet at the same degree of technological preparedness as the CE process or the manufacturing of gasoline from crude oil. The effect of biochar on reducing nitrous oxide emissions, for example,

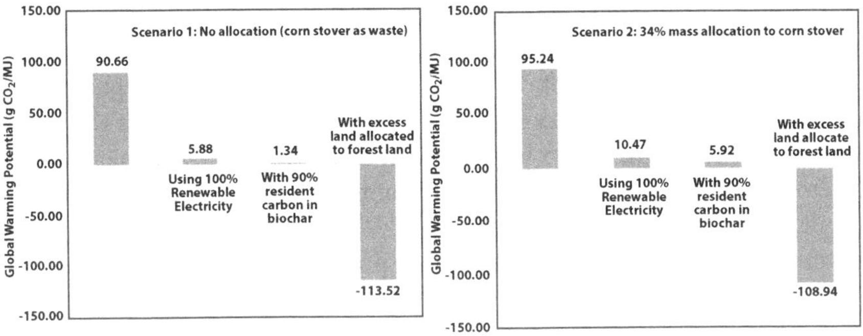

Figure 8.11 Waterfall chart showing the resultant global warming potential of stacking improvements in the Py-ECH system.

is not included in this analysis, which makes the advantages of the Py-ECH system inapplicable. Additionally, it has been established in the literature that the rate of biochar breakdown is extremely slow [18]. In reality, under ideal circumstances, Kuzyakov *et al.* measured an annual deterioration rate of 0.5%. In comparison to the analysis's estimated 82.5%, this translates to 90% of the carbon in the biochar being maintained after 20 years. Additionally, they calculate a half-life of 1,400 years and a mean residence time of 2,000 years under natural conditions, which are both noticeably longer than the time horizon taken into account in this research for biochar carbon breakdown. The need for more biomass feedstock for the CE system because to its lower energy production is another important factor that reduces Py-ECH's ability to mitigate climate change in comparison to CE. To increase carbon sequestration and lower the GWP of the Py-ECH system, additional land area that is spared might be set aside for forestland. According on the kind of forest, the yearly rate of carbon buildup in forests can range from 0.8 tons/ha/year to 5.1 tons/ha/year. The yearly carbon sequestration rates from American woods are estimated by the CCLUB model from GREET to be 2.4 tons C/ha/year, which is well within this range. It would be fascinating to learn how stacking these enhancements would affect the GWP of the Py-ECH system, even though it would make the eutrophication potential and water scarcity footprints of the two systems about similar. Figure 8.11 depicts this for both of the allocation possibilities taken into consideration during the investigation. The lower end of the range of values observed in the literature has been conservatively considered to represent the annual carbon sequestration rate for forests, which is 0.8 tons/ha/year. Higher levels will only further reduce the possibility of global warming.

The GWP of Py-ECH system may be reduced to levels that are practically carbon neutral for Scenario 1 by layering improvements in the source of energy and the grade of biochar generated, as shown in Figure 8.11. Additionally, the Py-ECH system can become carbon negative if the extra land space that is freed up because to its superior energy generation is given up to trees. There is still plenty of room for advancements that might result in carbon negative values for the Py-ECH system given the obvious advantages of a system run on renewable power over a system run on conventional electricity for the three environmental impacts studied in this research.

References

1. Sneha A. Akhade, Nirala Singh, Oliver Y. Gutiérrez, Juan Lopez-Ruiz, Huamin Wang, *et.al.* Electrocatalytic Hydrogenation of Biomass-Derived Organics: *Chem. Rev.* 2020 120 (20), 11370-11419, pp 27-47, 2016.
2. Bridgwater, A. V., Review of fast pyrolysis of biomass and product upgrading. *Biomass Bioenergy*, *38*, 68-94, 2012.
3. Wildschut, J.; Mahfud, F. H.; Venderbosch, R. H.; Heeres, H. J., Hydrotreatment of fast pyrolysis oil using heterogeneous noble-metal catalysts. *Ind. Eng. Chem. Res.*, *48* (23), 10324-10334, 2009.
4. Eranki, P. L.; Bals, B. D.; Dale, B. E., Advanced Regional Biomass Processing Depots: a key to the logistical challenges of the cellulosic biofuel industry. *Biofuels, Bioprod. Biorefin.*, *5* (6), 621-630, 2011.
5. Zhu, X.-G.; Long, S. P.; Ort, D. R., What is the maximum efficiency with which photosynthesis can convert solar energy into biomass? *Curr. Opin. Biotechnol.*, *19* (2), 153-159, 2008.
6. Lantz, E.; Wiser, R.; Hand, M., The past and future cost of wind energy. *National Renewable Energy Laboratory, Golden, CO, Report No. NREL/TP-6A20-53510*, 2012.
7. McKendry, P., Energy production from biomass (part 1): overview of biomass. *Bioresour. Technol.*, *83* (1), 37-46, 2002.
8. Bruun, E.; Müller-Stöver, D.; Ambus, P.; Hauggaard-Nielsen, H., Application of biochar to soil and N_2O emissions: potential effects of blending fast-pyrolysis biochar with anaerobically digested slurry. *Eur. J. Soil Sci.*, *62* (4), 581-589, 2011.
9. Bals, B. D.; Dale, B. E., Developing a model for assessing biomass processing technologies within a local biomass processing depot. *Bioresour. Technol.*, *106*, 161- 169, 2012.
10. Kim, S.; Zhang, X.; Reddy, A. D.; Dale, B. E.; Thelen, K. D.; Jones, C. D.; Izaurralde, R. C.; Runge, T.; Maravelias, C., Carbon-Negative Biofuel Production. *Environ. Sci. Technol.*, *54* (17), 10797-10807, 2020.

11. Humbird, D.; Davis, R.; Tao, L.; Kinchin, C.; Hsu, D.; Aden, A.; Schoen, P.; Lukas, J.; Olthof, B.; Worley, M. *Process design and economics for biochemical conversion of lignocellulosic biomass to ethanol: dilute-acid pretreatment and enzymatic hydrolysis of corn stover*; NREL/TP-5100-47764; National Renewable Energy Laboratory (NREL), Golden, CO.: 2011.
12. Lam, C. H.; Das, S.; Erickson, N. C.; Hyzer, C. D.; Garedew, M.; Anderson, J. E.; Wallington, T. J.; Tamor, M. A.; Jackson, J. E.; Saffron, C. M., Towards sustainable hydrocarbon fuels with biomass fast pyrolysis oil and electrocatalytic upgrading. *Sustain. Energy Fuels, 1* (2), 258-266, 2017.
13. Toffel, M. W.; Horvath, A., Environmental Implications of Wireless Technologies: News Delivery and Business Meetings. *Environ. Sci. Technol., 38* (11), 2961-2970, 2004.
14. Xu, H.; Sieverding, H.; Kwon, H.; Clay, D.; Stewart, C.; Johnson, J. M. F.; Qin, Z.; Karlen, D. L.; Wang, M., A global meta-analysis of soil organic carbon response to corn stover removal. *GCB Bioenergy, 11* (10), 1215-1233, 2019.
15. Smil, V., PHOSPHORUS IN THE ENVIRONMENT: Natural Flows and Human Interferences. *Annu. Rev. Energy Env., 25* (1), 53-88, 2000.
16. Emery, I.; Dunn, J. B.; Han, J.; Wang, M., Biomass Storage Options Influence Net Energy and Emissions of Cellulosic Ethanol. *Bioenergy Res., 8* (2), 590-604, 2015.
17. Dias Mayer, F.; Brondani, M.; Carrillo, M.; Hoffmann, R.; Lora, E., Revisiting Energy Efficiency, Renewability, And Sustainability Indicators In Biofuels Life Cycle: Analysis And Standardization Proposal. *J. Cleaner Prod., 252*, 119850, 2019.
18. Bai, M.; Wilske, B.; Buegger, F.; Esperschütz, J.; Kammann, C. I.; Eckhardt, C.; Koestler, M.; Kraft, P.; Bach, M.; Frede, H.-G.; Breuer, L., Degradation kinetics of biochar from pyrolysis and hydrothermal carbonization in temperate soils. *Plant Soil, 372* (1), 375-387, 2013.

9

Improving Energy Consumption in Smart Homes Based on the Internet of Things

Sanjeevikumar Padmanaban[1], Mostafa Azimi Nasab[1], Amritansh Sagar[2], Mohammad Zand[1], Mohammad Ali Dashtaki[3] and Morteza Azimi Nasab[1*]

[1]*Department of Electrical Engineering, IT and Cybernetic, University of South-Eastern Norway, Porsgrunn, Norway*
[2]*Department of Industrial Engineering, University of Padova, Padua, Italy*
[3]*Department of Electrical and Computer Engineering, Binghamton University-SUNY, Binghamton, NY, USA*

Abstract
The Internet of Things (IoT) revolutionizes the way that daily tools interact and share information by connecting to the internet. The integration of various technologies and communication solutions is a fundamental aspect of IoT. Key components of IoT include identification and tracking technologies, wired and wireless sensors, active networks, and protocols that enhance communication and object intelligence. This research aims to identify the specific concepts and technologies related to software networks based on IoT that can be utilized to create energy-efficient smart homes. Given that IoT network sensors have limited energy, the research focuses on extending the network's life span through the implementation of a sleep/wake mechanism.

Keywords: Energy management, smart city, smart homes, Internet of Things

9.1 Introduction

Connecting diverse gadgets through the web is the core of the idea behind the "Internet of Things" (IoT). The IoT is a self-configuring, globally distributed network that operates according to a set of agreed-upon protocols

**Corresponding author*: Morteza1368morteza@gmail.com

O.V. Gnana Swathika, K. Karthikeyan, Milind Shrinivas Dangate and Nicoletta Ravasio (eds.) Biomass and Solar-Powered Sustainable Digital Cities, (147–162) © 2024 Scrivener Publishing LLC

and standards for group and ad hoc information exchange. Intermediary "things" are smart objects that are consistently incorporated into a network of information and are given unique identifiers, physical features, and virtual qualities [1].

Two main features place different "things" as part of the "Internet of Things": internet connection and identity characteristics. Each smart object has a unique Internet Protocol address, enabling the device that sends or receives information to be identified. These smart objects can form systems that have the ability to communicate with each other. Such systems have the potential to create fundamental changes in homes, community places, companies, cities, and different economic sectors at domestic and international levels for each country [2]. Although all the details about the natural and predicted effects of the "Internet of Things" are associated with uncertainty, economic analysts have predicted that this technology will have a significant contribution to the economic growth of the next decade. Those who will be affected by this technology in the future include agriculture, energy, government management, healthcare, production and industry, and transportation. This technology can also be used in larger superstructures, especially in "smart cities," where rapid progress can be made in areas such as transportation and other urban services. Technological innovation like this makes it possible for computers and other gadgets to talk to one another and with people over the Internet. The inventory and expiration date of the food in a smart refrigerator that is connected to the internet, for instance. In reality, with the help of the internet's infrastructure, a person can manage and control their everyday objects remotely. There is a growing risk that disruptions, attacks, or theft of information will compromise IoT-driven systems and operations utilized on a massive scale. There are always two reasons for the ease of hacking devices based on the IoT: the first is that the aforementioned devices are connected through the internet, and the internet in general is at risk of hacking. The second reason is that many of the devices based on this technology were made by small companies that did not have enough resources and time to put powerful security structures in their devices. The IoT can become one of the great developments in the field of technology in the near future and change the way we live in many ways [3, 4].

9.1.1 The Role of Objects in the Internet of Things

Objects become intelligent when they are connected to the internet. However, smart objects require development in two major research areas: saving and harnessing energy and integration of intelligent components with materials. Developing solutions with the aim of creating a level of conflict that is as close as possible to zero is needed to consider energy consumption at all levels, from energy conservation and energy consumption to its consumption. Ordinary objects such as mobile phones should be able to harvest the required energy, whether through solar cells or through the conversion of vibrations and movement toward electrical energy [5]. The development of current technologies is not enough in this field, and the existing energy production capacity and processing power are very low to respond to future needs. While we are focused on the technologies and communication mechanisms of the IoT, we should not overlook the fact that energy forms the backbone of the IoT. The development of energy storage resources, in an effective and compact way, such as batteries, fuel cells, polymer printed batteries, as well as the design and construction of energy generating devices that have the ability to transmit, harness, or harvest energy using energy conversion methods, is a key factor and important in the development of intelligent wireless systems. The second axis of technological development in the field of smart objects is a step ahead. In the coming years, the integration of chips and antennas with non-standard substrates such as fabric and paper will lead to the emergence of special technologies [6, 7]. Metal sheets and new polymer-based substrates with conductive passages and paths and adhesive materials that are more compatible with harsh environments and environmentally friendly disposal, using today's materials such as silicone, will not be very efficient. In order to create a connection between the embedded chip on an integrated circuit and an antenna, instead of directly assembling the parts, we must use radio-frequency identification (RFID) inlays to produce tags with different shapes and sizes. Inductive or capacitive coupling of tape-like antennas with a special design prevents galvanic connection. This issue increases the level of confidence and enables faster processing of products. This work is done to physically integrate the RFID structure and the type of object that must be identified, in a way that gives the mentioned object the ability to act like an antenna from a physical point of view [8].

9.1.2 Bulk Data at the Heart of the Internet of Things Architecture

The IoT is characterized by the presence of vast amounts of data. The future of active IoT and sensors holds great promise, which naturally leads to significant commercial benefits. Although this idea may seem distant, these objects are actually very close to us. The term "Internet of Things" originally represented an incredible technological prospect that seemed unattainable in the past. However, the IoT has demonstrated its potential to reshape businesses. It signifies a unique communication mechanism between objects and the collection of a massive volume of data [9]. The IoT offers substantial commercial advantages to forward-thinking organizations that anticipate opportunities and leverage the IoT. Pioneers in the field have already launched or are implementing projects related to resource monitoring, consumption pattern tracking, and delivery of goods and services, which are expected to yield positive results. If your organization has embraced the IoT as part of its business strategy, then it is important to be aware of the technical and operational challenges that lie ahead [10].

9.1.3 Comprehensive Framework Based on the Internet of Things for Smart Homes

In the coming years, smart objects are expected to dominate the market, with their presence becoming ubiquitous in households. This trend necessitates the development of new and advanced services for smart homes. Consequently, the requirement for IoT-driven finding is becoming increasingly apparent. Recent publications propose the creation of a comprehensive framework for the IoT that is suitable for a wide range of applications. Radio-frequency identification (RFID) [11], wireless sensor networks, middleware, the cloud, and custom software development are all highlighted as essential IoT technologies. The researchers additionally divide business uses of the IoT into three groups: management and oversight, data collection and analysis, and internal communication and cooperation. Near-field communication (NFC), location-based services, and social networks are all added to the list of enabling technologies in a second article. Researchers offer a multi-tier architecture that includes sensors, networks, services, and user interfaces. Uncertainty exists about how services will be enabled, however, because the function of clouds is not well-defined [12]. The IoT has been investigated for potential use in the continual tracking of industrial manufacturing operations. The authors provide a holistic approach to energy management that can be used for both internal data mining and external cloud-based decision-making. However, the framework does not detail how external solution architects should be included in the process [13].

9.2 Smart Home

A home wireless sensor network is made up of all of the appliances and gadgets in the house that have wireless internet access. Every piece of information collected by your various gadgets is sent to a home hub or sink. Nodes in a domestic wireless sensor network are assumed to be reasonably intelligent terminals, capable of basic networking and data processing tasks. A home hub can be any device with data storage capacity, local processing capabilities, and the ability to communicate with devices outside the home wireless sensor network, such as a smart meter, computer, tablet, or smartphone. Both the home hub and its counterpart, the residential sink, are common features in residential smart or smart buildings. When compared to the main hub, which handles data from decentralized production sources, the residential hub needs an additional function. For the most part, consumers pool their use of renewable resources like a rooftop photovoltaic system in a residential building, which benefits everyone living there. Distributed renewable energy sources are technologically advanced in this sense [14, 15].

9.2.1 Cloud

The cloud collects information from numerous locations, such as personal logs, sensor readings from power lines and factories, and so on. The ability to store and process massive amounts of data must be a feature of the cloud. The cloud plays a key role in this system, which is why it is part of the most comprehensive component of the architecture [16].

9.2.2 Beneficial Program

Manufacturing, transmission, and distribution are all included in this tier of the smart network in addition to the smart home. Data from each subsection are uploaded to the cloud without any interference from other subsections. Common information that can be exchanged with this beneficial program includes electricity price, weather forecast, status of distribution/transmission lines, current and future consumption of the microgrid, and current and future generation of distributed sources of production related to the microgrid [17].

9.2.3 Third Party

Third-party applications have been developed using cloud data. Other terms used interchangeably are commercial software, industrial software, or IoT software. Developers of third-party application programs receive information from the cloud (private or public) and use these data to provide solutions in the form of web-based or mobile application programs [18].

9.2.4 User Interfaces

This level represents the relationships of users who provide data to end-users (announcements, recommendations, control of smart devices, etc.). Interpretation of the raw data in the table with reference to monthly (or even daily) consumption by individuals or families is a difficult task. A more complex visualization tool is needed not only to present the total consumption of family members but also consumption at the level of appliances/household appliances. This work is especially important because it allows customers to gain insight into the workings of the different devices in the homes they own, even the rigid devices that cannot be managed automatically so that users can exercise direct control over these devices in light of the nature of their consumption. Third-party apps commonly assess user relationships based on quality-of-service criteria, which requires the creation of explicit interactions among users [19, 20].

9.2.5 The Main Function of the Internet of Things

The IoT has three important functions: comprehensive collection, reliable transmission, and intelligent processing. Using RFID, sensors, two-dimensional codes, and barcodes, it is possible to collect information from objects anytime and anywhere. Providing information about objects through various secure and remote communication networks and the internet in real-time ensures reliable transmission. By using intelligent computing, such as cloud and fuzzy computing, it is possible to identify, analyze, and process large amounts of data and information to implement intelligent control of objects. In other words, the IoT can be considered a new form of network based on the internet, which is seen as a much larger network formed by computer devices and networks with advanced technologies [21–24].

9.2.6 Security in the Internet of Things

Despite the promises of a better life, the IoT faces several challenges, including immature technology, conflicting standards and protocols, heavy development costs, low public awareness, and security and privacy protection. The architecture is lagging behind in creating security and preventing information disclosure, and there is no general agreement to solve these issues in a complete and unified form in all fields. Secure architectures and structures have been introduced to prevent various security threats during the process of using the IoT to minimize various security risks [25]. For the security of the IoT, there is a need to integrate security in the field of physical equipment, information acquisition, information transmission, and information processing. The ultimate goal of security is to ensure confidentiality, integrity, authenticity of data, and information. Security models and architectures and frameworks should also consider three important factors of security in devices, security in communications, and security in management [25].

The security challenges for each layer are as follows [26]:

- The challenge of the security of the perceptual layer:
- Security of wireless sensor network nodes: challenges include security of nodes in the gateway, control of shared nodes, unauthorized access to the password key, unauthorized access to shared nodes, distributed denial of service (DDOS)/denial of service (DOS) attacks, physical damage to nodes, physical attacks channel blocking, document forgery attacks, copying attacks, broadcast attacks, and information manipulation.
- Node authentication security: Node security is one of the most important challenges.
- Privacy of object information: Object information is easily accessible in the IoT environment. Information on how to display and access the information of specified objects and also how to hide this information is one of the most important measures that should be taken. Preventing the forgery and fraud of object information is another big security challenge for the IoT.
- Network layer security challenge: The network layer is responsible for data transmission and is a channel that connects the cognitive layer and the application layer. Network security issues have not been fully resolved yet. The main threats are DOS/DDOS attacks, forgery, heterogeneous network attacks, attacks and risks of using IPv, WLAN application security challenges, and security threats of different communication networks.
- Application layer security challenge: The application layer includes the support of different sub-layers. The support program of the sub-layer requires intelligent access, calculations, and allocation of resources in all types of acquisition and selection, generation, and processing of data, in order to provide useful information for the sub-layer. In this process, in order to support the substrate, it is necessary to identify useful information, useless data, and even malicious data,

filtering and processing in time. Substrate security challenges include access, protecting property, verifying user's identity, maintaining privacy of information, and how to safely track data flow [26].

9.2.7 Operational Structure for Personal Medical Services Based on the Internet of Things

The operational structure for personal medical services based on the IoT requires consideration of the four major subsystems of VISA, information production, processing and storage, and consumer and startup.

9.3 Regulatory Applications and Their Use in Health Services

There is a dearth of research examining the impact of mobile management on healthcare. Research in this area has centered on the use of mobile cloud applications and Android mobile applications in smart homes to aid the elderly and the disabled in managing their day-to-day activities. Additionally, there is a subset of mobile management studies that focus on IoT-based health systems in smart homes. Control functions and applications for the smart home that are useful for older people and the disabled are additionally addressed. It is researched how older adults can benefit from using monitoring systems and medical reminders in smart houses [27].

9.3.1 Challenges

Modern home systems built on the IoT have many practical benefits, but they are not a replacement for a traditional communication infrastructure. The studied literature demonstrates that experts are worried about the pitfalls of IoT-driven applications for home automation [28].

9.3.2 Concerns About Healthcare

Medical infrastructures are under growing strain as the global population ages. About one-quarter of the elderly do not follow doctors' orders, leading to worse health and even death, according to reports. By 2060,

Europe's aged demographic is projected to grow from 25% to 53%, driving up demand for caretakers and clinicians and driving up healthcare expenses. Although they only account for 13% of the population, the elderly currently uses up more than 40% of the country's healthcare spending [28]. The percentage of Malaysians over 65 increased from 9.5% in 1991 to 2.6% (about 5.1 million people) in 2000 and is projected to rise to 5.9% (roughly 2.3 million people) in 10 February 2020. Software-based networks, also known as programmable networks, manage medium and large networks centrally based on software. In these networks, instead of hardware deciding where the passing packets should go, software treats each packet differently with the required definition conditions. Programmable systems are gaining popularity, and software-oriented networks are attempting to make them smarter by moving the data control function away from the physical router and switch and into the underlying virtual network software. The network is able to plan, scale, be flexible, be automated, be intelligent, and have network software developed by organizations thanks to a centralized software controller; this is being termed the biggest evolution of networking technologies in the last four decades [29, 30].

9.3.3 Architecture of Software-Oriented Networks

The SDN application controlling devices, which are responsible for the upkeep of the network as a whole, contain the intelligence of the network. Therefore, the network will appear to the application applications as a single entity and function as a logical switch. SDN allows telecom providers and carriers to manage and operate their networks from a centralized location, regardless of the hardware used or the manufacturer. The architecture of software-oriented networks includes the following layers [31]:

1. Application software layer: This layer includes end-user business applications used by SDN communication services. The boundary between the application layer and the control layer will be crossed by the northern APIs.
2. Control layer: A controller is a device that can see the whole network, knows how everything is connected, and makes decisions about how the network should be used. The network planning model is shifted from decentralized to centralized operation when an SDN controller is implemented. The controller isolates the physical layer components like routers, switches, firewalls, and load balancers from one another, much like a layer of network middleware might.

By using an SDN controller, it is possible to build unique networks to perform specific tasks based on criteria such as costs or needs in the security policy department.

9.3.4 SDN for Cloud Service Providers

Software-oriented networks play a special role in the cloud, addressing challenges faced by cloud service providers, such as multi-tenancy at the network level and resource level. Resources need to be shared in a safe and exclusive manner to ensure that one user's application programs do not affect another user's. SDN models have been designed to meet the needs of cloud computing. The first model focuses on multi-tenancy support, whereas the second model requires the cloud controller's support for network services, making SDN technology crucial for the cloud.

SDN enables cloud management platforms to coordinate network and virtualization services with computing and storage resources while ensuring security and application privacy protection. Cloud platforms can dynamically configure networks to provide necessary virtualization, firewalls, and traffic load balancers for each application. SDN also allows service providers to offer layered services to tenants, enabling fine-tuning of system performance for specific applications [32].

9.4 Previous Research

The article titled "Evaluation of Energy Consumption Reduction Methods in MPLS Networks Based on SDN" proposes a method based on SDN software to control and centralize network settings and disable network elements. The method involves identifying time periods with minimum traffic demand, ranking network elements based on their usage and importance, re-routing traffic to deactivate less-used elements, and ultimately deactivating node/link network elements. The proposed algorithm combines SDN controller capabilities with MPLS infrastructure to optimize performance and direct traffic from less-used elements in the network [32–35].

In another study, titled "Reducing Energy Consumption in Wireless Sensor Networks Using SDN_WISE Architecture," the focus is on the challenges faced by wireless sensor networks, such as limited power sources, limited memory, and low processing power. The research investigates the reduction of energy consumption and its impact on the lifetime of the wireless sensor network by implementing a wireless sensor network based on SDN_WISE architecture [36].

Similarly, in the research by Padmanaban, S. *et al.*, the research aims to reduce energy consumption in wireless sensor networks using clustering algorithms [37]. The limited power source of sensor nodes poses a challenge to the network's life span. Clustering protocols are considered effective methods to increase network lifetime, but they often impose high energy consumption on the cluster head. The research explores ways to reduce energy consumption and increase the life span of wireless sensor networks [38].

In the research by Padmanaban, S. *et al.*, the focus is on maximizing the overall lifetime of the network in wireless sensor networks. The selection of the cluster head, which consumes more energy than other nodes in the cluster, plays a crucial role in the life span of clusters. The research presents various protocols and algorithms for clustering and selecting cluster heads to extend the network's lifetime [39].

Several other studies have also contributed to the field of wireless sensor networks, addressing challenges such as energy consumption, network lifetime, and efficient data routing through clustering algorithms and distributed approximation algorithms [40].

9.5 Conclusion

This article discusses the potential for improving energy consumption in smart homes based on the IoT. The integration of various technologies and communication solutions is a fundamental aspect of IoT, and the article identifies specific concepts and technologies related to software networks based on IoT that can be utilized to create energy-efficient smart homes. The article also highlights the importance of energy management and the development of energy storage resources in the development of intelligent wireless systems. The article proposes a comprehensive framework for the IoT that is suitable for a wide range of applications, including monitoring and control, big data analytics and business, and information sharing and collaboration. The article also discusses the challenges of network layer security and application layer security in the context of smart homes. Finally, the article briefly touches on the potential for regulatory applications and their use in health services.

References

1. M. Zand, M. A. Nasab, A. Hatami, M. Kargar and H. R. Chamorro, "Using Adaptive Fuzzy Logic for Intelligent Energy Management in Hybrid Vehicles," *2020 28th ICEE,* , pp. 1-7, doi: 10.1109/ICEE50131.2020.9260941.IEEE Index

2. Ahmadi-Nezamabad, H., Zand, M., Alizadeh, A., Vosoogh, M., & Nojavan, S. (2019). Multi-objective optimization based robust scheduling of electric vehicles aggregator. *Sustain. Cities Soc.*, *47*, 101494.
3. Milani, O. H., Nguyen, T., Parekh, A., Cetin, A. E., & Prasad, B. (2023). 0537 Incident Hypertension Prediction in Obstructive Sleep Apnea using Machine Learning. *Sleep*, *46*(Supplement_1), A236-A237.
4. Zand, M., Nasab, M. A., Sanjeevikumar, P., Maroti, P. K., & Holm-Nielsen, J. B. (2020). Energy management strategy for solid-state transformer-based solar charging station for electric vehicles in smart grids. *IET Renew. Power Gener.*, *14*(18), 3843-3852.
5. Ghasemi, M., Akbari, E., Zand, M., Hadipour, M., Ghavidel, S., & Li, L. (2019). An efficient modified HPSO-TVAC-based dynamic economic dispatch of generating units. *Electr. Power Compon. Syst.*, *47*(19-20), 1826-1840.
6. Nasri, S., Nowdeh, S. A., Davoudkhani, I. F., Moghaddam, M. J. H., Kalam, A., Shahrokhi, S., & Zand, M. (2021). Maximum Power point tracking of Photovoltaic Renewable Energy System using a New method based on turbulent flow of water-based optimization (TFWO) under Partial shading conditions. In *Fundamentals and Innovations in Solar Energy* (pp. 285-310). Springer, Singapore.
7. Rohani, A., Joorabian, M., Abasi, M., & Zand, M. (2019). Three-phase amplitude adaptive notch filter control design of DSTATCOM under unbalanced/distorted utility voltage conditions. *J. Intell. Fuzzy Syst.*, *37*(1), 847-865.
8. Zand, M., Nasab, M. A., Neghabi, O., Khalili, M., & Goli, A. (2019, December). Fault locating transmission lines with thyristor-controlled series capacitors By fuzzy logic method. In *2020 14th International Conference on Protection and Automation of Power Systems (IPAPS)* (pp. 62-70). IEEE.
9. Z. Zand, M. Hayati and G. Karimi, "Short-Channel Effects Improvement of Carbon Nanotube Field Effect Transistors," *2020 28th Iranian Conference on Electrical Engineering (ICEE)*, Tabriz, Iran, 2020, pp. 1-6, doi: 10.1109/ICEE50131.2020.9260850.
10. Tightiz, L., Nasab, M. A., Yang, H., & Addeh, A. (2020). An intelligent system based on optimized ANFIS and association rules for power transformer fault diagnosis. *ISA Trans.*, *103*, 63-74.
11. Zand, M., Neghabi, O., Nasab, M. A., Eskandari, M., & Abedini, M. (2020, December). A hybrid scheme for fault locating in transmission lines compensated by the TCSC. In *2020 15th International Conference on Protection and Automation of Power Systems (IPAPS)* (pp. 130-135). IEEE.
12. Zand, M., Nasab, M. A., Khoobani, M., Jahangiri, A., Hosseinian, S. H., & Kimiai, A. H. (2021, February). Robust speed control for induction motor drives using STSM control. In *2021 12th Power Electronics, Drive Systems, and Technologies Conference (PEDSTC)* (pp. 1-6). IEEE.
13. Sanjeevikumar, P., Zand, M., Nasab, M. A., Hanif, M. A., & Bhaskar, M. S. (2021, May). Spider community optimization algorithm to determine UPFC optimal size and location for improve dynamic stability. In *2021 IEEE 12th*

Energy Conversion Congress & Exposition-Asia (ECCE-Asia) (pp. 2318-2323). IEEE.

14. Azimi Nasab, M., Zand, M., Eskandari, M., Sanjeevikumar, P., & Siano, P. (2021). Optimal planning of electrical appliance of residential units in a smart home network using cloud services. *Smart Cities*, *4*(3), 1173-1195.
15. Nasab, M. A., Zand, M., Padmanaban, S., Bhaskar, M. S., & Guerrero, J. M. (2022). An efficient, robust optimization model for the unit commitment considering renewable uncertainty and pumped-storage hydropower. *Comput. Electr. Eng.*, *100*, 107846.
16. Azimi Nasab, M., Zand, M., Padmanaban, S., & Khan, B. (2021). Simultaneous long-term planning of flexible electric vehicle photovoltaic charging stations in terms of load response and technical and economic indicators. *World Electric Vehicle Journal*, *12*(4), 190.
17. Zand, M., Nasab, M. A., Padmanaban, S., & Khoobani, M. (2022). Big data for SMART Sensor and intelligent Electronic devices–Building application. In *Smart buildings digitalization* (pp. 11-28). CRC Press.
18. Padmanaban, S., Khalili, M., Nasab, M. A., Zand, M., Shamim, A. G., & Khan, B. (2022). Determination of Power Transformers Health Index Using Parameters Affecting the Transformer's Life. *IETE J. Res.*, 1-22.
19. Sanjeevikumar, P., Samavat, T., Nasab, M. A., Zand, M., & Khoobani, M. (2022). Machine learning-based hybrid demand-side controller for renewable energy management. In *Sustainable Developments by Artificial Intelligence and Machine Learning for Renewable Energies* (pp. 291-307). Elsevier.
20. Khalili, M., Ali Dashtaki, M., Nasab, M. A., Reza Hanif, H., Padmanaban, S., & Khan, B. (2022). Optimal instantaneous prediction of voltage instability due to transient faults in power networks taking into account the dynamic effect of generators. *Cogent Eng.*, *9*(1), 2072568.
21. Asadi, A. H. K., Jahangiri, A., Zand, M., Eskandari, M., Nasab, M. A., & Meyar-Naimi, H. (2022, January). Optimal Design of High Density HTS-SMES Step-Shaped Cross-Sectional Solenoid to mechanical stress reduction. In *2022 International Conference on Protection and Automation of Power Systems (IPAPS)* (Vol. 16, pp. 1-6). IEEE.
22. Nasab, M. A., Zand, M., Hatami, A., Nikoukar, F., Padmanaban, S., & Kimiai, A. H. (2022, January). A Hybrid Scheme for Fault Locating for Transmission Lines with TCSC. In *2022 International Conference on Protection and Automation of Power Systems (IPAPS)* (Vol. 16, pp. 1-10). IEEE.
23. Dashtaki, A. A., Khaki, M., Zand, M., Nasab, M. A., Sanjeevikumar, P., Samavat, T., ... & Khan, B. (2022). A Day Ahead Electrical Appliance Planning of Residential Units in a Smart Home Network Using ITS-BF Algorithm. *Int. Trans. Electr. Energy Syst., 2022.*
24. Rastgoo, Samaneh, Zahra Mahdavi, Morteza Azimi Nasab, Mohammad Zand, and Sanjeevikumar Padmanaban. 2022. "Using an Intelligent Control Method for Electric Vehicle Charging in Microgrids" *World Electr. Veh. J.* 13, no. 12: 222. https://doi.org/10.3390/wevj13120222

25. Seresht, R. M., Miri, M., Zand, M., Azimi Nasab, M., Sanjeevikumar, P., & Khan, B. (2023). Frequency control scheme of an AC Islanded microgrid based on modified new self-organizing hierarchical PSO with jumping time-varying acceleration coefficients. *Cogent Eng.*, *10*(1), 2157982.
26. Mahdavi, Z., Samavat, T., Javanmardi, A. S. J., Dashtaki, M. A., Zand, M., Nasab, M. A., ... & Khan, B. (2024). Providing a Control System for Charging Electric Vehicles Using ANFIS. International Transactions on Electrical Energy Systems, 2024.
27. Padmanaban, S., Nasab, M. A., Shiri, M. E., Javadi, H. H. S., Nasab, M. A., Zand, M., & Samavat, T. (2023). The Role of Internet of Things in Smart Homes. *Artificial Intelligence-based Smart Power Systems*, 259-271.
28. Ahmadi, M., Rastgoo, S.,Mahdavi, Z., Nasab, M.A., Zand, M., Sanjeevikumar, P.,Khan, B.: Optimal allocation of EVs parking lots and DG in micro grid using two-stage GA-PSO. *J. Eng.* 2023, e12237 (2023).https://doi.org/10.1049/tje2.12237
29. Zand, M., Nasab, M. A., Padmanaban, S., Maroti, P. K., & Muyeen, S. M. (2023). Sensitivity analysis index to determine the optimal location of multi-objective UPFC for improvement of power quality parameters. *Energy Rep.*, *10*, 431-438.
30. Sagar, A., Kashyap, A., Nasab, M. A., Padmanaban, S., Bertoluzzo, M., Kumar, A., & Blaabjerg, F. (2023). A Comprehensive Review of the Recent Development of Wireless Power Transfer technologies for Electric Vehicle Charging Systems. *IEEE Access*.
31. Nasab, M. A., Zand, M., Dashtaki, A. A., Nasab, M. A., Padmanaban, S., & Blaabjerg, F. (2023). Uncertainty compensation with coordinated control of EVs and DER systems in smart grids. *Solar Energy*, *263*, 111920.
32. Nikoufard, M., & Hatami, M. (2016). Ultra-short novel transverse magnetic to transverse electric polarization rotator in hybrid integration of InGaAsP/silicon-on-insulator technologies. *Indian J. Phys.*, *90*, 211-217.
33. Padmanaban, S., Samavat, T., Nasab, M. A., Nasab, M. A., Zand, M., & Nikokar, F. (2023). Electric vehicles and IoT in smart cities. *Artificial Intelligence-based Smart Power Systems*, 273-290.
34. Samavat, T., Nazari, M., Ghalehnoie, M., Nasab, M. A., Zand, M., Sanjeevikumar, P., & Khan, B. (2023). A Comparative Analysis of the Mamdani and Sugeno Fuzzy Inference Systems for MPPT of an Islanded PV System. *Int. J. Energy Res., 2023*.
35. Padmanaban, S., Nasab, M. A., Shiri, M. E., Javadi, H. H. S., Nasab, M. A., Zand, M., & Samavat, T. (2023). The role of internet of things in smart homes. *Artificial Intelligence-based Smart Power Systems*, 259-271.
36. Zand, M., Nasab, M. A., Hanif, H. R., Ghahri, M. R., Padmanaban, S., & Khan, B. (2023). A hybrid scheme for optimal performance of photovoltaic system converters with multilayer structure. *The J. Eng.*, *2023*(1), e12218.

37. Padmanaban, S., Nasab, M. A., Samavat, T., Nasab, M. A., & Zand, M. Securing Smart Power Grids Against Cyber-Attacks. In *IoT and Analytics in Renewable Energy Systems (Volume 1)* (pp. 17-36). CRC Press.
38. Padmanaban, S., Nasab, M. A., Samavat, T., Zand, M., Nasab, M. A., & Hashemi, E. Cyber Security in Smart Energy Networks. In *IoT and Analytics in Renewable Energy Systems (Volume 2)* (pp. 309-325). CRC Press.
39. Padmanaban, S., Nasab, M. A., Samavat, T., Zand, M., & Nasab, M. A. Artificial Intelligence Techniques for Smart Power Systems. In *IoT and Analytics in Renewable Energy Systems (Volume 1)* (pp. 107-123). CRC Press.
40. Nikoufard, M., & Hatami, M. (2018). Analysis of ultra-compact TE to TM polarization rotator in InGaAsP and SOI technologies. *Optik-International Journal for Light and Electron Optics*, *153*, 9-15.

10

Performance Analysis of Supercapacitors in Renewable Energy System/ Microgrid/Electric Vehicle, and Comparative Study of Various Converters With Their Applications

Moovendan Murugu[1], Arul Rajagopalan[2*], V. T. Sreedevi[2], R. Sitharthan[2] and R. Kanimozhi[3]

[1]*School of Electrical Engineering, Vellore Institute of Technology, Chennai, Tamil Nadu, India*
[2]*Centre for Smart Grid Technologies, School of Electrical Engineering, Vellore Institute of Technology, Chennai, Tamil Nadu, India*
[3]*Department of Electronics and Communication Engineering, Annamalai University, Annamalainagar, Chidambaram, Tamil Nadu, India*

Abstract
Supercapacitors (SCs) are often used as energy storage devices in electric vehicle (EV) and microgrid (MG). In most cases, these SCs are used for power stabilizations to supply the loads in combination with batteries. The power levels and energy levels of SCs are higher than battery and conventional capacitor, respectively, that make them more suitable for several applications. Even though SCs act as good stabilizers in different application, converters are the most significant component that interfaces the energy storage systems and loads to meet the consumer power demand. The converters will efficiently increase the life span of energy storage devices. The appropriate selection of storage device by converters helps to supply the various types of loads for demand in the power. This review paper briefly discusses the various applications of SCs, their types, properties, and different converter topologies along with the comparisons. The characteristics of battery and SCs are also dealt in this review to have a clear idea toward EVs and grids.

**Corresponding author*: arulphd@yahoo.co.in

O.V. Gnana Swathika, K. Karthikeyan, Milind Shrinivas Dangate and Nicoletta Ravasio (eds.)
Biomass and Solar-Powered Sustainable Digital Cities, (163–196) © 2024 Scrivener Publishing LLC

***Keywords*:** Energy storage technologies, converter topologies, power electronic converters, electric vehicle, microgrid

10.1 Introduction

The requirement of high-performance energy storage device is one of the major concerns in the emerging technological world. Energy storage technologies (ESTs) that include battery and supercapacitors (SCs) can meet the energy requirements. SCs are dedicated to EST that provides appropriate energy and power density for midrange power applications [1]. SC is a kind of EST that stands in the middle level of conventional capacitor (CC) and battery [2]. Numerous alternative technologies have emerged to alleviate significant concerns about the energy problem. The technology's primary goal is to reduce greenhouse gas pollution [3]. SCs might have higher power densities, long operating spans, quicker charge, and discharge time intervals and are a clean and safe solution for storing electrochemical energy [4]. Currently, numerous types of capacitors for energy storage are available, classified based on the dielectric utilized or the physical state of the capacitor, as indicated in Figure 10.1 [5]. All capacitor that includes CC and SCs has a unique feature that ranges from small filter to high voltage power factor adjustment respectively [6]. In recent decades, research has exploded on SC to meet applications requiring improved qualities in the wheel of life, accuracy, and explicit energy. The configuration of SCs encountering electrochemical activity is like that of a battery.

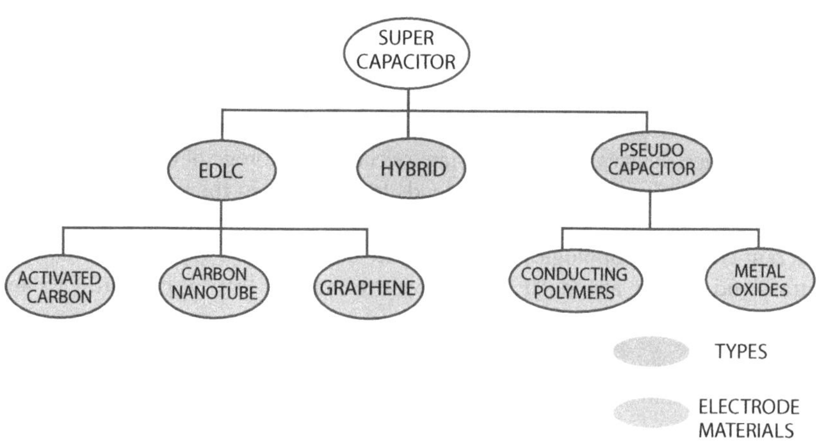

Figure 10.1 Types of supercapacitors.

The twin electrode immersed in an electrolyte and separated by a separator is the structure of an SC, as shown in Figure 10.2. These components determine the features of SCs. It stores energy by charging and discharging electrodes and electrolytes [7]. The SCs are like CCs in working as they have fast energy release and storage [8]. SCs, unlike ordinary capacitors, have included electrodes with a larger effective surface area, resulting in a tenfold increase in capacitance over CC [9]. While traditional capacitors store charge in the micro- to milli-farad range, SCs store charge in the 100-F to 1,000-F range [10]. SCs reach the pinnacle of power density compared to other energy storage devices. However, they have a significantly lower power density than typical capacitors, as demonstrated in the Rag one plot for various energy storage systems (ESSs) in Figure 10.3. Proper converter topologies improve the efficiency of SC applications. Converters are distinguished structurally by the type of isolation that they provide. Isolated or non-isolated converters are also possible options [11]. Previously published reviews focused on multi-input converters used in hybrid electric vehicles (HEVs) [12] and grid integration [13]. The review in [14] covered the major topologies (2015) when it was written. The review in [15] enclosed the topologies up to 2016. It has been a long time since novel topologies were proposed. As a result, a recent update in evaluation is required for the future. On the other hand, the authors did not thoroughly examine recently evolved topologies. The utilization of various types of converters in EST is purported in this review article.

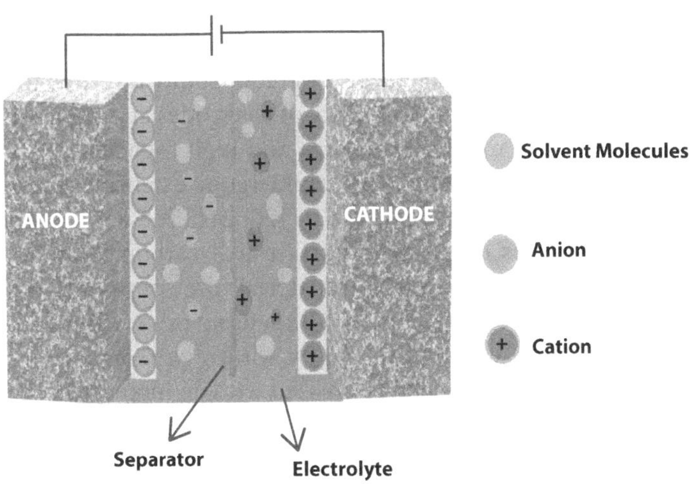

Figure 10.2 Structural classification of supercapacitor.

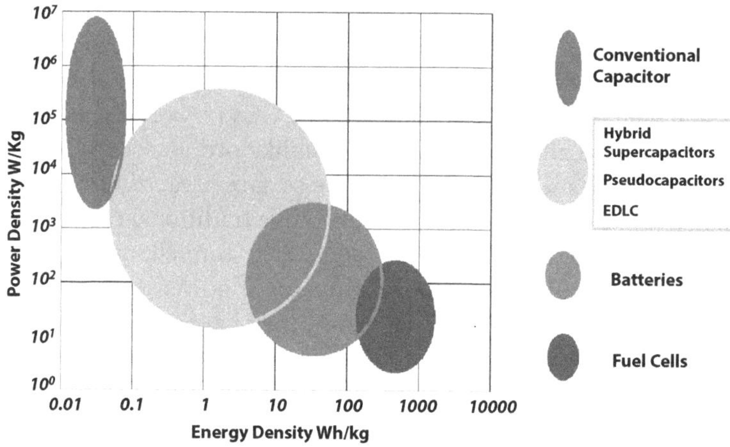

Figure 10.3 Rag one plot of energy storage devices.

10.2 Supercapacitor Fundamentals

Helmholtz initially described the electric double-layer phenomenon. Later, it was patented by Becker, who increased the surface area of SC by coating porous carbon on the electrodes forming a two-layer construction [16]. SC was authorized as a secondary storage device by Nippon Electric Company in 1971 [17]. The SCs are constructed of solvent molecules, anion, and cation present in electrolytes, two electrodes that work as anode and cathode, and the separators, which act as insulation between them. It impregnates the electrodes into the solution. The SCs store energy in the electrode-electrolyte interface. These types of EST will not include any chemical changes [18]. The high degree of recyclability is related to electrostatic charge transfer [19]. The contents of electrodes and their structures make the SCs to have more surface area and high capacitance compared to CC. These advanced structures have undergone extensive research, and the more recent updates are reported in [20, 21]. The key categories for SCs electrode materials are extended surface area, carbon compounds, solvents, and oxide layers [22]. Carbon materials are commonly utilized in commercialized SCs because of their inexpensive price, extended surface area, high conduction capability, stability with high temperature co-efficient [23]. The capacitance of SCs increased with an increase in the porosity of electrodes. The above facts impact the working area of the electrode that is exposed to the electrolyte solution. According to researchers, if the sizes of ions in the solution and the pores of the electrodes are equal, the capacitance will be peaked [24]. Another crucial component that influences SCs performance is an electrolyte. High voltage level, ionic level, and electrochemical balance as well as low resistance, viscidity, vaporization, and cost are all requirements for the electrolyte [25]. The aqueous electrolytes provide low voltage in contrast with their greater ionic level and low resistivity. The organic electrolytes have high voltage levels, which suits SC manufacturing, whereas ionic electrolytes are only desirable at specific temperature levels [26].

10.3 Applications and Functions of Supercapacitor

Although SCs are of different types according to their constructions, all of them are high in power density compared to battery; this makes them to work in conjunction with Energy storage devices and conventional capacitors in various applications such as EST, power electronic devices, UPS, renewable integration, traction, and stabilizers.

10.3.1 Energy Storage Technology

SCs function similarly to regular capacitors but have a significantly larger capacitance (kilo farads) and are packaged in smaller sizes [27]. SCs employ a high-permittivity dielectric with porous active carbon to enhance the electrode surface area, enabling huge amounts of energy to be stored at the electrode surface. Compared to organic electrolytes, the aqueous types are inexpensive have low resistance and a high-temperature range [28]. With aqueous and organic electrolytes, the voltage of one cell is around one volt and three volts, respectively. Because the electrodes are not chemically damaged, SCs can be employed in deep charge/discharge operations for hundreds of thousands of cycles. The SCs has a service life of 12 years and can be cycled over 500,000 times. They have a far better power density than batteries because the charges are stored on the electrodes. They are substantially faster than batteries at charging and discharging. Some advantages of SCs during working are good efficiency, low working temperature, and no hazardous substance release. SCs are limited for certain aspects, such as low energy density and fast discharging. As a result, SCs can charge and discharge energy for a short period. Although the SCs' life span can be affected by voltage and temperature changes. Thus, the design of a SC system should include an ageing model that considers these factors [29]. HEVs are a common application for SCs. Because of their quick charge and discharge capability, they are employed in them. SC hybrids with batteries combine the high-power density with the large energy density. The usage of SCs in HEVs extends the battery's life and reduces the size of EST [30–33]. SCs are also applicable in improving power quality and frequency stability, avoiding fluctuations in renewable energy systems. The usage of SCs in wind energy generation systems that use the doubly fed induction machine and permanent magnet synchronous machine was discussed in [34–36], respectively. The papers [37– 39] use only battery in EST.

However, in view of reducing the fluctuations, high power density storage devices are required like SCs. Hybridized energy storage systems, batteries, and SC for renewable power applications are presented in [40–42]. The goal of hybrid EST is for SCs to absorb higher level frequency fluctuations while batteries deal with lower-level frequency fluctuations. Thus, optimization for this two-level fluctuation control is to be made in the hybrid ESTs. Zhou *et al.* and Cheng [43, 44] discuss the use of SCs in solar applications. A solution for balancing the power generated by the SEAREV wave energy converter is obtained by controlling SCs. State of charge (SOC) is examined and compared [45]. Lower-level frequency or longer-term fluctuations and higher-level frequency or shorter-term fluctuations are available in renewable energy systems. SC-based EST looks to be one of the most appropriate alternatives for the above issue because of its high-power performance and cycling capability. However, SCs are not designed to stabilize the power that lasts for more than a minute. SCs should be hybridized with storage devices with higher energy levels for a universal treatment in MG and EV. Table 10.1 shows a clear view of characteristics comparison of SC and battery as energy storage device.

10.3.2 Uninterruptible Power Supplies (UPSs)

When there is a power outage, UPS is utilized for supplying the critical loads without any interruption by storing the energy in their energy storage device from the grid and, in turn, supplies the loads even during the failure of power outage from the grid. Because of their great energy density, battery is the first choice in their storage device. However, batteries have several drawbacks, such as lower power density, short life span, and increased losses by pulsing currents [46]. A UPS has EST that includes rechargeable batteries and SCs, such that the SCs are utilized for peak power absorption and delivery during backup and peak demand, respectively [47]. SCs would help extend the overall EST service life by suppressing the peak current transients implied on batteries and maintaining the battery charge at an appropriate level.

Table 10.1 Comparison of supercapacitor with batteries as energy storage device.

	Batteries							**Super capacitor**
	Lead acid	**Nickel cadmium**	**Nickel metal hydride**	**Lithium ion**	**Lithium polymer**	**Sodium sulfur**	**Vanadium redox**	**Electric double-layer capacitor**
Energy Efficiency (%)	70–80	60–90	50–80	70–85	70	70	80	95
Energy Density (Wh/Kg)	20–35	40–0	60–80	100–200	200	120	25	<50
Power Density (W/Kg)	25	140–180	220	360	250–1,000	120	80–150	4,000
Lifecycle (Cycles)	200–2,000	500–2,000	<3,000	500–2,000	>1,200	2,000	>16,000	>50,000
Self-Discharge	Less	Less	High	Moderate	Moderate	-	Negligble	Very high

10.3.3 Power Electronic Device

In the power system, the conversions from alternating to direct current and direct to direct current are required very often. Batteries are required in large backup storage, although DC links often employ CC for short storage. In power electronic converter applications, SCs can be a viable alternative. Because SCs have high capacitance, the size of the energy backup unit can be reduced, resulting in a higher power rating and lower cost, which makes them a viable alternative for battery and CC in power electronic converters. In [48], SC-based governor methods presented a converter with high efficiency. The power switching devices using bulky inductors can be replaced by SCs, which acts as an efficient voltage divider. In a multi-level inverter, the SCs were deployed as a direct current link was discussed [49]. The SCs' enormous capability aided in creating a strong direct current link with high stability, allowing the polluted load to be fed even when the voltage dropped.

10.3.4 Renewable Integration

In recent years, renewable energy systems have received much attention. However, due to the intermittent quality of many renewable energy systems, there are significant frequency–voltage swings during integration. As a result, an ESS is frequently required to balance electricity generation and consumption in a renewable energy system. Energy storage provides load optimization, energy trading, major frequency management, and supply peak demand [50]. To alleviate the effects of intermittency, SCs are used in the EST of wind energy generation systems [51]. The SCs boost the bus's low-voltage ride-through capabilities by reinforcing it during transients. In wind energy generation systems, the conversion in three-level inverters uses SCs to hike the capacitance by varying the voltage [52]. A basic power smoothing approach based on SC is used in a distributed network to condition renewable energy generation [53]. The goal of the technique was to manage the SCs SOC and generate the power required to smooth out the power changes of a renewable energy generation system. SC banks having a series of connections are utilized to implement the maximum power point tracking in photovoltaic generation system by their high charging and discharging characteristics for varying climatic conditions [54].

10.3.5 Stabilization and Traction

SCs can be used to mitigate the large transients in microgrids (MGs) that caused by constant power loads. SC module controller along with fuel cell is used as stabilizers. The controlling technique uses the Lyapunov theory model, which operates with the obtained local information. The simulations are carried out in MATLAB/Simulink software. In a decentralized model, the non-linear and robust SC-based PI controllers are applied, and a different analysis is done in the presence of single line-ground faults [55]. In hybrid energy storage system (HESS), the SC plays a vital role that counts on high-frequency components of power imbalance. Generally, the states of charge of HESS are concentrated. In a hybrid renewable energy system, the power balancing is done using HESS controllers. These stabilizers regulate DC-voltage under surplus and in deficit power mode [56]. In traction, due to the increase in population, the density of trains utilized increases. This weakens the direct-current substations. These lead to voltage sags and fluctuations in the alternating-current and direct-current buses by the nature of constant power loads. SC-based active stabilizers address issues such as direct-current voltage fluctuation and instability in the traction system. Large passive capacitors facilitate the rise of stability margin across the direct-current bus terminals. This leads to the addition of a bulky capacitor bank that degrades the efficiency and quick response of the system [57]. In such conditions, the SCs can attain the objective by overcoming the backdrops. The above-mentioned applications and functions of SCs are depicted in Tables 10.2 and 10.3.

Table 10.2 Applications and functions of SC analysis I.

Applications	System	Storage	Functions
Energy Storage	HEV	SC + battery	Extends battery life and reduces the EST size
Renewable Energy System	Wind energy, solar energy, and SEAREV wave energy	SC + battery	Reduces fluctuation, frequency stability, and two-level fluctuation control
Power Backup	UPS	SC + battery	Peak power absorption, and deliver peak demand during backup
Power Electronic Device	Power system	SC	SC-based converters, replacing bulky inductors in power switching devices, DC link in multilevel inverter, and creating strong DC link with high stability.

Table 10.3 Applications and functions of SC analysis II.

Applications	System	Storage	Functions
Integration	Renewable system	SC + Battery	Balances frequency–voltage swings during integration of RES, load optimization, energy trading, major frequency management, supply peak demand, and power smoothing, implementing MPPT in PV generation systems
Stabilization	Grid	SC + fuel cell	Stabilizer, regulate DC voltage, and HESS controllers used for power balancing
Fault Analysis	Grid	SC	SC-based PI controllers for fault analysis
Strengthening of DC Substations	Traction	SC	Large passive capacitors facilitate the rise of stability margin, and reducing the size of bulky capacitor bank

10.4 Power Converter Topologies for Energy Storage

HESS and MG can be linked using various topologies. Various topologies can combine highly efficient storage systems with huge energy density and power density [58–60]. In [61], a broad review was done on the methods of HESS that are conferred. Passive, semi-active, and active power converter topologies are depicted in Figure 10.4.

10.4.1 Passive Power Converter Topology

A less expensive, non-complex, and efficient method of connecting different storage devices with identical voltage is connected with the help of passive topology. Their inner resistance and VI characteristics decide the power sharing between huge energy density and power density storage. As a result, lower energy density types of storage devices are utilized as low pass filters for the HESS [62–65].

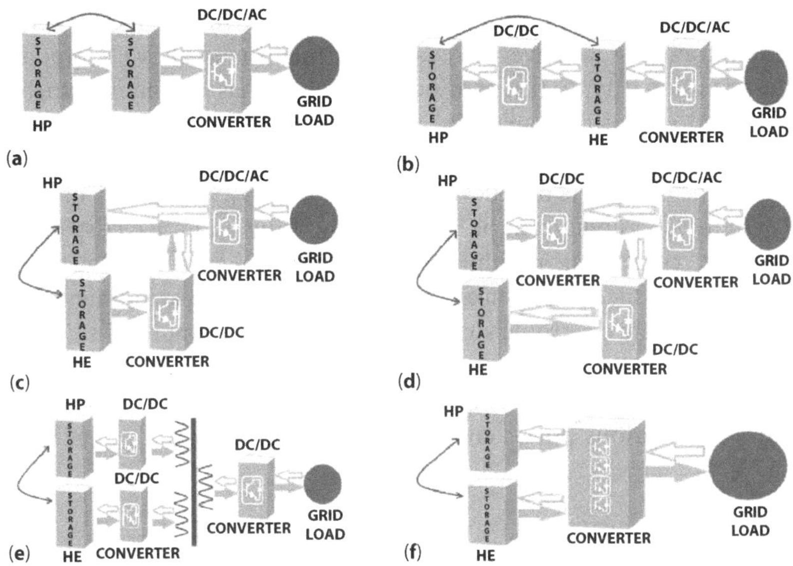

Figure 10.4 Topologies. **(a)** Passive, **(b)** semi-active, **(c)** series active, **(d)** parallel active, **(e)** isolated active, and **(f)** multilevel.

10.4.2 Semi-Active Power Converter Topology

Unlike the passive topology, a power converter is added in between the two different kinds of storage devices in the semi-active topology. The high energy storage devices are connected to the DC bus directly. Even though using a converter makes the system bulky and raises the cost, this type provides greater linearity and delivery capacity; such types are discussed in [66]. This type's additional usage of converters enhances the HESS's working range [67].

10.4.3 Active Power Converter Topology

These types are made of multiple backup power devices linked with the system through multiple power converters. Apart from the system's complexity, losses, and expensive cost, this topology can consciously control all the storage. This type gives an intelligent control strategy and is often used in a complete active HESS [68]. The concurrent active hybrid topology is depicted in Figure 10.4D. It uses multiple converters to control the power of both the storage. Conventional parallel topologies have multiple storage devices connected to a common DC bus through cascade converters, making the system inefficient. Momayyezan *et al.* [69] suggest a reconfigurable topology to solve the challenges mentioned above. Compared to the traditional type, the modifiable topology has advantages such as the reduced size of the capacitor, rise in efficiency during energy transfer modes. In some studies, the hybridized storage is controlled and connected to the bus by the multilevel converter [70–75]. The quality of power and system accessibility can be improved by using a multilevel structure. Meanwhile, combining multiple ESSs in a single converter lowers the expenses and reduces sophisticated control. This is because multilevel converters have many electronic switches, and control is complex. In [76–78], isolated-multiport adapters are utilized in multiple storage connected systems. Due to high capabilities, the active structure has been considered in recent years. Various power converters can be used to connect HESS to MG [79].

10.5 Comparative Study of Topologies

The hybridized energy storage topologies give more effect in the energy management systems. The straight control on the power storage system is not possible in passive types. In contrast, the power of one of the storages is controllable in the semi-active topology, and the voltage should be equal to the bus voltage. At the expense of poor performance, the active topology employs a reasonable control strategy to control the power of both storage facilities. Costs, efficiency, controllability, difficulty, and flexibility must all be viewed when selecting an appropriate topology. The passive configuration is simple and inexpensive, although difficult to manage. The active configuration outperforms the passive controllability and flexibility while still focusing on different constraints like SOC. Figure 10.5 gives a comparative study of different topologies [80]. However, the scale and risk are ridiculously expensive. Semi-active configuration offers minimal control at a cheaper price.

10.6 Converter Technologies in the Power Electronic Field

The topologies, abilities, and weak points of power electronic converters are described in this section. The structure of the console, the automated system, and the computational formalism are also proffered.

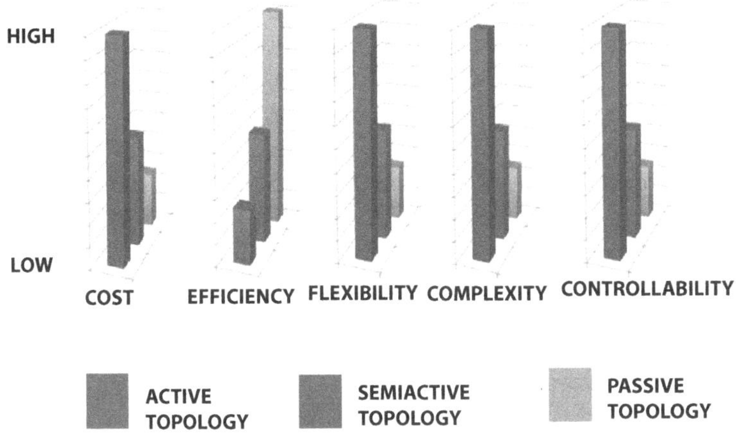

Figure 10.5 Comparison of different topologies.

Power electronic converters end up making use of semiconductor switches. These switches operate at a variable frequency up to 50 Hertz to 60 Hertz and, in some instances, up to 100-Megahertz shortwave radios. Alternating–direct current, direct–alternating current, direct–direct current, and alternating– alternating current are the types of power electronic converters [81]. The above converters are still being used in a variety of applications. In home appliances, an alternating–direct current converter is used to convert the incoming alternating current to a direct current. Direct–alternating current converters are used in grid-connected renewable energy sources, and adjustable speed drives to provide constant/variable AC electrical voltage. Direct–direct current converters are used in EST to achieve strictly controlled DC voltage for different loads. The direct–direct converter is also used to enhance the power output of photovoltaic systems and wind generators. Alternating–AC sources use alternating current converters for varying the voltage and frequency [82].

10.6.1 Converters for AC Source DC Applications

Alternating–direct current converters are being utilized to convert alternating voltage to a constant direct voltage output. Such converters are suitable for all types of power uses [83]. Uncontrolled rectifiers are made up of diodes and are also known as unidirectional alternating–direct current converters [84]. On the other hand, a diode rectifier produces quasi distortion in the input current. As a result, much work is required to create a filter that eschews AC grid distortion. Because diode rectifiers are incapable of meeting the IEEE standard of 5% of overall total harmonic distortion, numerous filtrations should be put in place at the input [85]. Bulky and heavy filters, on the other hand, may harm the circuit's active power, lowering transmission line effectiveness. Because of advancements in power semiconductor devices, unproductive diodes can now be replaced with convenient power factor correction (PFC) circuits that provide an appropriate incoming signal with a unity power factor and Total harmonic distortion just under 5% [86]. Figure 10.6 depicts two PFC circuits based on harmonic compensation concepts. Figures 10.6A and B show the harmonic injection by voltage source and current source, respectively.

The voltage injection structure consists of a filtration system of series H-bridge combinations linked by an inductance [87]. This active filter indemnifies the voltage ripple. On the other hand, the current injection method is a more effective way of compensating harmonics for PFC. Bidirectional series switches are used to infuse flux into main lines. This approach has the advantages of a reduced voltage rating, reference voltage swapping achievement, and low power dissipation, which are all advantages of this product. In both the harmonic injection compensation, the output voltage is uncontrollable.

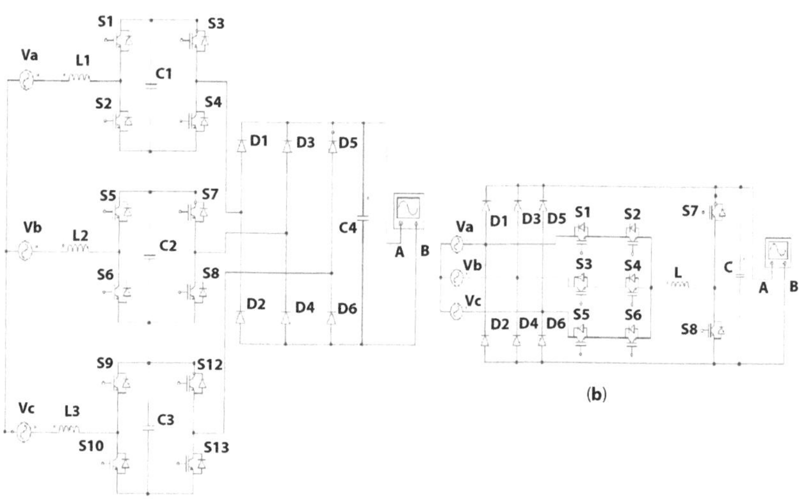

(a)

(b)

Figure 10.6 Harmonic injection methods. (a) Voltage source and (b) current source.

10.6.2 Converters for DC Source AC Applications

Direct–alternating current converters generate an alternating current output with specific parameters from a DC supply. Based on their load and source, the converter has been categorized into two categories: voltage source inverter (VSI) [88] and current source inverter (CSI) [89]. The alternating output voltage and current are controlled individually in VSI and CSI. VSI employ buck topology, whereas CSI employs voltage boost inversion. A three-phase direct–alternating current converter can perform well above 5-kilowatt power applications. Figures 10.7A and B depict three-level–three-phase VSI and CSI configuration, respectively. Each topology is connected to a three-phase load via three–half-bridge inverter legs. The circuit is built in the same way as a single-phase inverter but with an additional inverter leg. To overcome the challenges of high power-based devices, the number of phases can be increased. This method enhances output power and increases the inverter's regulating flexibility. A three-level line-to-line output signal can be produced using the proper control techniques. Switching losses, switching device stresses, harmonic distortion, and radio radiation can be lowered by employing the correct structures, govern, and cope ups. Multiple multilevel power converters with multiple components have been presented in the literature [90, 91].

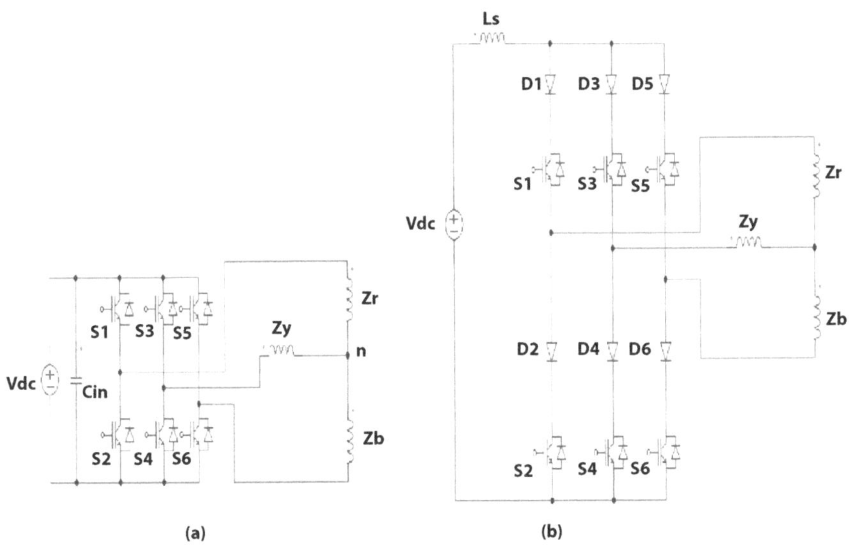

Figure 10.7 Three-level–three-phase (a) VSI and (b) CSI configuration.

10.6.2.1 Neutral Point Clamped Topologies

Large power users benefit more from neutral point clamped topologies (NPCTs). Figures 10.8A and B illustrate three-level and fie-level diode clamped NPCTs, respectively. Nonetheless, the diode clamped NPCT has unbalanced power transmission failures, leading to lower power output abilities. The three-level active NPCT addresses the shortcomings of the diode clamped NPCT by substituting anti-parallel diodes for the clamping diodes to provide a manageable route for the neutral current. As shown in Figures 10.8C and D, various active NPCTs such as five-level, seven-level, and nine-level are used in a variety of applications [92, 93].

10.6.2.2 Inverter Topologies With Multiple Levels

As depicted in Figures 10.9a, b, and c the output voltage requirement is met using multilevel inverter topologies that utilize various parameters of basic power cells. The multiple-input H-bridge has a separate power source for every basic compartment; meanwhile, the multi-level modular converter and flying capacitor have a shared input voltage. Multi-cell multi-level topologies benefit from being modular, resulting in low operating costs. Nevertheless, proactive and reactive components and peripheral devices

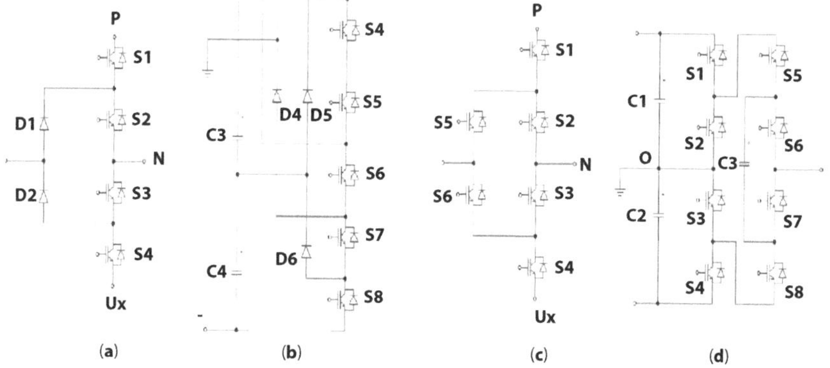

Figure 10.8 NPCT. (a) Three-level diode clamped, (b) five-level diode clamped, (c) three-level active, and (d) five-level active.

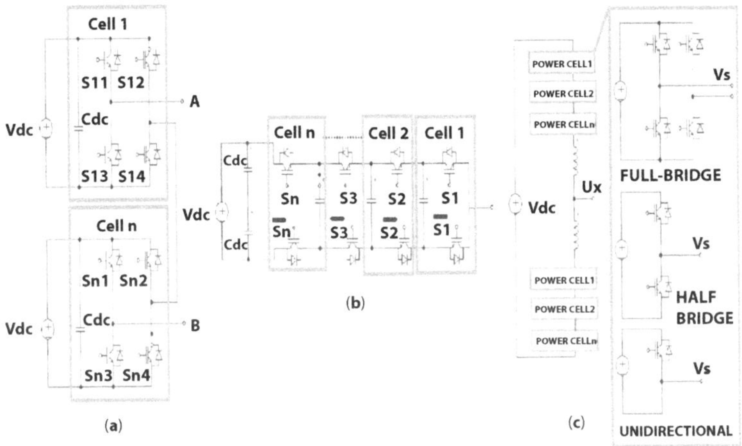

Figure 10.9 (a) Multiple-input H-bridge. (b) Flying capacitor. (c) Multi-level modular converter.

and enhanced controller design are involved in managing the potential difference in the capacitor [94, 95].

10.6.3 Converters for DC Source DC Applications

Direct–direct current converter reduces or increases the direct current voltage level. These can supply variable loads and regulate the speed of DC motors by pulse width modulation techniques. Complex structures can be built based on the application's needs by inductive devices between the input and output [96, 97]. These converters are classified into isolated and non-isolated.

10.6.3.1 Non-Isolated Converters

As shown in Figure 10.10, non-isolated converters are constructed with some active and passive devices such as inductors, capacitors, diodes, and a few switches. The magnetic characteristics of these converters make them bulkier and heavy in construction with low efficiency and power density. These converters can operate in two conduction modes such as continuous and discontinuous. Compared to continuous mode, the discontinuous mode has less efficiency and high ripple, although they are preferable for quick dynamic response [98]. These non-isolated converters will do

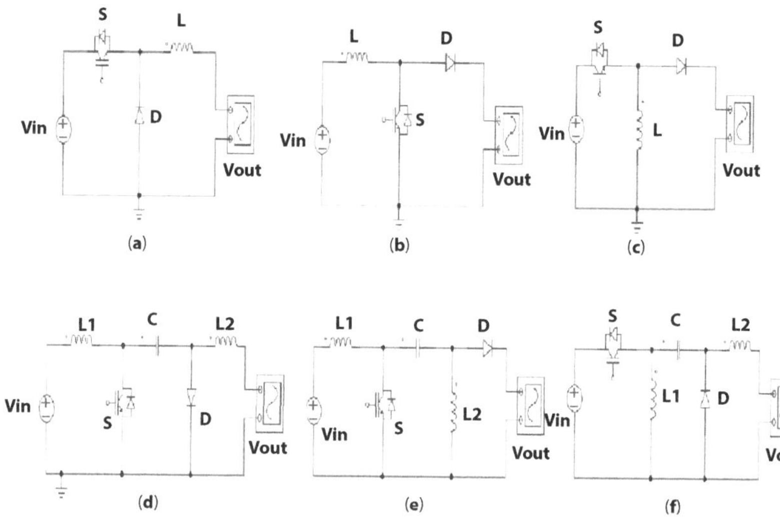

Figure 10.10 Non-isolated converters. (a) Buck. (b) Boost. (c) Buck–Boost. (d) CUK. (e) SEPIC. (f) Zeta.

different operations such as step up, step down, and combination of both, named buck, boost, and buck–boost converters, respectively [99]. A buck converter uses a diode that will allow the load current to free wheel around it while the supply is turned off and controls the output voltage [100, 101]. The boost converters are similar in characteristic to buck converters. Their inductors are charged magnetically when the switch is turned on and supplies the load through the diode when the switch is turned off. Thus, the addition of input voltage and the voltage supplied by the inductors gives the boosted voltage output [102, 103].

The buck–boost converters function both as a buck and boost according to the level of the duty cycle. If the duty cycle is less than 0.5, then buck conversion occurs, whereas, if it ranges from 0.5 to 1, then boost conversion will occur [104, 105]. The CUK converters are similar and superior to the buck–boost converter. The configuration makes the simplified drive circuits of CUK converters of a grounded source switch. The addition of a capacitor and a transformer makes them move to an isolated operating state. CUK converters can produce less ripple output by using coupled inductors and supply multiple outputs by utilizing secondary winding transformers [106, 107]. In such a way, the SEPIC converters will also do the buck and boost functions with a non-inverted output. The voltage levels are changed by the energy transfer between the passive components. SEPIC produces pulsating output current wave shape [108, 109].

ZETA converters work similarly to the single ended primary inductor converter (SEPIC). Its configuration is based on buck topology, and possession of floating switches makes them depend on additional components. Unlike CUK, these converters have input current with pulsating characteristics [110, 111]. Different non-isolated converters are shown in Figure 10.10. Tables 10.4 and 10.5 give a clear picture of comparative analysis of non-isolated converters.

10.6.3.2 Isolated Converters

Electrical isolation in direct–direct Current converters is achieved by using a transformer or a coupled inductor. They can provide satisfactory results by lowering the interferences and interruptions. Such converters operate at high frequencies, storing energy in inductors in one cycle and transferring

Table 10.4 Non-isolated converters comparison 1.

Converters	Loss	Stress level	Control
Buck	Negligible	Low	Complex
Boost	Negligible	Low	Complex
Buck–Boost	Negligible	Low	Complex
CUK	Low	Low	Complex
SEPIC	Low	Low	Complex
Zeta	Low	Low	Complex

Table 10.5 Non-isolated converters comparison 2.

Converters	Cost	Competence	Appearance	Handling difficulty
Buck	Less	High	Small	High
Boost	Moderate	High	Small	High
Buck–Boost	Moderate	High	Medium	High
CUK	Moderate	High	Medium	Moderate
SEPIC	Less	High	Small	High
Zeta	Moderate	High	Medium	High

it to the load in the next consequent cycles. As seen in Figure 10.11A, flyback converters are constructed with inductors instead of isolation transformers. The operation of a Flyback converter is identical to buck–boost converters in storing energy and transferring to load when the switch is in ON and OFF condition, respectively. These converters can supply applications that have low power [112]. A forward direct–direct current converter, as demonstrated in Figure 10.11B, is a type of buck converter that is built with a transformer that has no air space. This converter can supply power without storing it when the switch is turned on. The voltage levels of this converter could be adjusted by adjusting the transformers ratio. In moderate applications, this converter performs admirably [113]. As depicted in Figure 10.11C, the dual switch on the input side of the transformer is used to configure a push–pull converter. The transformer in this converter is bidirectionally excited. This converter has less interruption with a constant power supply [114]. As illustrated in Figure 10.11D, a half-bridge converter is also constructed with dual switches that are turned on and off alternately to energies the transformer. In this converter, voltage splitting can be possibly done using capacitors at the transformer primary. Because these converters are used in applications with power ranging less than 500 W, the switches present in them experience fewer voltage stresses [115]. As shown in Figure 10.11E, a full bridge converter is built with quadrupled

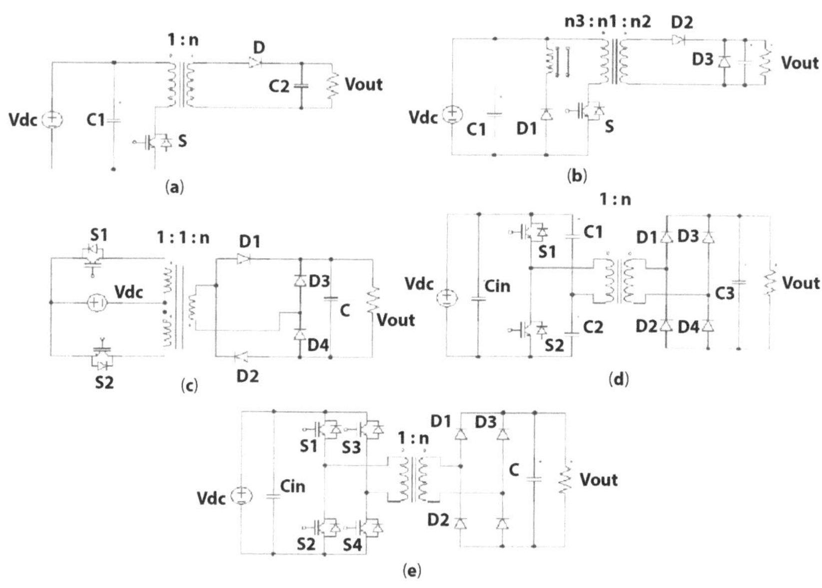

Figure 10.11 (a) Flyback. (b) Forward. (c) Push–pull. (d) Half-bridge. (e) Full bridge.

operated symmetrical switches on the input side of the transformer. For voltage conversion, this converter works similarly to a buck converter. These converters are suitable for power ranging above 1,000 W [116].

10.6.4 Converters for AC Source AC Applications

Alternating–alternating converters are commonly utilized in industries where the load types vary with different controlling method implementation. A DC-link voltage is always connected to a good quality alternating-to-alternating converter. Cycloconverters are the primary device for changing frequency from one level to another while avoiding DC-link [117]. It is constructed by both the half and full wave rectifiers. The rectifiers' triggering pulses can vary frequency and voltage autonomously and continuously [118]. Figures 10.12A and B show single- and three-phase cycloconverter used in electric locos and high-power systems. Usually, these converters utilize numerous pulses to maintain stability and linear output [119]. Figures 10.13A and B depict the 2 × 2 and 3 × 3 matrix converters. Matrix converters are built using bidirectional switches with no DC-link voltage. Matrix converters have a simple, compact structure and a power factor of one and deliver a stable sinusoidal output. Furthermore, this converter's maximum voltage is quick and equal to the maximum value of the input [120].

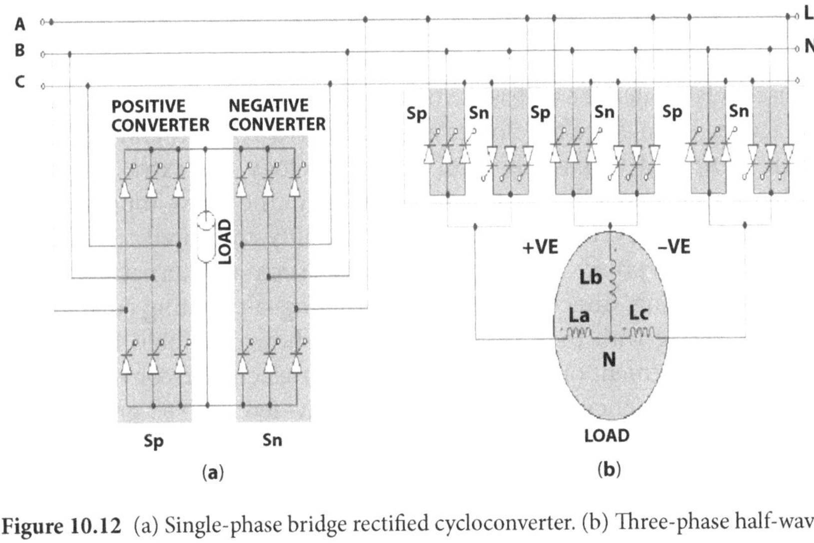

Figure 10.12 (a) Single-phase bridge rectified cycloconverter. (b) Three-phase half-wave cycloconverter.

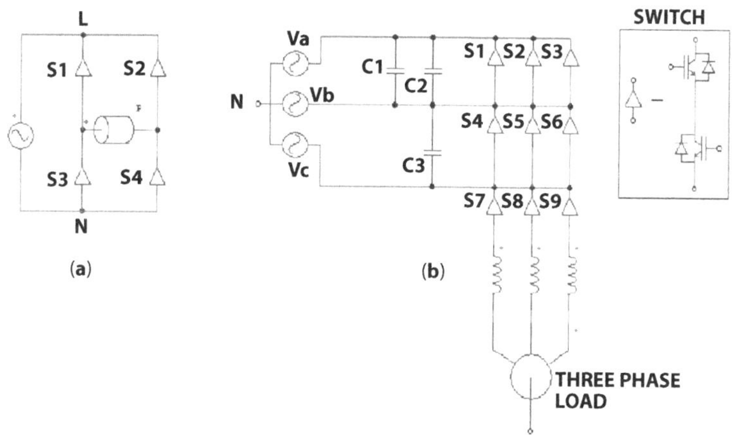

Figure 10.13 (a) 2 × 2 matrix converter. (b) 3 × 3 matrix converter.

10.7 Conclusion and Future Scope

The goal of this review is to analyze the applications of SCs and converters to increase the working efficiency and the comprehensive performance of an EST in AC/DC grids and EV applications. The usage of SC gives higher efficiency, better performance, reduction of size, cost, and pollution and, ultimately, increases the stabilization. SCs can improve the life cycles of several systems and devices such as ESS, UPS, power electronic devices, renewable energy integrations, stabilization, and tractions. Additionally, the analysis of different converter topologies gives the performance comparisons of different converters and their applications. The review paper precisely elaborates the different topologies such as passive, semi-active, and active. The passive topologies are productive, concise, and profitable. When compared to passive, semi-active topology has extra converters but has greater tunability and delivery capacity. The active topologies are higher in efficiency, flexibility, and controllability compared to the other topologies. The power electronic converters such as alternating–direct, direct–alternating, direct–direct, and alternating–alternating currents are discussed along with their pros and cons as shown in Table 10.6. Thus, the proper placement of SCs and proper power electronic converters may improve the efficiency of the renewable energy–fed AC/DC grids and in EVs. Even improvising converter topologies based on SCs can give better results than the conventional types.

This paper gives insight as follows: i) The proper usage of SC in the ESS and power electronic stabilizers gives a high-performance output of any

Table 10.6 Pros and cons of converters.

S. no.	Converter	Pros	Cons	Application
1	AC-DC Converter	Conduction losses are less; voltage rating are low; no voltage switching	Power factor is low; power supply is limited; harmonics distortions are high	Household electronic appliances; cell phone chargers; laptop chargers
2	DC-AC Converter	Switching losses are less; stresses are less; harmonics are less	Circuits are complex; conversion losses are heavy; costlier	Solar energy generation; wind energy generation
3	DC-DC Converter	Good dynamic responses; efficiency is high; cheaper in cost	Size is big; switching losses are high	EV; HEV; photovoltaic-based charging stations
4	AC-AC Converter	Efficiency is high; low maintenance; good flexibility control	Harmonics are high; size is big	Traction; high rated motors

systems, such as MG, EV, and applications that require high power density with a long-life cycle, etc. ii) Appropriate converter topology usage for various applications will give intelligent control, high efficiency, low cost, less complexity, high flexibility, and controllability. Overall, future research in active topologies such as multilevel converters will meet those objectives. iii) All converter topologies discussed above use CC for stabilization and boosting. Converters that are utilized in high power rating applications such as traction system are bulkier in nature, which can be overcome by proper utilization of SCs.

Acknowledgments

The authors are very grateful to the VIT University, Chennai, and Annamalai University, Annamalainagar, Chidambaram, Tamilnadu, India, for the extensive facilities in rendering this work.

Conflict of Interest

On behalf of all authors, the corresponding author states that there is no conflict of interest.

References

1. Kim BK, Sy S, Yu A, Zhang J, "Electrochemical supercapacitors for energy storage and conversion," *Handbook of clean energy systems*, pp.1-25, Jun. 2015.
2. Conway BE, "Electrochemical supercapacitors: scientific fundamentals and technological applications," *Springer Science & Business Media*, Apr 2013.
3. Chan, Candace K., Hailin Peng, Gao Liu, Kevin McIlwrath, Xiao Feng Zhang, Robert A. Huggins, and Yi Cui, "High-performance lithium battery anodes using silicon nanowires," *Nat. Nanotechnol.*, vol. 3, no. 1, pp. 31-35, 2008.
4. Simon P, Gogotsi Y, "Materials for electrochemical capacitors," *Nat. Mater.*, vol. 7, no. 11, pp. 845-54, Nov 2008.
5. Pandolfo, Tony, Vanessa Ruiz, Seepalakottai Sivakkumar, and Jawahr Nerkar, "General properties of electrochemical capacitors," *Supercapacitors: Materials, systems, and applications*, pp. 69-109, 2013.
6. Wu, Peng, Jingsong Huang, Vincent Meunier, Bobby G. Sumpter, and Rui Qiao, "Complex capacitance scaling in ionic liquids-filled nanopores," *ACS Nano*, vol. 5, no. 11 pp. 9044-9051, 2011.

7. Vivekchand, S. R. C., Chandra Sekhar Rout, K. S. Subrahmanyam, Achutharao Govindaraj, and Chintamani Nagesa Ramachandra Rao, "Graphene-based electrochemical supercapacitors," *J. Chem. Sci.* vol. 120, pp. 9-13, 2008.
8. Zhang, Li Li, and X. S. Zhao, "Carbon-based materials as supercapacitor electrodes," *Chem. Soc. Rev.* vol. 38, no. 9, pp. 2520-2531, 2009.
9. Huang, Jingsong, Bobby G. Sumpter, and Vincent Meunier, "Theoretical model for nanoporous carbon supercapacitors," *Angew. Chem.*, vol.120, no. 3, pp. 530-534, 2008.
10. Chmiola, John, G. Yushin, Yury Gogotsi, Christele Portet, Patrice Simon, and Pierre-Louis Taberna, "Anomalous increase in carbon capacitance at pore sizes less than 1 nanometer," *Science*, vol. 313, no. 5794, pp. 1760-1763, 2006.
11. Lavanya, A., J. Divya Navamani, K. Vijayakumar, and R. Rakesh, "Multi-input DC-DC converter topologies-a review," In *2016 International Conference on Electrical, Electronics, and Optimization Techniques (ICEEOT), IEEE*, pp. 2230-2233, 2016.
12. Reddy, K. Jyotheeswara, and Sudhakar Natarajan, "Energy sources and multi-input DC-DC converters used in hybrid electric vehicle applications–A review," *Int. J. Hydrogen Energy*, vol. 43, no. 36, pp. 17387-17408, 2018.
13. Yalamanchili, Krishna P., and Mehdi Ferdowsi, "Review of multiple input DC-DC converters for electric and hybrid vehicles," In *2005 IEEE vehicle power and propulsion conference, IEEE*, pp. 160-163, 2005.
14. Rehman, Zubair, Ibrahim Al-Bahadly, and Subhas Mukhopadhyay, "Multiinput DC–DC converters in renewable energy applications–An overview," *Renew. Sustain. Energy Rev.*, vol. 41, pp. 521-539, 2015.
15. Khosrogorji, Soheil, Mohammad Ahmadian, Hossein Torkaman, and S. Soori, "Multi-input DC/DC converters in connection with distributed generation units–A review," *Renew. Sustain. Energy Rev.*, vol. 66, pp. 360-379, 2016.
16. Becker, Howard I, "*Low voltage electrolytic capacitor*," U.S. Patent 2, 800, 616, issued July 23, 1957.
17. Sharma, Pawan, and T. S. Bhatti, "A review on electrochemical double-layer capacitors," *Energy Convers. Manage.*, vol. 51, no. 12, pp. 2901-2912, 2010.
18. Zhang, Lei, Xiaosong Hu, Zhenpo Wang, Fengchun Sun, and David G. Dorrell, "Experimental impedance investigation of an ultracapacitor at different conditions for electric vehicle applications," *J. Power Sources*, vol. 287, pp. 129-138, 2015.
19. Conway, Brian E, "Electrochemical supercapacitors: scientific fundamentals and technological applications" *Springer Science & Business Media*, 2013.
20. Frackowiak, Elzbieta, and Francois Beguin, "Carbon materials for the electrochemical storage of energy in capacitors," *Carbon*, vol. 39, no. 6, pp. 937-950, 2001.
21. Dhand, A. and Pullen, K, "Review of flywheel based internal combustion engine hybrid vehicles," *Int. J. Automot. Technol.*, vol. 14, pp. 797-804, 2013.

22. Lee, H., Cho, M.S., Kim, I.H., Do Nam, J., Lee, Y, "Ruox/polypyrrole nanocomposite electrode for electrochemical capacitors," *Synth. Met.*, vol. 160, no. 9-10, pp. 1055–1059, 2010.
23. Zhang, Y., Feng, H., Wu, X., Wang, L., Zhang, A., Xia, T., Dong, H., Li, X., Zhang, L, "Progress of electrochemical capacitor electrode materials: A review," *Int. J. Hydrogen Energy*, vol. 34, no. 11, pp. 4889–4899, 2009.
24. Largeot, C., Portet, C., Chmiola, J., Taberna, P.-L., Gogotsi, Y., Simon, P, "Relation between the ion size and pore size for an electric double-layer capacitor," *J. Am. Chem. Soc.*, vol. 130, no. 9, pp. 2730–2731, 2008.
25. Wang, G., Zhang, L., Zhang, J, "A review of electrode materials for electrochemical supercapacitors," *Chem. Soc. Rev.*, vol. 41, no. 2, pp. 797– 828, 2012.
26. Snook, G.A., Kao, P., Best, A.S, "Conducting-polymer-based supercapacitor devices and electrodes," *J. Power Sources*, vol. 196, no. 1, pp. 1–12, 2011.
27. Vazquez, S., Lukic, S.M., Galvan, E., Franquelo, L.G., Carrasco, J.M, "Energy storage systems for transport and grid applications" *IEEE Trans. Ind. Electron.*, vol. 57, no. 12, pp. 3881–3895, 2010.
28. Rabiee, A., Khorramdel, H., Aghaei, J, "RETRACTED: A review of energy storage systems in microgrids with wind turbines," *Elsevier*, 2013.
29. Chen, W., ˚Adnanses, A.K., Hansen, J.F., Lindtjørn, J.O., Tang, T, "Supercapacitors based hybrid converter in marine electric propulsion system," in *The XIX International Conference on Electrical Machines- ICEM 2010*. IEEE, pp. 1–6, 2010.
30. Khaligh, A., Li, Z, "Battery, ultracapacitor, fuel cell, and hybrid energy storage systems for electric, hybrid electric, fuel cell, and plug-in hybrid electric vehicles: State of the art," *IEEE Trans. Veh. Technol.*, vol. 59, no. 6, pp. 2806–2814, 2010.
31. Zhou, Z., Benbouzid, M., Charpentier, J.F., Scuiller, F., Tang, T, "A review of energy storage technologies for marine current energy systems," *Renew. Sustain. Energy Rev.*, vol. 18, pp. 390–400, 2013.
32. Thounthong, P., Chunkag, V., Sethakul, P., Davat, B., Hinaje, M, "Comparative study of fuel-cell vehicle hybridization with battery or supercapacitor storage device," *IEEE Trans. Veh. Technol.*, vol. 58, no. 8, pp. 3892–3904, 2009.
33. Schaltz, E., Khaligh, A., Rasmussen, P.O, "Influence of battery/ultracapacitor energy-storage sizing on battery lifetime in a fuel cell hybrid electric vehicle," *IEEE Trans. Veh. Technol.*, vol. 58, no. 8, pp. 3882–3891, 2009.
34. Dhand, A. and Pullen, K., "Review of flywheel based internal combustion engine hybrid vehicles," *Int. J. Automot. Technol.*, vol. 14, pp. 797-804, 2013.
35. Qu, L., Qiao, W., "Constant power control of DFIG wind turbines with supercapacitor energy storage," *IEEE Trans. Ind. Appl.*, vol. 47, no. 1, pp. 359–367, 2010.
36. Muyeen, S., Takahashi, R., Murata, T., Tamura, J, "Integration of an energy capacitor system with a variable-speed wind generator," *IEEE Trans. Energy Convers.*, vol. 24, no 3, pp. 740–749, 2009.

37. Goel, P.K., Singh, B., Murthy, S.S., Kishore, N, "Isolated wind-hydro hybrid system using cage generators and battery storage" *IEEE Trans. Ind. Electron.*, vol. 58, no. 4, pp. 1141–1153, 2009.
38. Abedini, A., Nikkhajoei, H, "Dynamic model and control of a wind turbine generator with energy storage," *IET Renew. Power Gener.*, vol. 5, no. 1, pp. 67–78, 2010.
39. Bragard, M., Soltau, N., Thomas, S., De Doncker, R.W, "The balance of renewable sources and user demands in grids: Power electronics for modular battery energy storage systems," *IEEE Trans. Power Electron.*, vol. 25, no. 12, pp. 3049–3056, 2010.
40. Li, W., Joos, G, "A power electronic interface for a battery supercapacitor hybrid energy storage system for wind applications," In: *2008 IEEE Power Electronics Specialists Conference*, IEEE, pp. 1762–1768, 2008.
41. Li, W., Joos, G., Belanger, J, "Real-time simulation of a wind turbine generator coupled with a battery supercapacitor energy storage system," *IEEE Trans. Ind. Electron.*, vol. 57, no. 4, pp. 1137–1145, 2009.
42. Jia, H., Fu, Y., Zhang, Y., He, W, "Design of hybrid energy storage control system for wind farms based on flow battery and electric double layer capacitor" In: *2010 Asia-Pacific Power and Energy Engineering Conference*, IEEE, pp. 1–6, 2010.
43. Zhou, H., Bhattacharya, T., Tran, D., Siew, T.S.T., Khambadkone, A.M, "Composite energy storage system involving battery and ultracapacitor with dynamic energy management in microgrid applications," *IEEE Trans. Power Electron.*, vol. 26, no. 3, pp. 923–930, 2010.
44. Cheng, Y, "Super capacitor applications for renewable energy generation and control in smart grids" In: *2011 IEEE International Symposium on Industrial Electronics*, IEEE, pp. 1131–1136, 2011.
45. Aubry, J., Bydlowski, P., Multon, B., Ahmed, H.B., Borgarino, B, "Energy storage system sizing for smoothing power generation of direct wave energy converters," In: *ICOE 2010*, pp. 978, 2010.
46. Dougal, R.A., Liu, S., White, R.E, "Power and life extension of battery ultracapacitor hybrids," *IEEE Trans. Compon. Packag. Technol.*, vol. 25, no. 1, pp. 120–131, 2002.
47. Lahyani, A., Venet, P., Guermazi, A., Troudi, A, "Battery/supercapacitors combination in uninterruptible power supply (ups)," *IEEE Trans. Power Electron.*, vol. 28, no. 4, pp. 1509–1522, 2012.
48. Kankanamge, K., Kularatna, N, "Improving the end-to-end efficiency of dc–dc converters based on a supercapacitor-assisted low-dropout regulator technique," *IEEE Trans. Ind. Electron.*, vol. 61, no. 1, pp. 223–230, 2013.
49. Ortuzar, M., Carmi, R., Dixon, J., Moran, L, "Voltage source active power filter, based on multi-stage converter and ultracapacitor dc-link," In: *IECON'03. 29th Annual Conference of the IEEE Industrial Electronics Society (IEEE Cat. No. 03CH37468)*, IEEE, vol. 3, pp. 2300–2305, 2003.

50. Vazquez, S., Lukic, S.M., Galvan, E., Franquelo, L.G., Carrasco, J.M, "Energy storage systems for transport and grid applications" *IEEE Trans. Ind. Electron.*, vol. 57, no. 12, pp. 3881–3895, 2010.
51. Abbey, C., Joos, G, "Supercapacitor energy storage for wind energy applications," *IEEE Trans. Ind. Appl.*, vol. 43, no. 3, pp. 769–776, 2007.
52. Jayasinghe, S.G., Vilathgamuwa, D.M, "Flying supercapacitors as power smoothing elements in wind generation," *IEEE Trans. Ind. Appl.*, vol. 60, no. 7, pp. 2909–2918, 2012.
53. Pegueroles-Queralt, J., Bianchi, F.D., Gomis-Bellmunt, O, "A power smoothing system based on supercapacitors for renewable distributed generation," *IEEE Trans. Ind. Electron.*, vol. 62, no. 1, pp. 343–350, 2014.
54. Mutoh, N., Inoue, T, "A control method to charge series-connected ultraelectric double-layer capacitors suitable for photovoltaic generation systems combining MPPT control method," *IEEE Trans. Ind. Electron.*, vol. 54, no. 1, pp. 374–383, 2007.
55. Azimi, S.M., Hamzeh, M, "Voltage/current large transient suppression in dc microgrids using local information and active stabilizing capability," *IEEE Syst. J.*, vol. 14, no. 1, pp. 1109–1116, 2019.
56. Krishan, O., Suhag, S, "Power management control strategy for hybrid energy storage system in a grid-independent hybrid renewable energy system: a hardware-in-loop real-time verification," *IET Renew. Power Gener*, vol. 14, no. 3, pp. 454–465, 2020.
57. Naseri, F., Farjah, E., Kazemi, Z., Schaltz, E., Ghanbari, T., Schanen, J.- L, "Dynamic stabilization of dc traction systems using a supercapacitor based active stabilizer with model predictive control," *IEEE Trans. Transp. Electrif.*, vol. 6, no. 1, pp. 228–240, 2020.
58. Pires, V.F., Romero-Cadaval, E., Vinnikov, D., Roasto, I., Martins, J, "Power converter interfaces for electrochemical energy storage systems–a review," *Energy Convers. Manage.*, vol. 86, pp. 453–475, 2014.
59. Ju, F., Zhang, Q., Deng, W., Li, J, "Review of structures and control of battery-supercapacitor hybrid energy storage system for electric vehicles," In: *2014 IEEE International Conference on Automation Science and Engineering (CASE)*, IEEE, pp. 143–148, 2014.
60. Ju, F., Zhang, Q., Deng, W., Li, J, "Review of structures and control of Battery-Supercapacitor Hybrid Energy Storage System For Electric Vehicles," *Taiwan: John Wiley Sons, Ltd*, 2016, ch. 13, pp. 303–318.
61. Zimmermann, T., Keil, P., Hofmann, M., Horsche, M.F., Pichlmaier, S., Jossen, A, "Review of system topologies for hybrid electrical energy storage systems," *J. Energy Storage*, vol. 8, pp. 78–90, 2016.
62. Jin, Z., Ouyang, M., Lu, Q. and Gao, D, "Development of fuel cell hybrid powertrain research platform based on dynamic test bed," *Int. J. Automot. Technol.*, vol. 9, pp. 365-372, 2008.

63. Zheng, J.P., Jow, T.R., Ding, M, "Hybrid power sources for pulsed current applications," *IEEE Trans. Aerosp. Electron. Syst.*, vol. 37, no. 1, pp. 288–292, 2001.
64. Dougal, R.A., Liu, S., White, R.E, "Power and life extension of battery ultra-capacitor hybrids," *IEEE Trans. Compon. Packag. Technol.*, vol. 25, no. 1, pp. 120–131, 2002.
65. Catherino, H.A., Burgel, J.F., Shi, P.L., Rusek, A., Zou, X, "Hybrid power supplies: A capacitor-assisted battery," *J. Power Sources*, vol. 162, no. 2, pp. 965–970, 2006.
66. Song, Z., Hofmann, H., Li, J., Han, X., Zhang, X., Ouyang, M, "A comparison study of different semi-active hybrid energy storage system topologies for electric vehicles," *J. Power Sources*, vol. 274, pp. 400–411, 2015.
67. Jian Cao, A, "new battery/ultracapacitor hybrid energy storage system for electric, hybrid, and plug-in hybrid electric vehicles/jian cao, a. emadi," *IEEE Trans. Power Electron.*, vol. 27, no. 1, pp. 122–132, 2012.
68. Cohen, I.J., Wetz, D.A., Heinzel, J.M., Dong, Q, "Design and characterization of an actively controlled hybrid energy storage module for high-rate directed energy applications," *IEEE Trans. Plasma Sci.*, vol. 43, no. 5, pp. 1427–1433, 2014.
69. Momayyezan, M., Abeywardana, D.B.W., Hredzak, B., Agelidis, V.G, "Integrated reconfigurable configuration for battery/ultracapacitor hybrid energy storage systems" *IEEE Trans. Energy Convers.*, vol. 31, no. 4, pp. 1583–1590, 2016.
70. Mo, R., Li, H, "Hybrid energy storage system with active filter function for shipboard mvdc system applications based on isolated modular multilevel dc/dc converter," *IEEE J. Emerging Sel. Top. Power Electron.*, vol. 5, no. 1, pp. 79–87, 2016.
71. Bharadwaj, C.A., Maiti, S, "Modular multilevel converter based hybrid energy storage system" In: *2017 IEEE PES Asia-Pacific Power and Energy Engineering Conference (APPEEC)*, IEEE, pp. 1–6, 2017.
72. Zhang, L., Tang, Y., Yang, S., Gao, F, "A modular multilevel converter based grid-tied battery-supercapacitor hybrid energy storage system with decoupled power control," In: *2016 IEEE 8th International Power Electronics and Motion Control Conference (IPEMC-ECCE Asia)*, IEEE, pp. 2964–2971, 2016.
73. Jiang, W., Xue, S., Zhang, L., Xu, W., Yu, K., Chen, W., Zhang, L, "Flexible power distribution control in an asymmetrical-cascaded-multilevel converter-based hybrid energy storage system," *IEEE Trans. Ind. Electron.*, vol. 65, no. 8, pp. 6150–6159, 2017.
74. Etxeberria, A., Vechiu, I., Baudoin, S., Camblong, H., Kreckelbergh, S, "Control of a vanadium redox battery and supercapacitor using a three level neutral point clamped converter," *J. Power Sources*, vol. 248, pp. 1170–1176, 2014.
75. Jayasinghe, S.G., Vilathgamuwa, D.M., Madawala, U.K, "Diode-clamped three-level inverter-based battery/supercapacitor direct integration scheme

for renewable energy systems," *IEEE Trans. Power Electron.*, vol.26, no. 12, pp. 3720–3729, 2011.
76. Piris-Botalla, L., Oggier, G.G., Garcia, G.O, "Extending the power transfer capability of a three-port dc–dc converter for hybrid energy storage systems," *IET Power Electronics*, vol. 10, no. 13, pp. 1687–1697, 2017.
77. Shreelekha, K., Arulmozhi, S, "Multiport isolated bidirectional dc-dc converter interfacing battery and supercapacitor for hybrid energy storage application," In: *2016 International Conference on Electrical, Electronics, and Optimization Techniques (ICEEOT)*, IEEE, pp. 2763–2768, 2016.
78. Ding, Z., Yang, C., Zhang, Z., Wang, C., Xie, S, "A novel soft-switching multiport bidirectional dc–dc converter for hybrid energy storage system," *IEEE Trans. Power Electron.*, vol. 29, vol. 4, pp. 1595–1609, 2013.
79. Anno, T., Koizumi, H, "Double-input bidirectional dc/dc converter using cell-voltage equalizer with flyback transformer," *IEEE Trans. Power Electron.*, vol. 30, no. 6, pp. 2923–2934, 2014.
80. Salman Hajiaghasi, Ahmad Salemnia, Mohsen Hamzeh, "Hybrid energy storage system for microgrids applications: A review," *J. Energy Storage*, vol. 21, pp. 543–570, 2019.
81. Zhang, N., Sutanto, D., Muttaqi, K.M, "A review of topologies of three port dc–dc converters for the integration of renewable energy and energy storage system," *Renew. Sustain. Energy Rev.*, vol. 56, pp. 388–401, 2016.
82. Zhang, G., Li, Z., Zhang, B., Halang, W.A, "Power electronics converters: Past, present and future," *Renew. Sustain. Energy Rev.*, vol. 81, pp. 2028–2044, 2018.
83. Wu, H., Wong, S.-C., Chi, K.T., Chen, Q, "Control and modulation of bidirectional single-phase ac–dc three-phase-legs PWM converters with active power decoupling and minimal storage capacitance," *IEEE Trans. Power Electron.*, vol. 31, no. 6, pp. 4226–4240, 2015.
84. Kolar, J.W., Friedli, T, "The essence of three-phase PFC rectifier systems," In: *2011 IEEE 33rd International Telecommunications Energy Conference (INTELEC)*, IEEE, pp. 1–27, 2011.
85. Duffey, C.K., Stratford, R.P, "Update of harmonic standard ieee-519: IEEE recommended practices and requirements for harmonic control in electric power systems," *IEEE Trans. Ind. Appl.*, vol. 25, no. 6, pp. 1025–1034, 1989.
86. Alam, M., Eberle, W., Gautam, D.S., Botting, C., Dohmeier, N., Musavi, F, "A hybrid resonant pulse-width modulation bridgeless ac–dc power factor correction converter," *IEEE Trans. Ind. Appl.*, vol. 53, no. 2, pp. 1406–1415, 2016.
87. Siwakoti, Y.P., Forouzesh, M., Pham, N.H, "Power electronics converters—an overview," *Control of Power Electronic Converters and Systems*, pp. 3–29, 2018.
88. Niewiara, L.J., Tarczewski, T., Grzesiak, L.M, "Application of dc/dc/ac dual voltage source inverter for current harmonics reduction in PMSM drive,"

In: *2017 19th European Conference on Power Electronics and Applications (EPE'17 ECCE Europe)*, IEEE, pp. 1, 2017.
89. Yamanaka, K., Hara, H., Kume, T., Ishii, S, "Bidirectional dc-ac current source inverter using coupled inductor," In: *2012 IEEE 13th Workshop on Control and Modeling for Power Electronics (COMPEL)*, IEEE, pp. 1–6, 2012.
90. Rodriguez, J., Bernet, S., Wu, B., Pontt, J.O., Kouro, S, "Multilevel voltage-source-converter topologies for industrial medium-voltage drives," *IEEE Trans. Ind. Electron.*, vol. 54, no. 6, pp. 2930–2945, 2007.
91. Kouro, S., Malinowski, M., Gopakumar, K., Pou, J., Franquelo, L.G., Wu, B., Rodriguez, J., Perez, M.A., Leon, J.I, "Recent advances and industrial applications of multilevel converters," *IEEE Trans. Ind. Electron.*, vol. 57, no. 8, pp. 2553–2580, 2010.
92. Marzoughi, A., Burgos, R., Boroyevich, D., Xue, Y, "Design and comparison of cascaded h-bridge, modular multilevel converter, and 5-l active neutral point clamped topologies for motor drive applications," *IEEE Trans. Ind. Appl.*, vol. 54, no. 2, pp. 1404–1413, 2017.
93. Shen, J., Schroder, S., Rosner, R., El-Barbari, S, "A comprehensive study of neutral-point self-balancing effect in neutral-point-clamped three-level inverters," *IEEE Trans. Power Electron.*, pp. 26, no. 11, pp. 3084–3095, 2011.
94. Melin, P.E., Rohten, J.A., Espinoza, J.R., Baier, C.R., Espinosa, E.E., Munoz, J.A., Riedemann, J.A, "Analysis and design of a multicell topology based on three-phase/single-phase current-source cells," *IEEE Trans. Power Electron.*, vol. 31, no. 9, pp. 6122–6133, 2015.
95. Dargahi, V., Sadigh, A.K., Corzine, K, "Enhanced double flying capacitor multicell power converter controlled with a new switching pattern," *IET Power Electron.*, vol. 8, no. 12, pp. 2386–2395, 2015.
96. Choi, B., Cho, B.H., Hong, S.-S, "Dynamics and control of dc-to-dc converters driving other converters downstream," *IEEE Transactions on Circuits and Systems I: Fundamental Theory and Applications*, vol. 46, no. 10, pp. 1240–1248, 1999.
97. Spiliotis, K., Gon¸calves, J.E., Van De Sande, W., Ravyts, S., Daenen, M., Saelens, D., Baert, K., Driesen, J, "Modeling and validation of a dc/dc power converter for building energy simulations: Application to bipv systems," *Appl. Energy*, vol. 240, pp. 646–665, 2019.
98. Forouzesh, M., Siwakoti, Y.P., Gorji, S.A., Blaabjerg, F., Lehman, B, "Step-up dc–dc converters: a comprehensive review of voltage-boosting techniques, topologies, and applications," *IEEE Trans. Power Electron.*, vol. 32, no. 12, pp. 9143–9178, 2017.
99. Singh, S.N, "Selection of non-isolated dc-dc converters for solar photovoltaic system," *Renew. Sustain. Energy Rev.*, vol. 76, pp. 1230–1247, 2017.
100. Wang, J.-M., Tzeng, L., Hsu, M.-T., Jian, H.-R, "A simple control scheme to avoid the sensing noise for the dc–dc buck converter with synchronous rectifier," *IEEE Trans. Ind. Electron.*, vol. 65, no. 6, pp. 5086–5091, 2017.

101. Veerachary, M, "Analysis of minimum-phase fourth-order buck dc–dc converter," *IEEE Trans. Ind. Electron.*, vol. 63, no. 1, pp. 144–154, 2015.
102. Wu, H.,Mu, T., Ge, H., Xing, Y, "Full-range soft-switching-isolated buck boost converters with integrated interleaved boost converter and phase shifted control," *IEEE Trans. Power Electron.*, vol. 31, no. 2, pp. 987–999, 2015.
103. Liu, H., Ji, Y.,Wang, L., Wheeler, P, "A family of improved magnetically coupled impedance network boost dc–dc converters," *IEEE Trans. Power Electron.*, vol. 33, no. 5, pp. 3697–3702, 2017.
104. Ahmed, H.F., Cha, H., Khan, A.A., Kim, H.-G, "A novel buck–boost ac–ac converter with both inverting and non inverting operations and without commutation problem," *IEEE Trans. Power Electron.*, vol. 31, no. 6, pp. 4241–4251, 2015.
105. Kang, F.-s., Park, S.-J., Cho, S.E., Kim, J.-M, "Photovoltaic power interface circuit incorporated with a buck-boost converter and a full-bridge inverter," *Appl. Energy*, vol. 82, no. 3, pp. 266–283, 2005.
106. Zhou, Q., Liu, C., Liu, X., Zhang, X., Xu, J., Cao, K, "Effect of intermediate capacitance on slow-scale instability of cuk power factor correction converter operating in discontinuous capacitor voltage mode," *IET Power Electron.*, vol. 10, no. 14, pp. 1975–1981, 2017.
107. Anand, A., Singh, B, "Design and implementation of PFC cuk converter fed srm drive," *IET Power Electron.*, vol. 10, no. 12, pp. 1539–1549, 2017.
108. Zhang, Z., Lin, J., Zhou, Y., Ren, X, "Analysis and decoupling design of a 30 mhz resonant sepic converter," *IEEE Trans. Power Electron.*, vol. 31, no. 6, pp. 4536–4548, 2015.
109. Moradpour, R., Ardi, H., Tavakoli, A, "Design and implementation of a new sepic-based high step-up dc/dc converter for renewable energy applications," *IEEE Trans. Ind. Electron.*, vol. 65, no. 2, pp. 1290– 1297, 2017.
110. Singh, A.K., Pathak, M.K, "Single-stage zeta-sepic-based multifunctional integrated converter for plug-in electric vehicles," *IET Electr. Syst. Transp.*, vol. 8, no. 2, pp. 101–111, 2018.
111. Andrade, A.M.S.S., da Silva Martins, M.L, "Quadratic-boost with stacked zeta converter for high voltage gain applications," *IEEE J. Emerging Sel. Top. Power Electron.*, vol. 5, no. 4, pp. 1787–1796, 2017.
112. Chen, H., Zheng, Z., Xiao, J, "Determining the number of transformer shielding winding turns for suppressing common-mode noise in flyback converters," *IEEE Trans. Electromagn. Compat.*, vol. 60, no. 5, pp. 1606–1609, 2017.
113. Roggia, L., Schuch, L., Baggio, J.E., Rech, C., Pinheiro, J.R, "Integrated full-bridge-forward dc–dc converter for a residential microgrid application," *IEEE Trans. Power Electron.*, vol. 28, no. 4, pp. 1728–1740, 2012.
114. Li, S., Xiangli, K., Smedley, K.M, "A control map for a bidirectional pwm plus phase-shift-modulated push–pull dc–dc converter," *IEEE Trans. Ind. Electron.*, vol. 64, no. 11, pp. 8514–8524, 2017.

115. Han, J.-K., Kim, J.-W., Moon, G.-W, "A high-efficiency asymmetrical half-bridge converter with integrated boost converter in secondary rectifier," *IEEE Trans. Power Electron.*, vol. 32, no. 11, pp. 8237–8242, 2017.
116. Zhifu, W., Yupu, W., Yinan, R, "Design of closed-loop control system for a bidirectional full bridge dc/dc converter" *Appl. Energy*, vol. 194, pp. 617–625, 2017.
117. Surapaneni, R.K., Yelaverthi, D.B., Rathore, A.K, "Cycloconverter based double-ended micro inverter topologies for solar photovoltaic ac module," *IEEE J. Emerging Sel. Top. Power Electron.*, vol. 4, no. 4, pp. 1354–1361, 2016.
118. Asad, R., Shoulaie, A, "Determination, simulation and implementation of a novel improved industrial cycloconverter," *Int. Trans. Electr. Energy Syst.*, vol. 23, no. 2, pp. 166–180, 2013.
119. Liu, Y., Heydt, G.T., Chu, R.F, "The power quality impact of cycloconverter control strategies," *IEEE Trans. Power Delivery*, vol. 20, no. 2, pp. 1711–1718, 2005.
120. Wheeler, P.W., Rodriguez, J., Clare, J.C., Empringham, L., Weinstein, A, "Matrix converters: a technology review," *IEEE Trans. Ind. Electron.*, vol. 49, no. 2, pp. 276–288, 2002.

11

Diesel Engines and Related Environmental Pollution Issues

Sathya Narayanan[1], Jayendra Kasture[2*] and Milind Shrinivas Dangate[1†]

[1]Chemistry Division, School of Advanced Sciences, Vellore Institute of Technology, Chennai, India
[2]VIT School of Law, Vellore Institute of Technology, Chennai, India

Abstract
Particulate matter emission emerges as a formidable challenge within diesel engine exhaust. Extensive research has gradually illuminated its intricate implications for the environment and human well-being. This includes a deeper understanding of its formation mechanisms within and beyond the diesel engine, along with critical strategies to regulate its production. This study focuses on addressing PM emissions reduction in diesel engines. The investigation initiates with the utilization of two interconnected diesel particulate filters (DPFs): an assisting DPF and a primary conventional honeycomb DPF. This innovative after-treatment approach supplants the prevalent diesel oxidation catalyst (DOC) + DPF system commonly employed in contemporary diesel engines. The novel system showcases potential not only for meeting stringent emission standards but also for substantial cost savings in the complete after-treatment setup. This cost reduction stems from downsizing and weight reduction of the main DPF, accompanied by alternative, less expensive component options. Notably, an issue of back pressure discovered in experimental trials is probed through one-dimensional model simulations, including its correlation with engine performance. Additionally, the study delves into fuel injection parameters, scrutinizing their effects on engine performance and emissions. The exploration further encompasses non-volatile particulate emissions under the influence of pilot injections, dissecting pivotal factors such as fuel quantity and injection timing. Through diverse engine operating conditions, the study involves the examination of concentrations of particulate numbers and the distribution of particle sizes, elucidating the intricate impacts of pilot injections

**Corresponding author*: dr.jakasture@gmail.com
†Corresponding author: milind.shrinivas@vit.ac.in

O.V. Gnana Swathika, K. Karthikeyan, Milind Shrinivas Dangate and Nicoletta Ravasio (eds.) Biomass and Solar-Powered Sustainable Digital Cities, (197–208) © 2024 Scrivener Publishing LLC

on particulate emissions. Moreover, the study ventures into the realm of biofuels, examining the potential of 10% alternative diesel fuel blends like rapeseed methyl ester (RME) and gas to liquid to curb particle emissions. The findings underscore that, even at low blend ratios, incorporation of selected alternative fuels can lead to substantial reductions in both total and non-volatile particle numbers. However, it is noteworthy that, in certain instances, nucleation mode particle count might rise. The study also highlights the phenomenon of the engine producing significant nucleation mode particles during the warm-up phase.

***Keywords*:** Methyl ester, gas-to-liquid, engine, environment, pollution, human health

11.1 Introduction

The diesel engine stands as an attractive power source with extensive applications, owing to its notable advantages over gasoline engines. These include superior fuel efficiency, reduced carbon monoxide emissions, and enhanced torque characteristics. Nevertheless, despite these merits, diesel engines also pose a significant challenge as major contributors to air pollution due to their substantial emission output, particularly in terms of particulate matter (PM) and nitrogen oxides (NOx). Addressing the mitigation of these pollutants has become a pivotal element in contemporary air quality management strategies.

11.2 Background

The origins of the diesel engine trace back to 1892, attributed to Rudolf Diesel's innovation featuring compression ignition. In a conventional diesel engine configuration, the cylinder intake is solely filled with air, with ignition occurring through the injection of a precise quantity of fuel directly into the combustion chamber. This injection takes place once the inducted air attains the requisite temperature and pressure conditions. This mechanism not only enables control over engine load by adjusting the fuel input but also eliminates the losses encountered in spark ignition engines. Unlike the latter, where fuel and air are mixed externally and subsequently throttled before entering the cylinder, this approach ensures more efficient combustion. This divergence in the initiation of combustion also plays a role in differentiating emissions between diesel and gasoline engines.

11.3 Diesel Engine Emissions

The diesel engine system, a cornerstone of transportation, industry, and power generation, encompasses a complex and efficient mechanism that converts diesel fuel into mechanical energy through combustion. This intricate system consists of various components, including the engine block, pistons, cylinders, fuel injection system, and exhaust system. The system's design and operation influence performance, fuel efficiency, and emissions.

The diesel engine's operation is characterized by the compression-ignition combustion process. Air is drawn into the cylinders, compressed to high pressures and temperatures, and, at the peak of compression, fuel is injected directly into the hot, pressurized air. The fuel–air mixture self-ignites due to the high temperatures, producing power through the expansion of the combustion gases. This process is renowned for its thermodynamic efficiency, yielding higher fuel efficiency compared to gasoline engines.

The fuel injection system plays a pivotal role in engine performance. Modern diesel engines employ common rail direct injection systems, which enhance fuel atomization, combustion efficiency, and emissions control. The precise control of injection timing and pressure optimizes combustion, reducing noise, NOx emissions, and PM formation. While diesel engines offer advantages in terms of efficiency and torque, they have been associated with environmental pollution issues, particularly NOx and PM emissions. To address this, emission reduction technologies such as exhaust gas recirculation (EGR) and diesel particulate filters (DPFs) are integrated into the system. EGR lowers combustion temperatures, reducing NOx formation, whereas DPFs capture PM from the exhaust gases.

11.3.1 Diesel Engine Emission Control Technologies

The after-treatment system for gas-diesel engines plays a pivotal role in mitigating emissions and ensuring compliance with stringent environmental regulations. This complex system is designed to reduce harmful pollutants generated during the combustion process, thereby minimizing their impact on air quality and public health. The development and implementation of effective after-treatment strategies are critical for achieving cleaner and more sustainable engine operations [1]. The after-treatment system typically consists of several components working in tandem to address specific pollutants. One key element is the selective catalytic reduction (SCR) system, which employs a catalyst to convert NOx into nitrogen and water vapor through a chemical reaction with a reducing agent, typically urea. This process significantly reduces the emission of NOx, a major contributor to smog and respiratory ailments [2]. Particulate matter (PM) is another major concern in diesel engine emissions. Diesel particulate filters (DPFs) are employed in the after-treatment system to capture and trap PM from the exhaust gases. These filters prevent fine particles from being released into the atmosphere, thus improving air quality and minimizing adverse health effects.

Furthermore, gas-diesel engines often integrate oxidation catalysts to facilitate the conversion of harmful carbon monoxide (CO) and unburned hydrocarbons into less harmful compounds. These catalysts accelerate oxidation reactions, breaking down these pollutants into carbon dioxide (CO_2) and water vapor. The effectiveness of the after-treatment system depends on various factors, including the quality of components, the engine's operating conditions, and the availability of the necessary reactants. Proper maintenance and monitoring are crucial to ensure the optimal performance of these systems over time.

11.3.2 Objectives

The primary goal of this study is to concentrate on the regulation of PM emissions from diesel engines, aligning with the central research domains outlined earlier. The investigation is structured into three key areas: after-treatment systems, injection strategies, and alternative fuels. The methodologies encompass the following steps:

- Establishing a comprehensive engine test setup with essential measurement equipment.
- Developing a one-dimensional engine model to facilitate simulations and analysis.
- Evaluating a series of novel after-treatment configurations to gauge their potential impact on future emissions and their influence on engine performance.
- Investigating various advanced fuel injection strategies that hold the potential to curtail both particulate and gaseous emissions.
- Exploring the emission characteristics of PM and gaseous pollutants resulting from new fuel blends.

By employing these methodologies, the study aims to contribute insights into innovative approaches for enhancing the control of PM emissions in diesel engines, ultimately aligning with the broader goals of emission reduction and environmental sustainability.

11.4 Particulates

Particulates, known as PM, encompass a complex amalgamation of minute solid fragments and liquid beads suspended in the atmosphere. Varying in dimension from coarse to fine, particulates are classified based on their size: PM10 (equal to or smaller than 10 micrometers) and PM2.5 (equal to or smaller than 2.5 micrometers). These diminutive entities emerge from diverse origins, both natural and human-caused, and wield significant influence over air quality, human well-being, and the ecosystem. Particulates emanate from natural sources like dust, pollen, and ocean mist, as well as human undertakings such as automobile discharges, industrial procedures, and building projects. The more minute PM2.5 particles present a more pronounced hazard due to their capacity to infiltrate deep into the respiratory tract, leading to an array of health complications. The unfavorable consequences of particulates on human health have propelled regulatory bodies and governments worldwide to formulate air quality benchmarks and recommendations. Monitoring and managing particulate levels have become imperative to alleviate their repercussions. Approaches like filtration systems, eco-friendly infrastructure, and emission-curbing technologies strive to curtail particulate discharges and protect public health [3].

Moreover, particulates exert environmental ramifications, including reduced clarity, soiling of surfaces, and modification of climatic patterns. Fine particulates can soak up and scatter sunlight, influencing the energy equilibrium of the atmosphere and potentially contributing to global warming.

11.4.1 Diesel Particulate Filter (DPF)

Diesel particulate filters (DPFs) are sophisticated emission control devices designed to capture and reduce PM emissions from diesel engines. As vehicular emissions play a pivotal role in air quality deterioration and adverse health effects, DPFs have gained prominence in mitigating these issues. DPFs operate by trapping PM, including fine particles like PM2.5, as exhaust gases pass through them. These filters consist of porous materials that enable the gas to flow while capturing the solid particles. Over time, the accumulated particles need to be removed to prevent filter clogging and maintain engine efficiency. This regeneration process can be passive or active, involving elevated exhaust temperatures to burn off the trapped particulates, transforming them into less harmful substances. Passive regeneration occurs naturally during sustained highway driving, where exhaust temperatures are high enough to incinerate trapped particulates. However, urban and stop-and-go driving conditions may hinder passive regeneration, necessitating active regeneration methods. Active regeneration involves altering engine parameters or injecting fuel into the exhaust system to elevate temperatures artificially, thus incinerating the particulates [4].

DPFs contribute significantly to reducing particulate emissions and, consequently, improving air quality and public health. However, it is important to note that, while DPFs are effective in curbing particulate emissions, they do not address other harmful components of diesel exhaust [5], such as NOx, which necessitate additional emission control technologies.

11.4.2 Exhaust Gas Recirculation (EGR)

Exhaust gas recirculation (EGR) is a vital emission control mechanism utilized in internal combustion engines to mitigate the creation of harmful NOx during the burning process. EGR achieves this by reintroducing a portion of exhaust gases back into the engine's intake air, effectively diluting the incoming fresh air and lowering the peak combustion temperature, which, in turn, curbs NOx formation. The EGR process involves diverting a fraction of exhaust gases from the exhaust manifold or exhaust pipe and directing it into the combustion chamber. This dilution of the air-fuel mixture decreases the oxygen concentration during combustion, reducing the

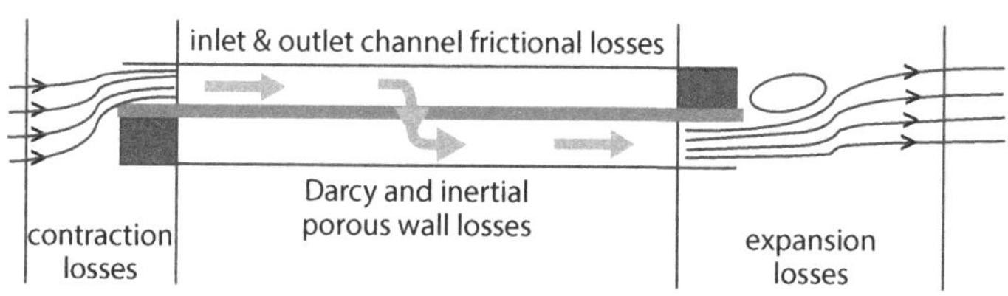

Figure 11.1 Illustration of flow path within a wall flow (DPF) channel, depicting pressure loss elements.

peak flame temperature that encourages NOx generation [6]. EGR systems can be pneumatic or electronically controlled, enabling precise regulation of the recirculation rate based on engine operating conditions. EGR offers several benefits beyond NOx reduction. It also improves engine efficiency by enhancing combustion stability and reducing knocking, leading to better fuel economy. Additionally, EGR contributes to lowering CO_2 emissions, a crucial factor in addressing climate change concerns.

While EGR effectively reduces NOx emissions, it presents certain challenges. One primary issue is the potential for increased soot accumulation in the engine due to reduced oxygen availability, necessitating complementary technologies like DPFs to capture these particles. Moreover, EGR systems demand careful calibration and maintenance to ensure optimal engine performance and emissions reduction.

11.4.3 Alternative Fuels for Diesel Engines

Alternative fuels fall into the category of advanced fuels, this classification encompasses a range of fuels including biodiesel, ethanol, hydrogen, and natural gas, all of which are commercially available for use in automobiles [7]. Additionally, there are fuels like biogas, ultra-low sulfur diesel, BTL, and GTL), among others, that are currently under investigation and development. These emerging fuels hold the potential to not only address the global energy supply in the face of limited conventional fuel resources but also offer the advantage of reduced emissions.

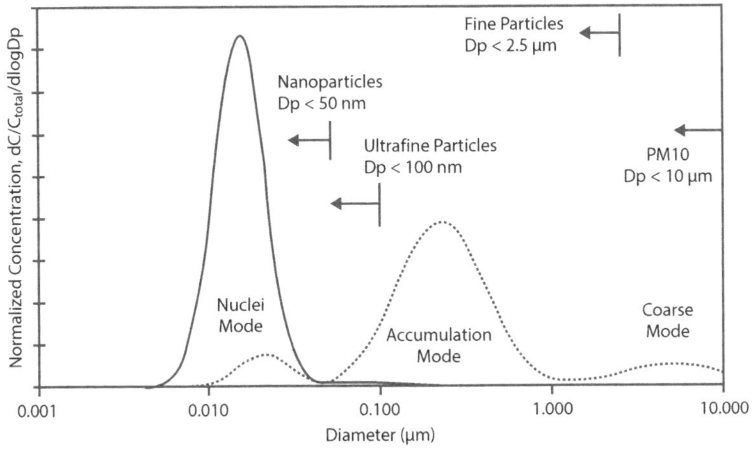

Figure 11.2 Standard particle size and distribution resulting from combustion engines.

11.5 Biodiesel

Biodiesel, a renewable alternative to conventional diesel fuel, is produced through a process called transesterification, playing a crucial role in sustainable energy solutions [5]. Transesterification involves the chemical transformation of triglycerides, typically sourced from plant oils or animal fats, into biodiesel and glycerol by reacting them with an alcohol, usually methanol or ethanol. The process of transesterification occurs in the presence of a catalyst, often NaOH or KOH, which accelerates the reaction and promotes the separation of the glycerol byproduct from the biodiesel. This chemical conversion results in the creation of biodiesel molecules with properties like those of conventional diesel fuel, making biodiesel compatible with existing diesel engines without necessitating significant modifications. Biodiesel offers several environmental advantages [8]. It emits fewer greenhouse gases and pollutants compared to conventional diesel, contributing to reduced air pollution and mitigating climate change. Furthermore, biodiesel is biodegradable and poses a lower risk of water and soil contamination in case of spills or leaks. Despite its benefits, biodiesel production via transesterification does present challenges. Feedstock availability, competition with food crops, and land use concerns are critical considerations in sustainable biodiesel production. Ensuring a steady supply of feedstock from non-food sources, such as algae or waste cooking oil, can address these issues.

11.5.1 Gas-to-Liquid (GTL) Diesel

Gas-to-liquid (GTL) diesel, a cutting-edge alternative to traditional diesel fuel, is synthesized through a groundbreaking conversion process that holds significant potential for the energy landscape. The GTL process involves the transformation of natural gas, primarily composed of methane, into high-quality liquid hydrocarbons resembling conventional diesel fuel in both chemical composition and performance [9]. The GTL technology entails a two-step procedure. First, natural gas is subjected to a reforming process, converting it into synthesis gas. In the subsequent step, a catalytic reaction, known as Fischer–Tropsch synthesis, transforms the syngas into a range of hydrocarbon molecules, including diesel, waxes, and even lubricants. The synthesized diesel, virtually free from impurities, possesses superior combustion properties, resulting in reduced emissions and enhanced engine efficiency. GTL diesel boasts several environmental and performance advantages. Its cleaner combustion leads to lower emissions of NOx, PM, and sulfur compounds, thus contributing to improved air quality and reduced impact on human health. Additionally, GTL diesel exhibits excellent cold-flow properties, making it suitable for extreme temperatures, which is especially advantageous in regions with harsh climates.

While GTL diesel shows promise, it does face challenges such as the requirement for large-scale capital investments due to the complexity of the conversion process and the need for advanced catalysts. However, ongoing research and development efforts aim to enhance process efficiency and address these challenges to make GTL diesel a more economically viable and widely accessible fuel option.

11.5.2 Combined Alternative Fuels with New Injection Strategies

The integration of alternative fuels with advanced injection systems has emerged as a promising avenue for enhancing combustion efficiency, reducing emissions, and diversifying the energy landscape. This synergy involves combining innovative fuel compositions, such as biofuels or synthetic fuels, with state-of-the-art injection technologies to optimize combustion processes in internal combustion engines [10].

The convergence of alternative fuels with modern injection systems holds the potential to address multiple challenges in the transportation sector. Direct injection systems, for instance, allow for precise control over fuel delivery and injection timing, facilitating more efficient combustion and minimizing wastage. When paired with alternative fuels, these systems can unlock synergies that lead to reduced emissions of greenhouse gases, NOx, and PM, contributing to improved air quality and meeting stringent emissions regulations. This combination can also contribute to enhanced energy security by utilizing diverse feedstock sources, such as algae, waste materials, or renewable electricity, to produce advanced fuels. Synthetic fuels, which can be tailored for specific combustion characteristics, enable higher compression ratios and leaner mixtures, translating to greater engine efficiency. Despite the potential benefits, challenges persist. Compatibility issues between alternative fuels and injection systems, differing combustion properties, and potential engine modifications need to be considered [9]. Rigorous research and development efforts are crucial to optimize these combinations and ensure reliable, efficient, and durable engine performance.

11.6 Summary

This study delves into the effect of emissions from diesel engines, particularly PM, on both human health and the environment. It explores the regulatory measures and methods adopted to control these emissions. The discussion begins with an overview of the harmful nature of these emissions, followed by an in-depth exploration of how PM is generated within and beyond the engine. The chapter also addresses the current methodologies for measuring PM, with a specific focus on particle number concentration, size, and dispersion, which are closely linked to human health concerns. The chapter emphasizes the importance of assessing non-volatile particle number concentration to enhance the accuracy and reliability of PM measurements. Furthermore, novel techniques for controlling PM emissions are examined, including the implementation of after-treatment systems for diesel engines. These systems have become indispensable for meeting present and future emission standards. The advancement of materials and designs for DOC and DPF has introduced innovative options for effectively reducing PM. The chapter also delves into the adoption of alternative fuels as a strategy to address fuel scarcity and lower emissions. In the realm of fuel injection, optimizing multiple injections alongside single injections demonstrates significant potential for further minimizing PM emissions. The careful adjustment of injection parameters may even give rise to alternative combustion modes like premixed charge compression ignition and homogeneous charge compression ignition, which offer reduced emissions of both PM and NOx simultaneously.

References

1. Agarwal AK (2007) Biofuels (alcohols and biodiesel) applications as fuels for internal combustion engines. *Prog. Energy Combust. Sci.*, 33:233–271
2. Burtscher H (2005) Physical characterization of particulate emissions from diesel engines: a review. *Aerosol Sci.*, 36:896–932
3. Englert N (2004) Fine particles and human health—a review of epidemiological studies. *Toxicol. Lett.*, 149:235–242.
4. Zheng M, Banerjee S (2009) Diesel oxidation catalyst and particulate filter modeling in active–flow configurations. *Appl. Therm. Eng.*, 29:3021–3035
5. Biswas S, Verma V, Schauer JJ, Sioutas C (2009) Chemical speciation of PM emissions from heavy-duty diesel vehicles equipped with diesel particulate filter (DPF) and selective catalytic reduction (SCR) retrofits. *Atmos. Environ.*, 43:1917–1925.

6. Lopez JM, Jimenez F, Aparicio F (2009) On-road emissions from urban buses with SCR + Urea and EGR + DPF systems using diesel and biodiesel. *Transp. Res. Part D.*, 14:1–5.
7. Dai A (2011) Drought under global warming: a review. *Wiley Interdiscip. Rev. Clim. Chang.*, 2(1):45–65.
8. Ng JH, Ng HK, Gan S (2010) Advances in biodiesel fuel for application in compression ignition engines. *Clean Technol. Environ. Policy*, 12:459–493.
9. Wiebenga MH, Kim CH, Schmieg SJ, Oh SH, Brown DB, Kim DH, Lee JH, Peden CHF (2012) Deactivation mechanisms of Pt/Pd-based diesel oxidation catalysts. *Catal. Today*, 184:197–204.
10. Channappagoudra, M., Ramesh, K., & Manavendra, G. (2020). Effect of injection timing on modified direct injection diesel engine performance operated with dairy scum biodiesel and Bio-CNG. *Renew. Energy*, 147, 1019–1032. https://doi.org/10.1016/j.renene.2019.09.070.

12

Production of Bio-Diesel and By-Product Processing

Sathya Narayanan[1], Jayendra Kasture[2*] and Milind Shrinivas Dangate[1]

[1]*Chemistry Division, School of Advanced Sciences, Vellore Institute of Technology, Chennai, India*

[2]*VIT School of Law, Vellore Institute of Technology, Chennai, India*

Abstract

In the current era, a significant dilemma confronting humanity revolves around the gradual depletion of Earth's fossil fuel reservoirs, an issue that cannot be avoided. The combustion of these fossil fuels, extensively employed for heating and transportation, stands as a pivotal contributor to the global warming crisis. Amidst this, an accessible and promising alternative energy source emerges in the form of biodiesel, poised to replace traditional petroleum-based diesel fuel. Derived from renewable biomass, biodiesel is primarily produced through alkali-catalyzed transesterification of plant oils. Its distinct advantages over conventional diesel fuel are compelling, particularly its capability to uphold a balanced carbon dioxide cycle due to its foundation on renewable organic materials. Biodiesel, either in its pure form or blended with petroleum diesel, seamlessly operates within standard diesel engines with minimal or no adjustments, utilizing existing infrastructure for distribution. The process kinetics of producing biodiesel through methanol-driven transesterification of plant oils are elucidated using an innovative solubility model that accounts for the interacting phases of the reaction mixture. This model unveils that the creation of methyl esters throughout the reaction course enhances the integration of oil into the methanol phase. The introduction of a new kinetic model incorporating ternary phase diagrams (oil/methanol/methyl esters) further enhances the understanding of product-facilitated oil integration, aligning well with experimental kinetics and suggesting prospects for future model enhancements. Furthermore, there are potential strategy to diminish reliance on methanol, thereby augmenting process economics by recycling what would otherwise become waste with the growth of biodiesel production. In-depth exploration

**Corresponding author*: dr.jakasture@gmail.com

O.V. Gnana Swathika, K. Karthikeyan, Milind Shrinivas Dangate and Nicoletta Ravasio (eds.) Biomass and Solar-Powered Sustainable Digital Cities, (209–220) © 2024 Scrivener Publishing LLC

using the bacteria strains *Pantoea agglomerans* and Clostridium past uranium in a controlled bioreactor showcases promising yields of alcohols from glycerol conversion. The investigation implies that refining fermentation conditions for *P. agglomerans* could potentially elevate alcohol yields.

The study culminates in the assessment of biodiesel fuels, both in their pure state and blended with mineral diesel. Findings indicate a rise in overall specific fuel consumption with higher biodiesel blends, juxtaposed with reduced emissions across all blends and oils used, except for an increase in NOx emissions.

Keywords: Renewable biomass, plant oil, glycerol, methanol, global warming

12.1 Introduction

Oil prices reached their highest point in 2004, exceeding $60 per barrel, a record for the past two decades. Surprisingly, even at this high cost, this growth was observed across all regions, with the Asia-Pacific region showing the most substantial rise, experiencing an increase of 860,000 b/d or 4%. As of the end of 2003, estimated proven oil reserves amounted to 1,147 million barrels, reflecting a 12% rise from the levels of 1993. However, this increase is attributed to variations in estimating quantities associated with oil wells rather than a standardized method. Nevertheless, considering current consumption patterns, these reserves are anticipated to sustain for approximately 40–60 years, without accounting for potential discoveries. Despite the ongoing significance of oil as a primary fuel source, particularly in the realm of transportation where suitable alternatives remain limited, there is a discernible shift in the scientific and engineering communities toward identifying sustainable substitutes. This shift acknowledges the finite nature of conventional oil reserves and the environmental considerations associated with their use. Consequently, the pursuit of alternatives aligns with the principles of sustainability.

12.2 Fossil Fuel Resources

In 2004, oil prices reached their highest point in two decades, surpassing the $60 per barrel mark. Despite the elevated cost, global oil consumption increased drastically marking a growth rate of 2.1%, with the Asia and Pacific regions experiencing the strongest boost, rising by 860,000 b/d or 4%. It is projected that existing fossil fuel will last for another five decades, excluding any new discoveries. While fossil fuel remains a crucial energy source, especially in the transportation sector where few viable alternatives exist, the scientific and engineering communities are undergoing a noticeable shift toward exploring alternative options that align with the principles of sustainability.

12.3 Environmental Issues

Fossil resources are essential for economic growth but their combustion releases CO_2, contributing to global warming. This effect is compounded by deforestation, which releases additional carbon. Encouraging plant growth becomes pivotal as plants store atmospheric carbon. Industrialized nations bear the responsibility of reducing emissions. Innovative strategies are necessary to prevent severe climate change. Agriculture also plays a role; cultivated areas like corn and rapeseed fields can annually store around 6 tons of carbon per hectare. Utilizing this stored carbon, for instance, in biomass power generation or vegetable oil-based fuels, can substantially lessen reliance on fossil fuels. The Kyoto Conference in 1997 highlighted the urgency of emission reduction. The European Union (EU) committed to an approximately 8.1% reduction in CO_2 emissions by 2010 compared to 1990 levels. Extensive research followed to determine effective strategies that minimize the economic impact on member states.

Renewable energy sources are key to achieving these goals. Biomass, along with wind, hydro, and solar energy, was identified as crucial by the EU. These efforts underscore a global commitment to address climate change by transitioning from fossil fuels to sustainable alternatives. Biodiesel offers several benefits compared to conventional petroleum-based diesel due to its renewable nature, ability to break down naturally, low emissions, non-toxic properties, and a higher flash point [1, 2].

12.4 Fueling the Future: Vegetable Oil as an Alternative for Diesel Engines

Oils that are derived have long been explored as sustainable alternatives to mineral oil-based fuels, with their potential recognized since the inception of internal combustion engines. Dr. Rudolf Diesel showcased the viability of vegetable-based oils as engine fuels back in 1900 when he presented an engine running entirely on oil taken from groundnut at the Paris Exposition. He foresaw their significance growing over time, comparable to the importance of petroleum and coal tar products in his era.

Experiments with various vegetable-based fuels, including castor oil, palm oil, and more, have been conducted to evaluate their potential as alternative energy sources. He underscored the versatility of his engine by demonstrating its capability to break fuel monopolies due to its adaptability to multiple fuel types.

12.5 Objectives of This Study

This research focuses on exploring various aspects of biodiesel production, encompassing kinetics, characteristics, efficiency, and emissions. Additionally, the study delves into methods for processing the by-product glycerol. The main objectives include the following:

1. Investigating the conversion process of waste oil and soya oil using MeOH and EtOH through transesterification.
2. Analyzing the power performance and emission levels of the produced biodiesels derived from different oil sources, as well as their blends with conventional mineral diesel. This investigation extends to include pure mineral diesel.
3. Evaluating potential applications for the by-product glycerol, with a particular emphasis on exploring the feasibility of achieving a "zero-waste" approach. This involves fermenting glycerol to produce alcohol, which can then be utilized within the biodiesel production process.

12.6 Requirements of Alternative Fuels to Diesel

Criteria for alternative fuels to replace diesel should ideally encompass the following prerequisites:

1. Compatibility: An alternative diesel fuel should have the capability to be employed in current diesel engines without necessitating expensive alterations to the engine configuration.
2. Local Origin: The fuel source should ideally be derived from materials accessible within the local area, ensuring a consistent and substantial supply.
3. Environmental Friendliness: The entire life cycle of the fuel, including its production, processing, and combustion,

should have minimal ecological impact, prioritizing environmental safety.

4. Comparable Properties: The fuel should exhibit energy density and qualities akin to traditional diesel, ensuring similar engine performance.
5. Durability: Utilizing the alternative fuel should not result in the reduction of engine lifespan or service intervals.
6. Accessibility and Cost-effectiveness: The fuel should be conveniently obtainable and economically feasible for end-users, promoting practical adoption.

12.6.1 Microbial Conversion of Glycerol to Alcohols: Enhancing Biodiesel Production and Sustainability

A comprehensive exploration of bacterial strains *Pantoea agglomerans* and *Clostridium pasteurianum* within a controlled bioreactor has revealed promising outcomes in terms of alcohol yields from glycerol conversion. This in-depth investigation sheds light on the potential of these bacteria to efficiently transform glycerol waste into valuable alcohols, contributing to the optimization of the biodiesel production process. The utilization of *Pantoea agglomerans* and *Clostridium pasteurianum* as bioconversion agents represents a significant advancement in sustainable biodiesel production. These bacteria exhibit a remarkable capability to convert glycerol, a by-product of the transesterification process, into valuable alcohols. The yields obtained from this conversion process underscore the potential economic and environmental benefits of harnessing waste materials for valuable outputs. The results of the investigation suggest that there exists room for further improvement through the refinement of fermentation conditions, particularly for *Pantoea agglomerans.* By fine-tuning the environmental parameters within the bioreactor, such as temperature, pH, and nutrient availability, researchers can potentially enhance the efficiency of alcohol production. This prospect holds promise for elevating alcohol yields and optimizing the overall glycerol conversion process. The significance of this research lies in its contribution to achieving a more sustainable and economically viable biodiesel production process. By effectively converting glycerol waste into alcohols through bacterial action, the research aligns with the principles of a circular economy, where waste materials are transformed into valuable resources. Moreover, the potential for improving alcohol yields by refining fermentation conditions highlights the ongoing commitment to innovation and optimization within the field of biodiesel production.

As we seek to reduce and mitigate environmental impacts, the utilization of bacterial strains for glycerol conversion represents a step forward in realizing these goals. The findings not only showcase the potential of bioconversion technologies but also emphasize the importance of continuous research and development in advancing sustainable energy solutions. Ultimately, this research contributes to the growing body of knowledge that is driving the transformation of the energy landscape toward a cleaner and more sustainable future.

12.6.2 Methods of Utilizing Vegetable Oils

1. The method of dilution combines substances, such as ethanol, with sunflower oil to decrease its viscosity. A study investigated the mixture of sunflower oil and diesel fuel in a ratio of 1:3 by volume, followed by evaluations of engines. However, employing this blend for extended periods in direct injection diesel engines was not advised due to potential complications like clogged injector nozzles and the thickening of lubricants. Even lower ratios resulted in the creation of carbon deposits on intake valves and noticeable wear on piston rings.
2. Microemulsification entails blending two liquids that do not naturally mix, facilitated by a surfactant. Fuel microemulsions involve combinations of vegetable oils with esters and dispersants, pure vegetable oils, or alcohol and a surfactant with or without diesel fuel. The presence of alcohol aids in cooling the combustion chamber, mitigating the risk of nozzle blockages due to their high heat of vaporization. Engines powered by microemulsions have been proposed to exhibit performance similar to traditional diesel fuel.
3. The pyrolysis method, a thermal degradation process without oxygen, offers a unique approach to producing biodiesel. This technique transforms oils into alkanes, alkenes, alkadienes, carboxylic acids, aromatics, and gases that can be further refined into biodiesel components. Provided insights into the pyrolysis process, while studies have explored the pyrolysis of triglycerides to obtain diesel-resembling products, highlighting its potential as an alternative fuel source. However, pyrolyzed vegetable oils may have elevated ash carbon residue despite acceptable sulfur, water, and sediment levels.

4. Transesterification is a vital process in the production of biodiesel, involving the conversion of vegetable oils or animal fats into biodiesel through the reaction with alcohol and a catalyst. This reaction results in the formation of fatty acid methyl esters (FAMEs), which are the main components of biodiesel. Transesterification is a key step due to its role in producing a cleaner-burning alternative fuel source [3].

12.6.3 Transesterification

Transesterification is a crucial chemical process within the realm of biodiesel production, involving the conversion of triglycerides found in feedstocks such as vegetable oils or animal fats into their corresponding fatty acid methyl or ethyl esters, along with glycerol as a by-product. This process is pivotal in transforming these renewable resources into usable biofuels that are compatible with existing diesel engines [4]. The transesterification reaction requires the use of a catalyst, often an alkali or acid, to facilitate the conversion of triglycerides and alcohol (usually methanol or ethanol) into esters and glycerol. The reaction occurs at moderate temperatures and atmospheric pressure, making it feasible for large-scale production [4]. Transesterification offers several advantages, including improving fuel properties such as viscosity and cold flow characteristics, thus addressing some of the challenges associated with using pure vegetable oils as fuels [4]. This process has been investigated extensively as a means to produce biodiesel on a commercial scale, aligning with the sustainability goals of reducing dependence on fossil fuels and mitigating greenhouse gas emissions [1]. The purity of the reactants, mainly water content and free fatty acids, is the crucial factor that determines the convenient route of transesterification [5, 6].

12.6.4 Catalysts

Transesterification utilizes two main categories of catalysts: acid and base catalysts. For acid catalysis, common choices include sulphuric acid and hydrochloric acid, whereas base catalysis involves the use of alkoxides. The quantity of catalyst employed in the reaction is typically expressed relative to the weight of the oil or fat being processed. The mechanisms underlying acid-catalyzed transesterification and base-catalyzed transesterification bear similarities to those of acid and base-catalyzed ester hydrolysis processes.

12.6.5 Acid-Catalyzed Transesterification

Acid-catalyzed transesterification proves particularly suitable when dealing with low-quality fats or materials abundant in free fatty acids. The presence of fatty acids tends to counteract catalysts and results in soap formation. Although base catalyst amounts can be tailored to account for substrate acidity, any soap produced may lead to heightened viscosity, gel creation, and hindered glycerol separation. An example of this technique is seen in the transesterification of soybean oil, using MeOH and EtOH and employing 1% weight conc. H_2SO_4 as a catalyst based on oil weight. Alcohols with fewer molecular weights display faster reaction times, whereas branched alcohols yield lower product yields. Fischer esterification's mechanism entails an acid-catalyzed substitution. Normally, alcohols lack the required nucleophilic strength to attack a carboxylic acid's carbonyl group. However, the presence of an acid catalyst leads to carbonyl group protonation, rendering it susceptible to nucleophilic attack. The subsequent loss of a proton yields an ester hydrate. Alkoxyl group protonation enables it to depart as water, resulting in a protonated ester that eventually loses a proton, yielding the product. The process of acid-catalyzed transesterification can be likened to a variation of Fischer esterification. In this context, the activation of the ester carbonyl occurs through protonation, enabling it to undergo nucleophilic attack by methanol. As methanol initiates this attack, it releases a proton, leading to the formation of an intermediate compound. Notably, conc. H_2SO_4 is a widely employed catalyst in both Fischer esterification and acid-catalyzed transesterification. Freedman *et al.* [7] conducted a study on the methanolysis of soybean oil, utilizing 1% weight conc. H_2SO_4 as a catalyst, along with a 30:1 methanol to oil molar ratio, and maintaining a temperature of 65°C. Over the course of 69 h, their experiment yielded a final conversion to methyl esters amounting to 93%.

12.6.6 Base-Catalyzed Transesterification

The base-catalyzed transesterification process for converting oils and fats into alkyl esters demonstrates a significantly higher reaction rate compared to its acid-catalyzed counterpart. Unlike acid-catalyzed reactions that often require elevated temperatures exceeding 100°C, base-catalyzed reactions proceed rapidly even at room temperature, making them more efficient [8]. This method proves especially effective when dealing with substrates low in free fatty acid content and nearly devoid of water. The cost-effectiveness of metal hydroxides and alkoxides makes them popular choices as proficient base catalysts. In the context of base-catalyzed alcoholysis often a metal hydroxide or alkoxide, in the chosen alcohol like methanol or ethanol, and then mixing it with the oil. The solubility of alcohol in oil varies based on factors such as the alkyl group size of the alcohol and the reaction temperature. To initiate the initial phase of the reaction, vigorous mixing is crucial due to the relatively low solubility of methanol in oil at room and moderate temperatures, like 60°C. As the conversion progresses, triglycerides undergo a sequential transformation into diglycerides, monoglycerides, and ultimately glycerol. The addition of alcohol triggers the formation of glycerol, which then settles at the bottom of the reaction vessel due to gravity. The glycerol-rich lower phase contains some alcohol and possibly monoglycerides. The upper phase, enriched with fatty esters, contains unreacted alcohol, glycerol, catalyst, mono-, di-, and triglycerides. The exothermic nature of base-catalyzed transesterification leads to a slight temperature rise upon adding the catalyst solution to the oil. This increase diminishes after a few minutes, maintaining a stable temperature throughout the reaction. However, using alkoxides can be cumbersome, often requiring the reaction of the metal with the alcohol. In contrast, the reaction mechanism remains consistent in the presence of metal hydroxides. While alkoxides offer efficiency, the practicality and cost-effectiveness of metal hydroxides, despite their slightly slower reaction rates, have led to their widespread industrial use. The base-catalyzed reaction surpasses the acid-catalyzed process significantly, permitting its use at ambient temperatures. Moreover, base catalysts are less corrosive to equipment compared to their acid counterparts, making them preferable for commercial applications.

12.7 Factors Affecting Kinetics

The transesterification process of vegetable oil follows a comprehensive mechanism consisting of three distinct reaction steps, the kinetics of which are intricate. Even when considering a simplified scenario involving a triglyceride with a single fatty acid (R1) reacting with alcohol, the detailed mechanism involves a minimum of 12 equilibrium reactions. These equilibrium reactions at each step of the transesterification process make the study of kinetics a complex endeavor. In line with Le Chatelier's principle, an excess of alcohol is essential within the reaction mixture to drive the equilibrium toward the desired product formation. The kinetics of transesterification display relatively sluggish behavior. This sluggishness can be attributed, in part, to the formation of a two-phase mixture resulting from the reactants, namely, vegetable oils and alcohol (typically methanol). This two-phase nature introduces either mass transfer limitations or solubility limitations, affecting the overall rate of the reaction. To counteract these limitations and expedite the reaction, intensive mechanical mixing becomes a necessity. By incorporating solubility data from the ternary system of oil/FAME/methanol, the concentrations of reactants were meticulously adjusted to closely align experimental findings with the kinetic models that were developed.

12.8 Legislation Governing Biodiesel

Taxation policies on biodiesel vary significantly across different countries, even within the same regions like the European Union. This divergence is evident when comparing taxation levels among various nations. For instance, in the United Kingdom, as of 2006, biodiesel faces a reduced tax of 20p per liter compared to regular petrol, yet it remains subject to a taxation rate of 27p per liter. Germany takes a more favorable approach, providing full tax exemption for biodiesel, also known as "801oon" or 100% biodiesel, until 2008, with potential extensions beyond that date. Austria shares a similar stance with a complete tax exemption for "80100" biodiesel, while biodiesel mixtures such as 8020 and 805 face standard duties. France's position on biodiesel blends appears less clear-cut. While a 100% tax is imposed on biodiesel, the country also offers subsidies to promote biodiesel production. The extent of these subsidies remains uncertain, and there are indications that a directive may discourage small-scale biodiesel production to evade taxation. Turning to the United States, biodiesel usage predominantly involves blended compositions, with popular blends like 8020 and 805 biodiesel. The U.S. taxation system for biodiesel is intricate and multifaceted. National policy dictates a proportional reduction in taxation for increasing percentages of biodiesel in blends, up to 8020 or 20% biodiesel. However, individual states retain the authority to formulate their policies. In Illinois, for instance, the national law is upheld, but a complete tax exemption is granted for "80100" biodiesel. Moreover, biodiesel plants with an annual production exceeding 30 million U.S. gallons have the opportunity to access grants of up to $15 million.

The discrepancies in biodiesel taxation policies across countries underline the complexity of aligning economic incentives with environmental goals. These variations highlight the diverse approaches governments take to promote or regulate the production and consumption of biodiesel, ultimately shaping the growth and development of this alternative fuel source in the global energy landscape.

12.9 Conclusion

The current global energy landscape demands innovative solutions to address the challenges posed by depleting fossil fuel reserves, environmental concerns, and the imperative for sustainable energy sources. Biodiesel emerges as a promising alternative that aligns with these goals. Derived from renewable biomass through alkali-catalyzed transesterification, biodiesel offers distinct advantages over conventional diesel fuel, including its potential to mitigate carbon emissions and utilize existing infrastructure seamlessly. The research explored various facets of biodiesel production, from its kinetics to the utilization of by-product glycerol. The innovative solubility model sheds light on the complex interplay of phases during transesterification, whereas the incorporation of ternary phase diagrams enhanced our understanding of product-facilitated oil integration. This research opens avenues for refining production processes and models to optimize biodiesel yield and quality. The investigation into the microbial conversion of glycerol into alcohols showcases the potential to enhance process economics and sustainability. By leveraging microbial strains within controlled bioreactors, researchers demonstrated the feasibility of turning glycerol waste into valuable alcohols, reducing the reliance on methanol and minimizing waste. These findings underline the versatility of biodiesel production processes and the potential to create a closed-loop system. The evaluation of biodiesel's performance within a diesel engine facility offers valuable insights into its practical application. The study's focus on engine operation, emissions, and combustion characteristics unveils a nuanced picture of biodiesel's behavior under various conditions. Although higher biodiesel blends show increased fuel consumption, they present a trade-off with reduced emissions, except for NOx emissions. These findings inform the ongoing development of biodiesel formulations and engine optimization strategies. In conclusion, biodiesel presents a viable pathway toward sustainable energy production and consumption. With its renewable origins, compatibility with existing infrastructure, and potential to mitigate environmental impacts, biodiesel holds promise as a significant player in the transition away from conventional fossil fuels. Continued research, innovation, and collaboration between scientific, engineering, and policy communities are essential to realize the full potential of biodiesel and address the challenges posed by our evolving energy landscape.

References

1. Imtenan, S.; Varman, M.; Masjuki, H.H.; Kalam, M.A.; Sajjad, H.; Arbab, M.I.; Rizwanul Fattah, I.M. Impact of low temperature combustion attaining strategies on diesel engine emissions for diesel and biodiesels: A review. *Energy Convers. Manag.* 2014, 80, 329–356.
2. Palash, S.M.; Kalam, M.A.; Masjuki, H.H.; Masum, B.M.; Rizwanul Fattah, I.M.; Mofijur, M. Impacts of biodiesel combustion on NOx emissions and their reduction approaches. *Renew. Sustain. Energy Rev.* 2013, 23, 473–490.
3. Demirbas, A. (2009) Progress and Recent Trends in Biodiesel Fuels. *Energy Convers. Manage.*, 50, 14-34.
4. Marchetti, J. M., *et al.* "Technological platforms for the production of biofuels from algae." *Renew. Sustain. Energy Rev.* 14.2 (2010): 172-180.
5. Okwundu, O.S.; El-Shazly, A.H.; Elkady, M. Comparative effect of reaction time on biodiesel production from low free fatty acid beef tallow: A definition of product yield. *SN Appl. Sci.* 2019, 1, 140.
6. Singh, S.P.; Singh, D. Biodiesel production through the use of different sources and characterization of oils and their esters as the substitute of diesel: A review. *Renew. Sustain. Energy Rev.* 2010, 14, 200–216.
7. Freedman, B.; Pryde, E.H.; Mounts, T.L. Variables affecting the yields of fatty esters from transesterified vegetable oils. *J. Am. Oil Chem. Soc.* 1984, *61*, 1638–1643.
8. Rashid, Umer, *et al.* "A critical review on the production of bioenergy from microalgae." *Renew. Sustain. Energy Rev.* 100 (2019): 324-337.

13

Effective Use of Biofuel in a Diesel Engine

Sathya Narayanan[1], Nasrin I. Shaikh[2*] and Milind Shrinivas Dangate[1]

[1]Chemistry Division, School of Advanced Sciences, Vellore Institute of Technology, Chennai, India

[2]Department of Chemistry, College of Military Engineering, Old Mumbai Pune Hwy, Dapodi, Pimpri-Chinchwad, Maharashtra, India

Abstract

The substitution of conventional diesel fuel with ultracarbofluids has emerged as a prospective strategy to mitigate the emission of pollutants like CO_2, CO, HCs, and smoke. These innovative ultracarbofluids consist of a blend of coal suspended in fuel oil and water, typically in a proportion of 5:3:2, complemented by a minor stabilizing additive. A comprehensive analysis of existing literature pertaining to the economic dimensions of coal and fuel oil production, as well as the characteristics of charcoal and vegetable oils, was undertaken. The feasibility of replacing coal and fuel oil with charcoal and vegetable oils was critically examined. An empirical investigation was carried out using novel blends of bio-ultracarbofluids, setting them apart from conventional ultracarbofluids by incorporating renewable charcoal and vegetable oil instead of coal and fuel oil. Exhaust emissions were quantified using Mexa-32 1E and Mexa-21 1E analyzers, covering the speed range from 1,000 to 3000 rpm in 200-rpm intervals. The outcomes were juxtaposed against those derived from a standard reference diesel fuel.

The formulations of bio-ultracarbofluids exhibited diminished brake power outputs, ranging from 5.6% to 20.7% lower, yet marked enhancements in CO_2, CO, HC, and smoke emissions were observed. Among the influencing factors within the formulations, the type and quantity of charcoal emerged as paramount. Charcoal characterized by higher volatile content at a mass ratio of 30% yielded the most favorable results, displaying marginally decreased brake power output while significantly curbing exhaust pollutants. In summation, bio-ultracarbofluids manifest promising potential as viable diesel fuel substitutes due to their substantially reduced harmful exhaust emissions compared to conventional diesel.

*Corresponding author: dr.nasrinshaikh@yahoo.com

O.V. Gnana Swathika, K. Karthikeyan, Milind Shrinivas Dangate and Nicoletta Ravasio (eds.) Biomass and Solar-Powered Sustainable Digital Cities, (221–230) © 2024 Scrivener Publishing LLC

The findings lay the foundation for devising an optimal performance-driven specification for bio-ultracarbofluids. Furthermore, avenues for additional valuable research and potential engine adjustments are highlighted. However, it remains crucial to undertake long-term testing to assess the impact of bio-ultracarbofluids on engine components. Although the prevailing economic conditions are currently unfavorable, they remain subject to fluctuations in crude oil prices and geopolitical considerations, such as economic and environmental policies implemented by entities like the European Union.

Keywords: Bio-ultracarbofluids, diesel fuel replacement, hydrocarbon, stabilising additive, environmental pollution

13.1 Energy Resources

In the 1970s, the rapid depletion of non-renewable global resources, which could be exhausted soon, became evident. This concern, coupled with apprehensions about harmful emissions from vehicles and power plants, particularly greenhouse gases like CO_2 and pollutants like NOx and SO2, led technologically advanced nations to shift their focus to renewable energy sources. Notable technological progress has been made in areas such as solar heating and photovoltaics. The drive to replace petroleum-based diesel fuels, partially or fully, arises from the need to conserve finite resources and address the significant pollutants produced by their combustion [1]. Many developing countries contribute up to 14% of global energy needs. Despite the availability of arable land for energy crops, liquid fuels dominate transportation. Liquid fuels' popularity stems from the easy extraction, transportation, and conversion of hydrocarbon (HC)-rich petroleum into various forms. However, liquid fuels lack routine ash generation and facilitate engine distribution. The world consumes energy equivalent to 80 billion tonnes of oil annually, with over 90% derived from finite fossil deposits that will deplete over time.

Energy is essential for heating, electricity generation, and transportation. The global population relies on energy, with 400 million registered passenger cars worldwide, doubling in two decades. Car ownership varies, with North America at 50 cars per hundred people, Europe at 40, and Japan at 30. Worldwide averages stand at 10 cars per hundred people, whereas countries like India and China have less than one car per hundred people. Cars primarily use liquid fuels like petrol or diesel, raising the possibility of a renewable source for a significant portion. Environmentally beneficial exhaust emissions with reduced pollution potential would also be advantageous.

13.2 Rationale and Significance of Biofuels

The environmental and public health consequences of diesel engines have spurred an urgent need for innovative solutions. The detrimental effects of diesel emissions, ranging from nitrogen oxides (NOx) and particulate matter to carbon monoxide (CO) and HCs, have underscored the need to shift away from conventional fossil fuels. Biofuels offer an alternative with the potential to reduce these harmful emissions and thus mitigate environmental damage. As global efforts to tackle climate change intensify, the exploration of biofuels gains greater significance due to their potential to align transportation systems with sustainable development goals [2].

13.3 Types of Biofuels and Their Characteristics

Biofuels encompass a diverse range of options, each with unique properties and potential benefits. Ethanol, derived from plant-based feedstocks, offers the advantage of reduced CO_2 emissions and is often blended with gasoline. On the other hand, biodiesel, typically derived from vegetable oils, presents an attractive alternative for diesel engines due to its compatibility with existing infrastructure and engines [3]. Furthermore, bio-ultra-carbofluids, a novel formulation comprising coal, fuel oil, and water, have been investigated for their potential to enhance combustion efficiency and decrease emissions. The distinct characteristics of these biofuel types and their implications for engine performance warrant a detailed exploration [4].

13.4 Advantages and Challenges of Biofuels

The advantages of biofuels extend beyond emissions reduction. Their renewable nature addresses concerns about fossil fuel depletion and provides a mechanism for sustainable resource management. Biofuels can also play a pivotal role in rural development by creating new opportunities for agriculture and bio-energy production. However, challenges such as land-use competition, feedstock availability, and potential indirect land-use change effects must also be acknowledged. The economic feasibility of large-scale biofuel production and distribution is another critical consideration, as are the energy inputs required for cultivation, processing, and transportation [5].

13.5 Ultracarbofluids: Pioneering Advances in Biodiesel Production

In the realm of sustainable energy, a groundbreaking innovation emerges with the integration of ultracarbofluids in the realm of biodiesel production. These fluids, characterized by their exceptional thermal and electrical conductivity due to the inclusion of carbon-based nanoparticles, hold immense promise for revolutionizing biodiesel processing. This delves into the potential application of ultracarbofluids in the realm of biodiesel production, shedding light on their unique attributes and their potential to enhance the efficiency and sustainability of this renewable energy source. This discussion is grounded in recent research strides and technological breakthroughs.

13.6 The Potential of Ultracarbofluids in Biodiesel Production

Ultracarbofluids, a subset of nanofluids, are colloidal suspensions containing carbon-based nanoparticles such as graphene and carbon nanotubes, dispersed within a base fluid. This distinctive composition imparts unparalleled thermal and electrical conductivity to these fluids, making them intriguing candidates for enhancing the biodiesel production process. One of the key challenges in biodiesel production is the need for efficient heat transfer during the transesterification process, where triglycerides are converted into biodiesel through a reaction with alcohol and a catalyst. Ultracarbofluids, with their exceptional thermal conductivity, can play a pivotal role in rapidly and uniformly conducting heat within the reaction mixture [6]. This could lead to accelerated reaction rates, reduced energy consumption, and improved yield in the biodiesel production process. Moreover, the electrical conductivity of ultracarbofluids holds potential for optimizing the separation and purification stages of biodiesel production. Electrostatic techniques, empowered by the electrical properties of these fluids, could facilitate the separation of glycerol and biodiesel, streamlining the overall process and reducing the need for extensive chemical additives [7].

13.7 Incorporating Ultracarbofluids: Challenges and Future Prospects

While the integration of ultracarbofluids in biodiesel production offers exciting prospects, several challenges must be addressed. These include the stability and compatibility of the nanoparticles within the fluid matrix, potential interactions with catalysts, and the scalability of the production process. As ongoing research addresses these challenges, the utilization of ultracarbofluids in biodiesel production could potentially lead to a more sustainable and efficient pathway toward renewable energy generation.

13.8 Experimental Methodology

The assessment of biofuels' efficacy in diesel engines necessitates a robust experimental methodology. A comprehensive experimental setup involving a Lister-Petter 600cc 2-cylinder, four stroke diesel engine equipped with a Heenan-Froude DPX I water brake dynamometer facilitates accurate measurements of brake power output and engine performance. The utilization of Mexa-32 1E and Mexa-21 1E analyzers enables the quantification of CO_2, CO, HC, and smoke emissions, offering insights into the environmental impact of the biofuel blends. The choice of engine parameters, operating conditions, and data collection techniques are essential for obtaining reliable and actionable results [8].

13.8.1 Performance Evaluation of Bio-Ultracarbofluid Blends

The core of this study lies in the evaluation of bio-ultracarbofluid blends as substitutes for conventional diesel fuels. The collected experimental data reveal the intricate relationship between engine performance and emissions reduction. The observed reduction in brake power output across various speed intervals is attributed to the differences in combustion characteristics and energy content between the bio-ultracarbofluid blends and the reference diesel fuel. This trade-off between performance and emissions underscores the importance of a nuanced understanding of the fuel's impact on engine behavior.

13.8.2 Bio-Ultracarbofluids: A Paradigm Shift in Biodiesel Evolution

In the landscape of sustainable energy, a pioneering evolution unfolds as bio-ultracarbofluids enter the realm of biodiesel production. These innovative fluids, characterized by their exceptional thermal and electrical conductivity owing to the infusion of carbon-based nanoparticles, stand poised to redefine the biodiesel landscape. This delves into the potential of bio-ultracarbofluids within the biodiesel context, shedding light on their distinctive attributes and the transformative impact they can have on enhancing the efficiency and sustainability of renewable energy production. This discussion draws from recent strides in research and technological advancements.

13.8.3 The Potential of Bio-Ultracarbofluids in Biodiesel Production

Bio-ultracarbofluids represent a pioneering subset of nanofluids, where carbon-based nanoparticles such as graphene and carbon nanotubes are expertly integrated into the fluid matrix. This unique composition confers unmatched thermal and electrical conductivity to these fluids, rendering them not only intriguing but also exceptionally promising for elevating biodiesel production processes. The heart of biodiesel production lies in the transesterification process, wherein triglycerides undergo conversion into biodiesel through interaction with alcohol and a catalyst [9]. Herein lies a challenge: efficient heat transfer. The exceptional thermal conductivity of bio-ultracarbofluids can revolutionize this aspect by facilitating rapid, uniform heat conduction within the reaction mixture [10]. This has the potential to accelerate reaction rates, diminish energy consumption, and enhance overall biodiesel yield.

Furthermore, the electrical conductivity intrinsic to these fluids presents an avenue for refining separation and purification stages in biodiesel production and harnessing their electrical properties, electrostatic techniques could streamline the separation of glycerol and biodiesel, streamlining processes and mitigating the need for extensive chemical additives.

13.8.4 Navigating Challenges and Future Prospects

Despite the promise they hold, the integration of bio-ultracarbofluids into biodiesel production is not without challenges. Factors such as nanoparticle stability, compatibility within the fluid matrix, potential interactions with catalysts, and scalability demand careful consideration. As ongoing research addresses these hurdles, bio-ultracarbofluids could emerge as a pivotal force propelling biodiesel production toward greater sustainability and efficiency.

13.8.5 Emissions Analysis and Environmental Impact

The emissions analysis provides a comprehensive overview of the environmental impact of bio-ultracarbofluid blends. The consistent reduction in CO_2, CO, and HC across the speed range suggests the potential of these biofuel formulations to substantially curtail the release of pollutants that contribute to air quality deterioration and environmental degradation. The implications of emissions reduction extend beyond immediate environmental benefits; they align with global efforts to mitigate climate change and reduce the overall carbon footprint of transportation systems.

13.8.6 Influence of Charcoal Properties on Performance

A significant aspect of the study pertains to the role of charcoal properties on engine performance and emissions [7]. The type and volatile content of charcoal emerged as pivotal factors influencing the outcomes of the bio-ultracarbofluid blends. The selection of charcoal with higher volatile content, approximately 27.2%, led to more favorable results in terms of emissions reduction and engine performance. Furthermore, the optimal mass ratio of charcoal within the blend, determined to be 30%, highlights the intricate balance between achieving reduced brake power output and significantly diminished exhaust pollutants.

13.8.7 Economic Considerations and Policy Implications

While the technical feasibility of biofuels as substitutes for conventional diesel fuels is demonstrated in this study, economic viability remains a crucial consideration. Fluctuations in crude oil prices and evolving environmental policies, such as carbon pricing mechanisms, play a significant role in determining the cost-effectiveness of biofuels. Policy frameworks that incentivize biofuel adoption and create a level playing field for alternative fuels are instrumental in accelerating the transition toward greener transportation systems. Moreover, the potential for biofuels to contribute to rural development, job creation, and energy security adds another layer to their socio-economic significance.

13.9 Conclusion and Future Directions

The comprehensive evaluation of bio-ultracarbofluids as alternatives to conventional diesel fuels underscores their potential as a viable solution to mitigate the environmental impact of internal combustion engines. The intricate interplay between engine performance, emissions reduction, and fuel properties highlights the complexity of transitioning to biofuels. The findings pave the way for establishing performance-oriented specifications for bio-ultracarbofluids and offer insights into engine adjustments that optimize fuel compatibility. The long-term effects of bio-ultracarbofluids on engine components necessitate further exploration, emphasizing the need for extended testing to ensure durability and reliability.

In a broader context, the journey toward effective biofuel utilization is intertwined with evolving global dynamics. As attitudes toward sustainability evolve, the economic considerations surrounding biofuels may witness shifts that foster their adoption. Collaborative efforts among researchers, industry stakeholders, and policymakers are imperative for realizing the full potential of biofuels and integrating them into a greener and more sustainable transportation ecosystem.

References

1. Jones, M. *et al.*, "*Biomass for Energy and Industry-Policy, Environment, Production and Harvesting*", Elsevier, London, pp 228-234,1989
2. Knothe, G., *et al.*, "*Vegetable oils as Alternative diesel fuels*", SAE Technical paper No. 920194, Society of Automotive Engineers, Warrendale, Pennsylvania, U. S. A.
3. Alengebawy, A.; Mohamed, B.A.; Ghimire, N.; Jin, K.; Liu, T.; Samer, M.; Ai, P. Understanding the environmental impacts of biogas utilization for energy production through life cycle assessment: An action towards reducing emissions. *Environ. Res.* 2022, 213, 113632.
4. Shell, "*Bio-fuels in the European Community*", May 1992.
5. Mehmood, M.A.; Ibrahim, M.; Rashid, U.; Nawaz, M.; Ali, S.; Hussain, A.; Gull, M. Biomass production for bioenergy using marginal lands. *Sustain. Prod. Consum.* 2017, 9, 3–21.
6. Antonini, OR *et al.*, "The Ultracarbofluids: An Oil Agglomeration and Slurryation Process for Clean Slurry Production", pp. 293-312, *V EEC Biomass Energy Conference*, 1989.
7. Poole, A., "*Charcoal Pyrolysis in Africa*" Academic Press, 1982.
8. Heywood J. B. *Internal Combustion Engine fundamentals*, pp. 598, 1995.

9. Schwab, A. W., Bagby, M. O. and Freeman, B. "Preparation and Properties of Diesel Fuels from Vegetable Oils" *Fuel*, Vol. 66, pp. 1373-1378,1987.
10. Singh, A.R.; Singh, S.K.; Jain, S. A review on bioenergy and biofuel production. *Mater. Today Proc.*, 2022, 49, 510–516.

14

Enhancing Bio-Oil Stability and Fuel Quality Through Additive Integration

Sathya Narayanan[1], Nasrin I. Shaikh[2*] and Milind Shrinivas Dangate[1]

[1]*Chemistry Division, School of Advanced Sciences, Vellore Institute of Technology, Chennai, India*

[2]*Department of Chemistry, College of Military Engineering, Old Mumbai Pune Hwy, Dapodi, Pimpri-Chinchwad, Maharashtra, India*

Abstract

Being one of the globe's prominent palm oil producers, Malaysia has encountered a significant challenge in managing the substantial volume of waste from oil palm biomass. A solution to this issue involves transforming these biomass wastes into biofuel through the use of rapid pyrolysis technology. However, to address the unfavorable characteristics of the resulting fast pyrolysis bio-oil, additional refinement via direct solvent incorporation is necessary. Additionally, the relatively high production expenses associated with biofuels often impede their commercial viability. Therefore, the formulated blend of solvent and oil must meet both functional fuel requirements and economic benchmarks in order to effectively compete with traditional diesel fuel. In this particular section, a multi-phase (computer-aided molecular design) approach was applied to the task of designing a solvent for bio-oil. A model based on rough-set theory and machine learning was proposed to establish correlations between feedstock characteristics, pyrolysis conditions, and pyrolysis bio-oil properties by generating decision rules. These generated decision rules were analyzed scientifically to uncover underlying patterns while ensuring their logical coherence. The resultant decision rules can serve to guide the selection of optimal feedstock compositions and pyrolysis conditions for producing pyrolysis bio-oil with desired fuel properties. To validate the outcomes derived from computational methodologies, experimental investigations were conducted. The resulting pyrolysis bio-oils were combined with identified solvents in varying ratios. Additionally, the blending of the solvent-oil mixture with diesel was facilitated through surfactants, enabling emulsification.

***Keywords*:** Biofuel, pyrolysis, bio-oil, fuel additive, economy

**Corresponding author*: dr.nasrinshaikh@yahoo.com

O.V. Gnana Swathika, K. Karthikeyan, Milind Shrinivas Dangate and Nicoletta Ravasio (eds.) Biomass and Solar-Powered Sustainable Digital Cities, (231–242) © 2024 Scrivener Publishing LLC

14.1 Background Problems

The historical focus of research has been on physically mixing pyrolysis bio-oil with solvents, but the detailed connections between bio-oil components and solvents have been overlooked. Traditional methods for solvent selection involved time-consuming trial-and-error experiments. Computer-aided molecular design (CAMD) emerged as a solution, predicting solvents that meet property targets. CAMD uses reverse engineering to build molecules for specific properties. Past solvent design attempts relied on property prediction models like the group contribution (GC) method, but complex computations limited their use. Topological indices offer an alternative, but combining them with GC models was challenging. Molecular signature descriptors capture structural details in a 2D formula and can manage complex molecular structures. However, managing the complexity of the CAMD problem is crucial. Biofuel commercialization faces challenges like high production costs and evolving regulations. Incorporating solvents can enhance biofuel properties, but optimizing performance may increase expenses. Economic considerations are essential to compete with conventional diesel. Multi-objective optimization (MOO) is vital to balance bio-oil quality and blend profitability. Pyrolysis bio-oil properties vary based on biomass sources and conditions, impacting solvent selection. The higher heating value relates to elemental composition, but estimating it remains uncommon. Bio-oil's low pH and instability affect engine fuel use. Machine learning (ML) tools aid thermochemical biomass processing, but some lack transparency. The rough-set theory and ML (RSML) approach offers understandable rules while incorporating domain knowledge. While CAMD identifies solvent candidates, model accuracy needs verification. Experimental validation confirms real-world feasibility. Stable emulsification strategies are needed for solvent-oil-diesel blends, considering varying properties like low heating value and corrosiveness. Research historically focused on physically mixing pyrolysis bio-oil with solvents, but their complex relationships were often neglected. Traditional solvent selection methods were time-consuming, leading to the emergence of CAMD for predicting solvents with desired properties. Past attempts at solvent design relied on property prediction models, but challenges arose from computational complexities. Molecular signature descriptors offer a way to manage complex molecular structures in CAMD. Challenges in biofuel commercialization include high production costs and evolving regulations, making economic considerations crucial. Balancing bio-oil quality and blend profitability requires MOO. Bio-oil properties, influenced by biomass and conditions, impact solvent selection. Machine learning tools aid in biomass processing, with the rough-set approach providing transparency. Verification of CAMD results through experimental validation is vital, as is the development of stable emulsification strategies for solvent-oil-diesel blends.

14.2 Biomass Source Materials

Biomass source materials comprise natural elements, discarded items, or leftover remnants with the potential for energy generation. These organic materials stem from agricultural origins, encompassing both plant and animal-derived by-products, alongside residues from aquatic and industrial sources. Virtually, any variety of biomass source materials can be harnessed for energy production through either biochemical processes.

14.3 Dedicated Energy Crops

The term "dedicated energy crops" refers to plants that are deliberately cultivated with the primary goal of producing energy, requiring minimal maintenance and expenses. These crops can be broadly categorized into three main groups: grassy, woody, and oil-bearing crops, as illustrated in Figure 14.1. Among these, rapidly growing grassy plant varieties like poplar and willow are typically cultivated in perennial plantations [1]. Woody plants usually follow growth cycles spanning from three to 7 years, whereas

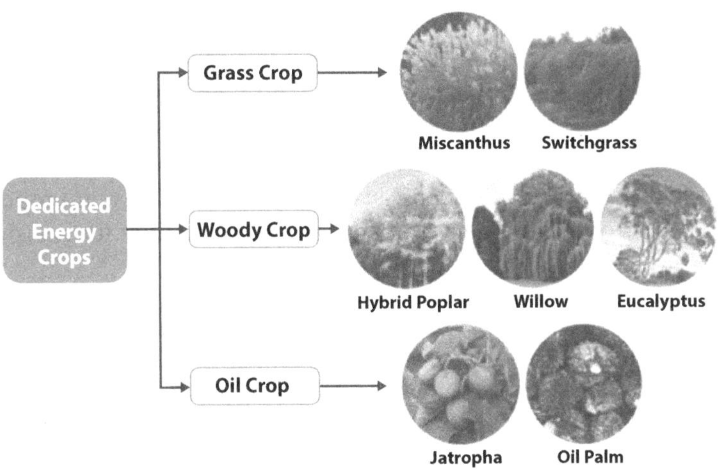

Figure 14.1 Categories of dedicated energy crops.

grassy plant species have a shorter 1-year cycle. Switchgrass shows promise as a valuable biofuel source due to its adaptability to less fertile agricultural land and diverse climatic conditions [2]. In contrast, Miscanthus, a perennial grass from Asia, exhibits rapid growth during warm seasons and demands minimal application of fertilizers and pesticides [3].

14.4 Biomass Thermal Conversion

Biomass thermal conversion involves subjecting biomass feedstock to heat under an inert atmosphere, resulting in the production of biochar, bio-oil. This process draws parallels with traditional charcoal-making methods that have been employed for centuries. The final composition of the products generated through pyrolysis is influenced by factors such as temperature, vapor residence time, heating rate, and feedstock size [4]. Among these factors, temperature has the most profound impact on the composition of the end products of pyrolysis. Lower temperatures (<300°C) and slower heating rates yield biochar and gaseous by-products like carbon dioxide and water. Temperatures exceeding 650°C trigger biomass decomposition, resulting in gaseous outputs like carbon monoxide, hydrogen, and methane taking precedence

In the case of fast and flash pyrolysis, high temperatures and rapid heating rates bypass the initial reaction stage and directly engage the second reaction pathway, predominantly yielding liquid products.

14.4.1 Slow Pyrolysis

Slow pyrolysis is a thermal conversion process that transforms biomass into biochar, bio-oil, and syngas through controlled heating in an oxygen-limited environment [5]. This process occurs at relatively lower temperatures and longer residence times compared to fast pyrolysis, resulting in a higher yield of biochar and a more stable bio-oil with different properties. Slow pyrolysis offers unique advantages and opportunities for sustainable resource utilization.

In slow pyrolysis, biomass feedstock is heated to temperatures typically ranging from 350°C to 700°C in the absence of O_2 or with limited O_2 supply.

14.4.2 Fast Pyrolysis

Fast pyrolysis is a process that involves the rapid heating of biomass in the absence of oxygen to produce valuable bio-oil, as well as biochar and syngas. This process typically operates at high temperatures and short residence times, leading to the rapid breakdown of complex biomass molecules into smaller, liquid-rich fractions. Fast pyrolysis holds great promise for converting biomass into renewable fuels and chemicals, making it a key technology in the pursuit of sustainable energy solutions. During fast pyrolysis, biomass is quickly heated to temperatures ranging from 400°C to 600°C in a reactor with limited or no oxygen supply. This rapid heating causes the breakdown of biomass components through a complex series of thermal degradation and chemical reactions [6]. The primary product of interest is bio-oil, a viscous liquid mixture rich in oxygenated compounds. This bio-oil can be further refined into various valuable products, such as transportation fuels, chemicals, and materials. In recent years, research efforts have focused on optimizing the fast pyrolysis process to improve bio-oil quality, yield, and stability. Additionally, the effects of different operating parameters on bio-oil yield and composition contribute to a better understanding of the process optimization.

14.4.3 Flash Pyrolysis

Flash pyrolysis stands as an accelerated procedure characterized by ultra-short reaction periods, sometimes mere seconds, demanding swift elevation of temperature. This method hinges on specialized arrangements of reactors like entrained flow and fluidized bed reactors, engineered to achieve exceedingly brief intervals during which biomass is present, typically a matter of seconds [6]. To enable efficient flash pyrolysis, biomass particles need to be minute, facilitating rapid heating. Nevertheless, this technique does introduce specific technological complexities. The corrosive nature of the resulting bio-oil and the presence of solids within it emerge as noteworthy issues.

14.4.4 Enhancing the Properties of Pyrolysis Bio-Oil through Physical Upgrading

Enhancing the properties and usability of pyrolysis bio-oil through physical upgrading methods has gained traction as a strategy to overcome its inherent limitations and make it a more versatile and valuable product. Pyrolysis bio-oil, obtained from the thermal degradation of biomass, is characterized by high oxygen content, water content, and instability, which can hinder its direct utilization. Physical upgrading techniques focus on improving the bio-oil's quality by removing undesirable components, reducing its water content, and increasing its energy density. One of the primary methods for physical upgrading is vacuum distillation, which involves the separation of bio-oil into different fractions based on their boiling points. This process helps to reduce the water content and remove volatile compounds from the bio-oil, resulting in a more stable and energy-dense product. Another physical upgrading technique is filtration, which can be used to remove solid particles, char, and ash from the bio-oil. Filtration not only enhances the bio-oil's cleanliness but also prevents clogging and fouling of combustion systems. In addition, centrifugation is employed to separate the bio-oil into distinct layers based on density, allowing the removal of heavy phases that can contribute to instability and degradation [7]. These physical upgrading methods can be combined to achieve synergistic effects. For example, a combination of vacuum distillation and centrifugation can help produce fractions of bio-oil with improved stability and energy content.

14.4.5 Enhancing Pyrolysis Bio-Oil Through Chemical Upgrading

Pyrolysis bio-oil, a complex mixture derived from the thermal degradation of biomass, has gained significant attention as a potential renewable energy source and feedstock for various applications. However, its chemical composition is often characterized by high oxygen content, water content, acidity, and instability, which limits its direct use. To overcome these challenges and enhance the quality of pyrolysis bio-oil, chemical upgrading processes have emerged as promising strategies. Chemical upgrading involves the modification of the bio-oil's composition and properties through various chemical reactions. These reactions can include deoxygenation, hydrodeoxygenation, hydrodeoxygenation, hydrodenitrogenation, and hydrodesulfurization. Through these transformations, the undesirable components such as oxygen, nitrogen, and sulfur compounds can be selectively removed, resulting in a bio-oil with improved stability, lower acidity, and enhanced energy content. The process of chemical upgrading typically involves the use of catalysts to facilitate the desired reactions. Catalysts can help in breaking down complex molecules, promoting the rearrangement of functional groups, and facilitating the removal of heteroatoms. The choice of catalyst and reaction conditions plays a crucial role in determining the extent and selectivity of the upgrading process. Researchers have explored various catalysts for chemical upgrading, including supported metal catalysts, zeolites, and transition metal sulfides. These catalysts have shown effectiveness in reducing the oxygen and nitrogen content of bio-oil and improving its energy content. For instance, the use of various catalysts for the hydrodeoxygenation of pyrolysis bio-oil and found that a combination of Ni-Mo/γ-Al2O3 catalyst exhibited superior performance in terms of oxygen removal and energy content enhancement. Furthermore, integrated processes that combine chemical upgrading with other techniques such as hydrotreating and fractional distillation have been proposed to maximize the benefits of bio-oil enhancement. These integrated approaches aim to achieve higher yields of valuable liquid fuels and chemicals while minimizing waste and by-products.

14.5 Employing Ultrasound for Bio-Oil and Diesel Emulsification

Ultrasound-assisted emulsification has emerged as a promising technique for producing stable emulsions of bio-oil and diesel fuel, addressing the challenges of immiscibility and facilitating the integration of renewable bio-oil into conventional fuels. This process involves the application of high-frequency sound waves to induce cavitation, resulting in the breakup of immiscible liquids into finely dispersed droplets. The resulting emulsions exhibit improved stability, enhanced combustion characteristics, and reduced environmental impact.

The use of ultrasound for emulsifying bio-oil and diesel offers several advantages. Firstly, it allows for the creation of emulsions with uniform droplet size distributions. The cavitation forces generated during ultrasound treatment ensure that larger bio-oil droplets are broken down into smaller, more uniform droplets, which, in turn, leads to better fuel atomization and more efficient combustion. Secondly, ultrasound-assisted emulsification enhances the kinetic stability of the emulsions. The finely dispersed bio-oil droplets are less prone to coalescence and separation, resulting in emulsions that remain stable over extended periods, even in the presence of temperature fluctuations and mechanical stress. Furthermore, the improved combustion properties of emulsified bio-oil and diesel blends contribute to reduced emissions and enhanced fuel efficiency. The increased surface area of the finely dispersed bio-oil droplets promotes more complete combustion, leading to lower levels of particulate matter and other pollutants.

The successful application of ultrasound-assisted emulsification depends on various factors, including the frequency and intensity of ultrasound, treatment time, and the characteristics of the liquids being emulsified. Proper optimization of these parameters is crucial to achieving the desired emulsion properties and performance benefits. Several research studies have investigated the potential of ultrasound-assisted emulsification for bio-oil and diesel blends. The effects of ultrasound parameters on emulsion stability and droplet size distribution highlight the importance of optimizing ultrasound conditions for achieving desired emulsion properties. The combustion performance of emulsified bio-oil and diesel blends demonstrates improved combustion efficiency and reduced emissions compared to pure diesel fuel. In conclusion, ultrasound-assisted emulsification presents a promising approach to enhance the compatibility and properties of bio-oil and diesel blends. By creating stable and uniform emulsions, this technique offers a pathway to effectively integrate renewable bio-oil into conventional diesel fuels, contributing to more sustainable energy solutions.

14.6 Emulsifying Pyrolysis Bio-Oil and Diesel

Emulsifying pyrolysis bio-oil with diesel fuel has gained attention as a strategy to address the challenges associated with the utilization of bio-oil as a renewable energy source. Pyrolysis bio-oil, derived from the thermal decomposition of biomass, possesses a complex composition with high oxygen content and polar compounds, making it incompatible with hydrocarbon-based diesel fuel [1]. Emulsification, a process that involves the dispersion of one liquid phase within another in the form of tiny droplets, offers a means to improve the compatibility of these two immiscible liquids. The process of emulsifying pyrolysis bio-oil and diesel fuel involves mixing the two liquids and subjecting them to mechanical energy, such as stirring or ultrasonication. This mechanical energy leads to the formation of emulsions, where the bio-oil is dispersed as tiny droplets within the diesel matrix. Emulsification improves the miscibility of bio-oil and diesel, making it easier to handle, transport, and utilize in combustion systems.

Emulsification offers several advantages when blending pyrolysis bio-oil and diesel:

1. **Homogeneous Mixture:** Emulsification results in a uniform mixture of bio-oil and diesel, preventing phase separation and ensuring consistent fuel properties.
2. **Enhanced Combustion:** The finely dispersed bio-oil droplets in the emulsion promote better mixing with air during combustion, leading to improved combustion efficiency and reduced emissions.

3. **Stability:** Emulsification can enhance the long-term stability of the bio-oil and diesel blend, reducing the risk of droplet coalescence and phase separation.
4. **Reduced Carbon Footprint:** By incorporating renewable bio-oil into diesel, emulsification contributes to lowering the carbon footprint of diesel fuel and supports sustainable energy initiatives.
5. **Compatibility:** Emulsified bio-oil and diesel blends can be used in existing diesel engines and infrastructure without significant modifications.

However, achieving stable and well-dispersed emulsions requires careful selection of emulsification techniques, emulsifiers, and processing conditions. The properties of the bio-oil, such as viscosity, water content, and chemical composition, also influence the emulsion formation and stability. Researchers have investigated various aspects of emulsifying pyrolysis bio-oil and diesel. The effect of different emulsification methods on emulsion stability and combustion performance highlights the importance of proper emulsifier selection and optimization. Additionally, the influence of bio-oil composition on emulsion formation and stability sheds light on the factors affecting emulsion properties. In conclusion, emulsifying pyrolysis bio-oil with diesel offers a promising approach to enhance the compatibility and usability of these two distinct liquids. By creating stable and homogeneous emulsions, this technique contributes to the integration of renewable bio-oil into conventional diesel fuels, supporting the transition toward more sustainable energy sources.

14.7 Utilizing RST

Utilizing Reactive Separation Technology (RST) for biodiesel production represents an innovative and efficient approach to simultaneously convert feedstock oils and remove unwanted by-products in a single process step. Biodiesel, a renewable and sustainable alternative to fossil fuels, is typically produced through transesterification, a chemical reaction between triglycerides in feedstock oils and alcohol. Reactive Separation Technology integrates this reaction with separation processes, enhancing reaction efficiency, reducing reaction time, and improving overall biodiesel yield [8].

In RST, feedstock oils, such as vegetable oils or animal fats, are mixed with alcohol and a catalyst, and the mixture is exposed to a combination of heat and vacuum conditions. This innovative approach promotes the transesterification reaction while simultaneously driving the removal of the glycerol by-product. The application of heat and vacuum reduces the boiling point of glycerol, facilitating its removal as vapor [9]. This continuous removal of glycerol helps shift the equilibrium of the reaction toward biodiesel formation, resulting in higher yields.

The advantages of utilizing RST for biodiesel production include the following:

1. **Enhanced Reaction Rates:** RST accelerates the transesterification reaction, reducing the time required to achieve high conversion rates. The continuous removal of glycerol helps maintain favourable reaction conditions and minimize the inhibition of the reaction.
2. **Higher Biodiesel Yields:** By shifting the equilibrium toward biodiesel formation through continuous glycerol removal, RST enhances overall biodiesel yield, leading to increased production efficiency.
3. **Simplified Process:** RST integrates the reaction and separation processes, simplifying the overall biodiesel production scheme. This can lead to reduced capital and operating costs compared to traditional methods.
4. **Reduced Catalyst Usage:** The continuous removal of glycerol prevents catalyst deactivation, allowing for the use of lower catalyst concentrations and minimizing the need for post-reaction separation steps.
5. **Flexibility in Feedstocks:** RST can be applied to a wide range of feedstock oils, including low-quality oils, waste cooking oils, and animal fats, further enhancing the versatility of biodiesel production.

Researchers have explored the application of RST for biodiesel production and highlighted its potential benefits. For instance, a comprehensive study on the utilization of RST for the production of biodiesel from various feedstocks, demonstrating the improved yields and reaction rates achieved through this technology.

In conclusion, utilizing RST for biodiesel production offers a promising approach to enhance the efficiency and yield of the transesterification reaction while simultaneously removing glycerol by-product. By integrating the reaction and separation processes, RST streamlines biodiesel production, making it a viable and sustainable alternative to conventional fossil fuels.

14.8 Conclusion

In conclusion, Malaysia's position as a prominent palm oil producer has led to a significant challenge in managing the substantial waste generated from oil palm biomass. Transforming these biomass wastes into biofuel through rapid pyrolysis technology presents a solution, although challenges remain in addressing the unfavorable characteristics of the resulting fast pyrolysis bio-oil. Additional refinement through solvent incorporation is essential, and economic viability is a key consideration in this endeavor. This section has explored a multi-phase approach, involving CAMD and RSML, to enhance its effective utilization. CAMD has proven useful in designing solvents for bio-oil, with molecular signature descriptors accommodating property prediction models. A consistency rule was developed to manage the complexity of the CAMD challenge. Economic benchmarks were integrated into the approach through fuzzy MOO. RSML emerged as a valuable tool for correlating feedstock characteristics, pyrolysis conditions, and bio-oil properties, offering decision rules that guide optimal production strategies. Experimental validation was conducted to verify computational outcomes, highlighting the feasibility of solvent-oil blends. Emulsification strategies were employed to blend pyrolysis bio-oil with diesel using surfactants, allowing for enhanced stability and improved fuel properties. This comprehensive approach addresses both the technological challenges of bio-oil enhancement and the economic considerations crucial for competing with traditional diesel fuel. Furthermore, the application of rough set theory in diverse fields, such as pattern recognition, information retrieval, and fragrance molecule design, underscores its practical versatility and potential for solving complex problems. Overall, this section demonstrates the integration of advanced computational methodologies, theoretical frameworks, and experimental validation to tackle the multifaceted challenges associated with pyrolysis bio-oil enhancement and utilization.

References

1. Singh, D.; Sharma, D.; Soni, S.; Sharma, S.; Sharma, P.K.; Jhalani, A. *A review on feedstocks, production processes, and yield for different generations of biodiesel.*
2. Aransiola, E.F.; Ojumu, T.V.; Oyekola, O.; Madzimbamuto, T.; Ikhu-Omoregbe, D. A review of current technology for biodiesel production: State of the art. *Biomass Bioenergy,* 2014, 61, 276–297.
3. Salehi Jouzani, G.; Aghbashlo, M.; Tabatabaei, M. Biofuels: Types, Promises, Challenges, and Role of Fungi. In *Fungi in Fuel Biotechnology*; Springer: Berlin/Heidelberg, Germany, 2020; pp. 1–14.

4. Jeswani, H.K.; Chilvers, A.; Azapagic, A. Environmental sustainability of biofuels: A review. *Proc. R. Soc. A*, 2020, 476, 20200351.
5. Akram, F.; ul Haq, I.; Raja, S.I.; Mir, A.S.; Qureshi, S.S.; Aqeel, A.; Shah, F.I. Current trends in biodiesel production technologies and future progressions: A possible displacement of the petro-diesel. *J. Clean. Prod.*, 2022, 133479.
6. Bashir, M.A.; Wu, S.; Zhu, J.; Krosuri, A.; Khan, M.U.; Aka, R.J.N. Recent development of advanced processing technologies for biodiesel production: A critical review. *Fuel Process. Technol.*, 2022, 227, 107120.
7. Karmakar, B.; Halder, G. Progress and future of biodiesel synthesis: Advancements in oil extraction and conversion technologies. *Energy Convers. Manag.*, 2019, 182, 307–339.
8. Kulprathipanja. (2019). Reactive Separation Processes. *CRC Press EBooks*, 1-17, https://doi.org/10.1201/9780203741238-1.
9. Paul Kt Liu, Muhammad Sahimi, & Tsotsis, T. T. (2012). Process intensification in hydrogen production from coal and biomass via the use of membrane-based reactive separations. *Curr. Opin. Chem. Eng.*, 1(3), 342–351.

15

Biofuels from Microbes and Waste Resources

Sathya Narayanan[1], Nasrin I. Shaikh[2*] and Milind Shrinivas Dangate[1]

[1]*Chemistry Division, School of Advanced Sciences, Vellore Institute of Technology, Chennai, Tamil Nadu, India*
[2]*Department of Chemistry, College of Military Engineering, Old Mumbai Pune Hwy, Dapodi, Pimpri-Chinchwad, Maharashtra, India*

Abstract

The global energy landscape is at a critical point, necessitating a shift from fossil fuels to sustainable alternatives, spurring extensive research into liquid biofuels. This comprehensive review explores diverse strategies, technologies, and goals in developing sustainable liquid biofuels, highlighting the need to balance energy security, environmental sustainability, and resource efficiency. The paper begins by addressing the energy challenge, emphasizing that reliance on a single alternative fuel technology is inadequate. It notes the limitations of current biofuels from sugar and oil crops and stresses the importance of sourcing sustainable liquid fuels from marginal land biomass, renewable plant materials, and waste.

A significant challenge with conventional liquid biofuels is their suboptimal physical properties, which limit their applicability. The variations in chemical structure among biofuels, especially biodiesel, lead to inconsistent properties and engine performance, underscoring the need for drop-in fuels with predictable and improved characteristics. The review then details biodiesel production from coffee ground waste, discussing its composition, lipid extraction methods, and the transesterification process. It highlights the environmental benefits of using coffee waste for biodiesel but also acknowledges the challenges of feedstock variability, energy-intensive extraction, and economic feasibility.

The discussion then shifts to pyrolysis, a thermal decomposition technique that offers precise control over fuel properties. The versatility of pyrolysis in producing various fuel products with tailored characteristics is explored, along with the

**Corresponding author*: dr.nasrinshaikh@yahoo.com

O.V. Gnana Swathika, K. Karthikeyan, Milind Shrinivas Dangate and Nicoletta Ravasio (eds.)
Biomass and Solar-Powered Sustainable Digital Cities, (243–254) © 2024 Scrivener Publishing LLC

potential for upgrading bio-oil to ensure its stability for integration into existing infrastructure. The paper also introduces cross metathesis in organic chemistry, emphasizing its impact on synthesizing diverse compounds through alkene exchange. It explains the role of transition metal-based catalysts, particularly Grubbs-type catalysts, in this process, highlighting cross metathesis's potential in organic synthesis.

The review outlines key research objectives for enhancing sustainable liquid biofuel production, including advanced feedstock selection, integrated biorefineries, innovative conversion technologies, catalytic upgrading, microbial engineering, carbon capture and utilization, waste valorization, life cycle analysis, and policy considerations. Each objective is discussed in detail, illustrating its role in a comprehensive and sustainable biofuel production approach. The synthesis concludes by emphasizing the collaborative and multidisciplinary nature of this endeavor, highlighting the need to synchronize efforts across diverse research domains to achieve a future with cleaner, more resilient, and energy-secure paradigms. By integrating various technologies to harness synergies and mitigate limitations, the research community aims to replace fossil fuels with sustainable alternatives.

***Keywords*:** Biodiesel, lipid, pyrolysis, biomolecule, ignition engines

15.1 Background

There are many alternative fuel technologies, but none of them can replace transport fuels by themselves. We need to use different technologies to meet the growing energy and fuel demand of the world. We cannot use enough sugar and oil crops to make enough biofuels to replace fossil fuels. We need to use biomass from renewable plant materials, or waste materials to make sustainable liquid fuels. Current liquid biofuels have bad physical properties that limit their use. For example, bioethanol causes corrosion and has lower energy density, which makes it unsuitable for some vehicles and aviation. Biodiesel has low oxidative stability and high pour and cloud points, which makes it hard to store and use in cold temperatures and aviation. This causes variation in the fuel properties, which is not good because it can affect the engine performance. To make a good drop-in fuel, we need to reduce the variation in the structure and make the fuel properties more predictable. This can be done by making a single-molecule fuel or by using methods that can control the product distribution.

These are the criteria that any new fuel or technology must meet or exceed to be a good replacement for fossil fuels.

15.2 Biodiesel

Biodiesel is a biofuel that can be used instead of diesel and comes from biological sources like plant oils and animal fats. It is one of the main biofuels produced globally, with a production of 28.3 billion liters in 2013.The most common and cheapest way to make biodiesel is by reacting triglycerides with three times the amount of alcohol (usually methanol because it is cheap) in the presence of an alkali catalyst (usually sodium or potassium hydroxide). This reaction is called transesterification. Most biodiesel plants use a little more methanol than needed (usually 4.5–6 times the amount of triglyceride) to make the reaction go faster, according to Le Chatelier's principle. This method produces biodiesel with a 98% yield, but it also produces crude glycerol as a by-product, which is hard to purify and requires a lot of energy. Therefore, many researchers are looking for other ways to do transesterification, such as using different catalysts or enzymes that have lower energy consumption and less soap formation.

The quality of biodiesel as a fuel depends on the fatty acid methyl ester (FAME) profile. Different FAMEs have very different physical properties depending on how saturated they are and how long their chains are. Generally, saturated FAMEs with long chains have not only high viscosities and melting points but also high cetane numbers. These FAMEs are not good for diesel replacements because they would make the fuel too thick at low temperatures. Higher saturation also means lower oxidative stability. Oxidation of biodiesel can change its fuel properties, such as increasing its viscosity, which can affect the engine performance. Unsaturated FAMEs can also affect the exhaust emissions, usually increasing the amount of NOx. This may be because unsaturated FAMEs have lower cetane numbers, which means they burn at higher temperatures that favour NOx formation.

15.3 Research Objectives

The primary goal of this project is to create and evaluate potential eco-friendly liquid fuels sourced from microorganisms and waste materials. The focus is on enhancing the physical attributes and selectivity of the resulting products. Given the variety of technologies capable of generating sustainable liquid fuels, various competitive approaches will be investigated. This approach enables a direct comparison of these technologies and an understanding of how they can be synergistically integrated. The specific objectives of this study include the following:

1. Evaluating the viability of utilizing spent coffee grounds as an environmentally sustainable source for producing biodiesel, specifically in terms of its physical characteristics.
2. Exploring the process of cross-metathesis between triglycerides, a key biodiesel component, and ethene, aiming to produce two fuels with improved properties and controlled structure.
3. Investigating the conversion of representative ketonic compounds present in the vapor resulting from the pyrolysis of lignocellulosic biomass into bio-oil, using metal-supported zeolite catalysts.

15.4 Biodiesel From Coffee Ground Waste

Producing Biodiesel from Coffee Grounds

The pursuit of sustainable and renewable energy sources has led researchers to explore unconventional feedstocks for biodiesel production, such as coffee ground waste. Coffee is one of the most widely consumed beverages globally, and its production generates significant amounts of waste in the form of used coffee grounds. This waste, which is often discarded, holds potential as a feedstock for biodiesel production, offering an innovative solution for both waste management and biofuel generation.

1. **Coffee Ground Waste Composition:** Coffee grounds are rich in lipids (fats and oils), which are the primary raw materials for biodiesel production. The lipid content in coffee grounds can vary depending on factors such as the coffee variety and roasting process. The presence of these lipids makes coffee grounds a promising feedstock for biodiesel synthesis through transesterification.
2. **Lipid Extraction:** To convert coffee ground waste into biodiesel, the lipids must be extracted. Various methods, including solvent extraction and mechanical pressing, can be employed to separate the lipids from the solid coffee grounds. Once extracted, the lipids can undergo further processing to produce biodiesel through transesterification reactions.
3. **Transesterification Process:** Transesterification is the key chemical process that converts lipids into biodiesel. In this process, the lipids are reacted with an alcohol (typically

methanol or ethanol) in the presence of a catalyst to produce biodiesel and glycerol as a by-product. Proper optimization of the transesterification conditions, such as temperature, catalyst type, and molar ratio of alcohol to lipids, is crucial to achieve high biodiesel yields [1].

4. **Environmental Benefits:** Biodiesel derived from coffee ground waste offers several environmental benefits. By utilizing waste that would otherwise end up in landfills, this approach contributes to waste reduction and resource recovery. Additionally, the use of biodiesel reduces greenhouse gas emissions compared to fossil diesel, as biodiesel combustion releases fewer net carbon dioxide emissions.
5. **Challenges and Considerations:** While the concept of producing biodiesel from coffee ground waste is promising, several challenges must be addressed. These include the variability in lipid content among different coffee varieties, the energy-intensive nature of lipid extraction, and competition with other uses for coffee waste, such as composting. Economic feasibility and scalability are also important considerations for commercial implementation.

In conclusion, the production of biodiesel from coffee ground waste represents a novel approach that aligns with the principles of sustainability and circular economy. By valorizing coffee waste to generate a renewable fuel source, this innovative solution addresses waste management challenges and contributes to the transition to a more sustainable energy landscape.

15.5 Pyrolysis for Controlled Fuel Properties

Pyrolysis, a thermal decomposition process in the absence of oxygen, has garnered substantial interest as a versatile method to tailor the properties of solid and liquid fuels. By subjecting biomass, plastics, or other carbonaceous feedstocks to controlled pyrolytic conditions, researchers have been able to engineer fuel properties such as calorific value, viscosity, and stability, thereby addressing energy and environmental challenges.

One of the key advantages of pyrolysis is its ability to produce a spectrum of fuel products with varying properties. For instance, fast pyrolysis at higher temperatures and shorter residence times predominantly yields bio-oil, a liquid fuel with potential applications akin to conventional petroleum products. This bio-oil can be further upgraded through processes like hydrotreating to enhance its stability and compatibility with existing infrastructure [2].

15.6 Cross Metathesis

Cross metathesis is a powerful synthetic technique in organic chemistry that involves the exchange of alkene moieties between two different olefinic compounds. In the context of ethene, cross metathesis entails the reaction between ethene and another olefinic substrate, typically a more complex alkene or alkynyl compound, to generate structurally diverse products. This reaction, catalyzed by transition metal-based catalysts, has garnered significant attention due to its capacity to form complex carbon-carbon double bonds, which are integral to the construction of various organic molecules. Cross metathesis is often utilized for the creation of small- to medium-sized rings, as well as for the modification of naturally occurring products and the synthesis of pharmaceuticals and agrochemicals. The mechanism of cross metathesis involves the coordination of the transition metal catalyst to both participating olefins, followed by the exchange of the alkene fragments to yield the desired products. Grubbs-type catalysts, named after Nobel laureate Robert H. Grubbs, are frequently employed in this context due to their high activity and compatibility with a wide range of functional groups. In a seminal study by Nguyen and Schrock, they demonstrated the application of cross metathesis with ethene in the synthesis of functionalized cyclopentenes. The reaction involved the coupling of ethene with a strained cyclic alkene, yielding a complex trisubstituted cyclopentene product with high efficiency. This work highlighted the potential of ethene cross metathesis as a valuable tool in organic synthesis.

15.7 Toward Enhanced Sustainable Liquid Biofuel Production

The growing demand for sustainable and renewable energy sources has prompted increased research into the production of liquid biofuels as alternatives to fossil fuels. Liquid biofuels, such as biodiesel and bioethanol, offer the advantage of being compatible with existing infrastructure and internal combustion engines, making them viable options for reducing greenhouse gas emissions and dependence on non-renewable resources.

To further enhance the sustainability and efficiency of liquid biofuel production, researchers are exploring innovative technologies and strategies across various aspects of feedstock selection, conversion processes, and end-use applications.

1. **Advanced Feedstock Selection:** Efficient liquid biofuel production starts with the careful selection of feedstocks. Biomass feedstocks that do not compete with food production, such as non-edible plant materials and agricultural residues, are preferred to avoid negative impacts on food security and land use. Lignocellulosic materials, such as agricultural residues and dedicated energy crops, are being investigated for their potential to yield high quantities of fermentable sugars and lignocellulosic hydrocarbons suitable for bioethanol and bio-oil production.
2. **Integrated Biorefineries:** The concept of integrated biorefineries involves the conversion of various feedstock components into multiple value-added products. Beyond liquid biofuels, these biorefineries can produce biochemicals, bioplastics, and other bio-based products. Integrating various conversion processes, such as fermentation, pyrolysis, and enzymatic hydrolysis, within a single facility optimizes resource utilization and minimizes waste [3].
3. **Advanced Conversion Technologies:** Innovative conversion technologies are essential for improving the efficiency of biofuel production. For instance, enzymatic and microbial approaches are being developed to efficiently break down complex carbohydrates in lignocellulosic feedstocks into fermentable sugars. Additionally, thermochemical processes like pyrolysis and hydrothermal liquefaction can convert a wider range of feedstocks into bio-oil, which can serve as a precursor for various liquid biofuels [2].
4. **Catalytic Upgrading:** Catalytic processes play a crucial role in upgrading bio-oil and other intermediates to yield high-quality liquid biofuels. Catalytic hydrotreating and hydrodeoxygenation can reduce the oxygen content of bio-oil, improving its stability and energy density. Moreover, the use of heterogeneous catalysts can enhance selectivity and minimize undesirable by-products [4].
5. **End-Use Applications and Blending:** Liquid biofuels can be integrated into the existing fuel infrastructure by blending

them with conventional fuels. However, it is essential to ensure compatibility, engine performance, and emission characteristics. Research is ongoing to optimize blending ratios and develop advanced engine technologies capable of utilizing higher percentages of biofuels without compromising efficiency or emissions [5].

6. **Pre-treatment and Feedstock Diversity:** The effectiveness of biofuel production hinges on the efficient breakdown of complex biomass structures into fermentable sugars or bio-oil precursors. Various pre-treatment methods, such as steam explosion and acid hydrolysis, are being researched to improve the accessibility of cellulose and hemicellulose in lignocellulosic feedstocks. Furthermore, the exploration of diverse feedstock sources, including algae and waste streams, enhances resource utilization and reduces competition with food crops [6].
7. **Microbial Engineering and Synthetic Biology:** Advancements in microbial engineering and synthetic biology are enabling the design of tailored microorganisms capable of efficiently converting sugars into bioethanol or other biofuels. Genetic modifications that enhance metabolic pathways, tolerance to fermentation inhibitors, and yield improvements are at the forefront of research to create more productive and robust biocatalysts [7].
8. **Carbon Capture and Utilization (CCU):** To achieve a closed carbon cycle, researchers are investigating carbon capture and utilization strategies in biofuel production. This involves capturing CO_2 emissions released during biofuel combustion and utilizing them to enhance feedstock growth. Algae, for example, can be cultivated using CO_2 emissions, leading to higher biomass yields for subsequent biofuel conversion [8].
9. **Waste Valorization:** The concept of waste valorization involves utilizing organic waste streams as feedstocks for biofuel production. By converting agricultural residues, food waste, and other organic materials into biofuels, this approach not only addresses waste management challenges but also contributes to a circular economy by generating value from materials that would otherwise be discarded.
10. **Life Cycle Analysis and Sustainability Metrics:** A holistic assessment of liquid biofuel production's environmental

impact requires life cycle analysis (LCA) that considers factors from feedstock cultivation to end-use emissions. Integrating LCA and sustainability metrics helps identify potential trade-offs and provides insights into the most environmentally and economically viable pathways [9].

11. **Policy and Socioeconomic Considerations:** Effective biofuel production strategies must consider the broader policy and socioeconomic context. Supportive policies, incentives, and regulations play a critical role in encouraging investments in biofuel research, development, and commercialization. Additionally, ensuring that biofuel production aligns with social and economic goals is vital for sustainable growth [10].

15.8 Conclusion

The quest for sustainable liquid fuels is a multifaceted journey that requires a comprehensive approach to address the pressing challenges of energy security, environmental sustainability, and resource efficiency. The current landscape of alternative fuel technologies highlights the need for diverse strategies that collectively contribute to meeting the world's growing energy demands. The limitations of individual fuel sources emphasize the importance of combining different technologies that can complement each other's strengths and mitigate weaknesses. The drive to replace fossil fuels with viable alternatives has led to the exploration of innovative solutions, such as producing liquid fuels from biomass, waste materials, and microbial sources. This pursuit necessitates a deep understanding of the intricate relationships between fuel properties, chemical composition, and production methods. The evolution of biodiesel production, for instance, reflects the ongoing efforts to improve fuel quality, enhance physical characteristics, and minimize environmental impacts. The journey toward enhanced sustainable liquid biofuel production encompasses a multitude of research objectives. Evaluating unconventional feedstocks like spent coffee grounds, harnessing the potential of cross metathesis for controlled fuel properties, and leveraging pyrolysis for tailored outcomes demonstrate the innovative spirit that underpins biofuel research. These endeavors not only strive to create more efficient and eco-friendly fuel sources but also embrace the principles of circular economy and waste valorization.

As the research objectives outlined here highlight, a harmonious interplay of advanced feedstock selection, integrated biorefineries, innovative conversion technologies, catalytic upgrading, and meticulous attention to end-use applications is essential. Embracing microbial engineering, pre-treatment techniques, carbon capture and utilization, and waste valorization adds depth to the arsenal of strategies aimed at producing sustainable liquid fuels. The path forward is illuminated by considerations of life cycle analysis, sustainability metrics, policy frameworks, and socioeconomic factors. Achieving a balance between environmental benefits, economic feasibility, and societal acceptance is paramount for the successful integration of sustainable liquid fuels into the global energy landscape. Through a combination of innovative technologies, rigorous scientific exploration, and a commitment to environmental stewardship, the aspiration of replacing fossil fuels with sustainable alternatives is within reach. By aligning our efforts and perspectives, we can pave the way for a cleaner, more resilient, and energy-secure future.

References

1. Freedman, B., *et al.* "Transesterification kinetics of soybean oil." *J. Am. Oil Chem. Soc.*, 1984.
2. Bridgwater, A. (2012, March). Review of fast pyrolysis of biomass and product upgrading. *Biomass Bioenergy*, *38*, 68–94. https://doi.org/10.1016/j.biombioe.2011.01.048
3. Cherubini, F. (2010, July). The biorefinery concept: Using biomass instead of oil for producing energy and chemicals. *Energy Convers. Manage.*, *51*(7), 1412–1421. https://doi.org/10.1016/j.enconman.2010.01.015
4. Oasmaa, A., Kuoppala, E., Ardiyanti, A., Venderbosch, R. H., & Heeres, H. J. (2010, September 1). Characterization of Hydrotreated Fast Pyrolysis Liquids. *Energy & Fuels*, *24*(9), 5264–5272. https://doi.org/10.1021/ef100573q
5. Graboski, M. S., & McCormick, R. L. (1998, January). Combustion of fat and vegetable oil derived fuels in diesel engines. *Prog. Energy Combust. Sci.*, *24*(2), 125–164. https://doi.org/10.1016/s0360-1285(97)00034-8
6. Ragauskas, A. J., Williams, C. K., Davison, B. H., Britovsek, G., Cairney, J., Eckert, C. A., Frederick, W. J., Hallett, J. P., Leak, D. J., Liotta, C. L., Mielenz, J. R., Murphy, R., Templer, R., & Tschaplinski, T. (2006, January 27). The Path Forward for Biofuels and Biomaterials. *Science*, *311*(5760), 484–489. https://doi.org/10.1126/science.1114736
7. Lee, S. K., Chou, H., Ham, T. S., Lee, T. S., & Keasling, J. D. (2008, December). Metabolic engineering of microorganisms for biofuels production: from bugs to synthetic biology to fuels. *Curr. Opin. Biotechnol.*, *19*(6), 556–563. https://doi.org/10.1016/j.copbio.2008.10.014

8. Brennan, L., & Owende, P. (2010, February). Biofuels from microalgae—A review of technologies for production, processing, and extractions of biofuels and co-products. *Renew. Sustain. Energy Rev.*, *14*(2), 557–577. https://doi.org/10.1016/j.rser.2009.10.009
9. Gnansounou, E., Dauriat, A., Villegas, J., & Panichelli, L. (2009, November). Life cycle assessment of biofuels: Energy and greenhouse gas balances. *Bioresour. Technol.*, *100*(21), 4919–4930. https://doi.org/10.1016/j.biortech.2009.05.067
10. Börjesson, P., & Tufvesson, L. M. (2011, January). Agricultural crop-based biofuels - resource efficiency and environmental performance including direct land use changes. *J. Cleaner Prod.*, *19*(2–3), 108–120. https://doi.org/10.1016/j.jclepro.2010.01.001

16

Optimal Scheduling of a Microgrid Incorporating Renewables and Demand Response Using a New Heuristic Optimization Technique

Karthik Nagarajan[1], Arul Rajagopalan[2*], Krishna Kumba[3], R. Sitharthan[2], Sowmmiya Uthayakumar[4], Rajkumar Murugesan[5] and R. Kanimozhi[6]

[1]Department of Electrical & Electronics Engineering, Hindustan Institute of Technology & Science, Chennai, Tamil Nadu, India
[2]Centre for Smart Grid Technologies, School of Electrical Engineering, Vellore Institute of Technology, Chennai, Tamil Nadu, India
[3]School of Electrical Engineering, Vellore Institute of Technology, Chennai, Tamil Nadu, India
[4]Department of Electrical Engineering, SRM Institute of Science and Technology, Kattankulathur, Tamil Nadu, India
[5]VIT Business School, Vellore Institute of Technology, Chennai, Tamil Nadu, India
[6]Department of Electronics and Communication Engineering, Annamalai University, Annamalainagar, Chidambaram, Tamil Nadu, India

Abstract
Customers of electric utilities that participate in demand response are encouraged to use less energy than they typically do in order to better balance the supply and demand for energy. It includes adjustments to consumption when power system operators need them to maintain grid stability. These adjustments are made by giving consumers incentives through tariffs or other means to encourage them to shift their demand peaks to times when more energy is produced. The generated power from wind turbines and photovoltaic systems is highly uncertain because of the stochastic nature of wind speed and solar radiation. This study examines the best microgrid scheduling for both conventional and renewable energy sources. The optimal schedule of microgrid with the incorporation of demand response is determined using the enhanced pelican optimization algorithm (EPOA), a new

**Corresponding author*: arulphd@yahoo.co.in

O.V. Gnana Swathika, K. Karthikeyan, Milind Shrinivas Dangate and Nicoletta Ravasio (eds.)
Biomass and Solar-Powered Sustainable Digital Cities, (255–284) © 2024 Scrivener Publishing LLC

technique for optimization inspired by nature. The operating process includes the implementation of an incentive-based demand response program. By maximizing the demand response program's profit and reducing the cost of the generators and trading, the microgrid can reach the best possible economic situation. Weighted sum is utilized to find the optimal Pareto solutions for this multi-objective model, and fuzzy satisfaction is used to choose the trade-off solution. The application outcomes are compared to EPOA and other optimization techniques. The findings of the optimization show that both test systems taken into account by the demand response program consume less unused energy.

Keywords: Demand response, demand side management, optimal scheduling, microgrid, enhanced pelican optimization algorithm, operation cost

List of Symbols and Abbreviations

I, i	Number and index of generators
J, j	Number and index of customers
N_P, n	Number and index of objective functions
S, s	Number and index of scenarios
T, t	Number and index of timeslots in the scheduling horizon
A	Wind turbine rotor swept area
A_C	PV array's area
a_i, and b_i	Coefficients of the ith generator's cost function
CM_j	Interruptible power's daily limit for customer j
Cp	Coefficient of the wind turbine's output
DR_i	Maximum ramp down rate of generator i
D_t	Total demand of the system at time t
hhub	Desired height
href	Reference height
Ipv_t	Solar irradiation of the PV (kWh/m^2)
K_1 and K_2	Cost coefficients
θ	Customer type between $[0 \leq \theta \leq 1]$
$P_{i,max}$	Maximum output of generator i
$P_{i,min}$	Minimum output of generator i
Pt_{max}	Maximum transferable power between the main grid and MG
S_p	Scenario possibility
S_t	Forecast PV power

UB	Total budget of the utility
UR_i	Maximum ramp up rate of generator i
V	Wind velocity at hub height
$Vhub_t$	Hourly wind speed
$Vref_t$	Hourly wind speed at href
W_t	Forecast wind power
β	Power law exponent β =1/7
η_{PV}	Efficiency of the PV
η_W	Efficiency of the wind
ρ_{air}	air density
σ_t	Locational marginal price
$C(\theta,x)$	Cost sustained by a customer
C_i	Conventional generator *i*'s cost
$U_{j,s}$	Customer benefit
$U_{o,s}$	Utility benefit
$P_{i,t}$	Conventional generator i generated power from at time t
Ps_t	Hourly generated power of the PV
Pt_t	MG and the main grid's transferable power of time t
Pw_t	Wind generator's hourly energy output
W_1, W_2	Weighted sum technique coefficients
x	Power consumption
y	Customer gained value of monetary compensation
$\lambda_{j,t}$	Power interruptibility value
POA	Pelican optimization algorithm
EPOA	Enhanced pelican optimization algorithm
OPF	Optimal power flow
LMP	Locational marginal pricing
MG	Microgrid
PV	Photovoltaic
DRNN-LSTM	Deep recurrent neural network with long short-term memory units
ESS	Energy storage systems
DMO	Distribution market operator
DR	Demand response
WT	Wind turbine
EV	Electric vehicle
TL	Transferable load

EDP	Extreme day pricing
MMG	Multi-microgrid
GAMS	General algebraic modeling system
RES	Renewable energy sources
HBA	Honey badger algorithm
INFO	Weighted mean of vectors
LINFO	Leader-based mutation-selection INFO

16.1 Introduction

As the world moves toward a sustainable energy future, microgrids (MGs) have emerged as a promising solution to integrate renewable energy sources (RESs), enhance energy efficiency, and improve grid resilience. A MG is a localized and autonomous energy system that can generate, distribute, and manage electricity independently or in conjunction with the main grid [1, 2]. To achieve optimal and efficient operation, MGs face the challenge of balancing multiple conflicting objectives, such as maximizing renewable energy utilization, minimizing operational costs, ensuring grid stability, and incorporating demand response (DR) strategies [3]. This concept of multi-objective optimal dispatch of a renewable energy-based MG with the incorporation of DR addresses the complexities of managing a diverse set of energy resources while considering the dynamic nature of renewable energy generation and varying energy demands. This optimization problem seeks to strike a balance between economic efficiency, environmental sustainability, and reliability in the MG's operation. With the growing concern over climate change and environmental degradation, the integration of RESs, such as solar PV, wind, and biomass, becomes critical [4]. Renewable energy not only reduces greenhouse gas emissions but also offers a path toward energy independence and resilience. However, their intermittent nature and fluctuating energy outputs pose challenges for reliable and stable MG operation. Microgrids encompass a variety of energy resources with different cost profiles and operating characteristics [5, 6]. Managing the interactions among these diverse resources is challenging. Additionally, energy demands can vary throughout the day, making it crucial to optimize energy generation and distribution in real time. The incorporation of DR strategies, which involve engaging consumers to modify their energy consumption patterns, adds another layer of complexity to the dispatch problem. The multi-objective optimization approach recognizes the existence of several competing objectives. Unlike traditional single-objective optimization, which focuses on a single goal, this approach seeks to find a set of Pareto-optimal solutions that represent trade-offs among different objectives [7]. In our case, we aim to maximize renewable energy utilization, minimize operating costs, ensure grid stability, and effectively implement DR actions. To tackle this complex problem, various techniques and algorithms, such as mathematical optimization, genetic algorithms, reinforcement learning, and fuzzy logic, can be employed. These methods help in exploring the vast solution space and finding near-optimal solutions that strike the right balance between the conflicting objectives [8]. Achieving multi-objective optimal dispatch offers numerous benefits [9]. It reduces greenhouse gas emissions and fossil fuel dependency, enhances energy cost savings, improves grid reliability, and strengthens the resilience of local energy systems. The MG may also involve users, conserve energy, and stabilize the system using DR. Jin *et al.* [10] achieved optimum MG dispatch with DR. A DR scheme with real-time pricing was used to model a large MG with intermittent renewables and variable building demands. The thermoelectric linked MG dispatch technique considered renewable energy uncertainty and customer satisfaction [11]. Microgrid modeling was also given. We previously designed and optimized an islanded hybrid MG for various load dispatch techniques by examining component size, power system responses, and cost analyses [12]. In [13], a freestanding MG with wind turbines, photovoltaics, diesel engine unit, load, and battery energy storage system created a multi-objective optimum dispatch model. In [14], the enhancement of operating cost and safety for multi-MG systems, alongside improvements in the convergence performance of solution algorithms, was explored within the context of considering uncertainties related to existing loads, renewable energy, and electric vehicle usage through the utilization of a novel double-layer robust optimization dispatch model. In [15], CPLEX solver in GAMS developed dynamic economic load dispatch in a grid-connected and heat network-connected MG. DC multi-MGs use a completely distributed fixed-time dual-layer controller for load power-sharing and voltage management [16].

The distributed tertiary controllers for multi-MGs modify the voltage references of the MGs to optimize the power flow between them and obtain the lowest possible global total production cost. In our earlier study [17], the economic energy management of a water-energy MG and an optimization of mixed integer nonlinear programming were investigated. The model controlled summer energy supplies and loads. Joint active and reactive power dispatch in isolated MGs was proposed using a distributed technique [18]. Our prior study used a deep recurrent neural network with long short-term memory units (DRNN-LSTM) model to estimate community MG aggregated power load and PV power output [19]. Meanwhile, an optimum load dispatch model for a grid-connected community MG with residential power load, PV arrays, electric vehicles (EVs), and energy storage devices was created under three scheduling scenarios. This study suggested four dispatching instances and a multi-objective economic dispatching model for MG accounting for EVs and TL [20]. It has been determined through simulation calculations and comparison analysis that EVs' orderly charging and discharging, along with the involvement of TL, can cause the operation of MG systems to achieve the best results in terms of financial costs, efficiency, and security. In [21], a two-layer interdependent network for EDPs of an MMG system was provided. This network was made up of the MMG system as the bottom layer and an information network formed of agents as the top layer. In order to reduce operating costs and emissions, an integrated optimal operation strategy was proposed for a grid-connected MG in our earlier study conducted [22]. A stochastic technique had been used to predict the uncertainties associated with load demands and RESs. Additionally, a DR program with incentives had been put in place. In [23], a novel market-based scheduling approach was proposed, taking into account the mathematical framework's limitations for risk, distribution market operator (DMO), and linear approximation of AC power flow. The proposed model was created to provide an in-depth viewpoint on reducing the day-ahead operation cost. Dispatching units, ESS, wind turbines, and DR resources were all part of the MG.

16.2 Problem Formulation

Demand response formulas are employed on the consumer side of the MG, which is made up of traditional generators and RES on the supply side. The RES is made up of a wind and solar energy system. An example of a PV generator's hourly energy output is given in [24]:

$$S_t = n_{pv} A_C lpv_t \tag{16.1}$$

where n_{pv} is the solar PV generator's efficiency, A_C is the PV array's surface area, and S_t is the solar generator's hourly energy output [24]. lpv_t is the solar irradiation incident on the PV array per hour in kilowatt-hours per square meter. Wind speed is provided as a significant impact on the hourly output of a wind generator.

$$vhub_t = vref_t(\frac{hhub}{href})^{\beta} \tag{16.2}$$

$$W_t = 0.5n_w \rho_{air} C_p A V^3 \tag{16.3}$$

where V is the wind speed at the turbine's hub, ρ_{air} is the air density, C_p is the power coefficient of the wind turbine (this varies by design), A is the swept area of the rotor, n_w is the efficiency of the wind generator, and W_t is the hourly energy output of the turbine.

In the sections that follow, we will discuss the mathematical models that underpin both the supply-side MG and the demand-side DR.

16.2.1 Grid-Connected Microgrid

In this study, we make the assumption that there is a trading mechanism in place that enables the transfer or sale of electricity between the main grid and the MG in both directions. The purpose of this trading strategy is to address the sporadic character of RESs. Therefore, let W_t represent the projected maximum wind power output from the wind generator, and S_t represent the projected maximum solar power output from the solar generator. In this context, we define Pw_t as the power generated by the wind generator and Ps_t as the power generated by the solar generator within the MG at time t. In the event that the supply of the MG fails to meet its need, it becomes necessary to get electricity from the primary grid. Conversely, if the supply of the MG surpasses its demand, then the surplus power may be sold to the primary grid. Therefore, we designate Pr_t as the transferable power between the MG and the main grid at time t. If we assume that locational marginal prices (LMPs) [25] are used for procuring power between the main and micro grid over a designated interface bus (denoted as γ_t), then the overall cost of transactions for exchanging transferrable power may be represented as $C_r(Pr_t)$ and is provided in ref. [24].

$$C_r(Pr_t) = \begin{cases} \gamma_t \times Pr_t & Pr_t > 0 \\ 0 & Pr_t = 0 \\ -\gamma_t \times Pr_t & Pr_t < 0 \end{cases} \tag{16.4}$$

The primary goal in the grid linked mode is to reduce the fuel cost of the traditional generators and the transaction costs associated with the transferable electricity. This objective function may be expressed as follows:

$$min \sum_{t=1}^{T} \sum_{i=1}^{I} C_i(P_{i,t}) + \sum_{t=1}^{T} C_r(Pr_t) \tag{16.5}$$

Subject to

$$\sum_{i=1}^{I} P_{i,t} + Pw_t + Ps_t + Pr_t = D_t - \sum_{j=1}^{J} X_{j,t} \tag{16.6}$$

$$P_{i,min} \leq P_{i,t} \leq P_{i,max} \tag{16.7}$$

$$0 \leq Pw_t \leq W_t \tag{16.8}$$

$$0 \leq Ps_t \leq S_t \tag{16.9}$$

$$-Pr_{max} \leq Pr_t \leq Pr_{max} \tag{16.10}$$

$$-DR_i \leq P_{i,t+1} - P_{i,t} \leq UR_i \tag{16.11}$$

where Pr_t represents the transferable power between the main grid and the MG at time t; the variable $C_r(Pr_t)$ represents the transaction cost associated with trading transferable power at time t. W_t denotes the forecasted maximum wind power that can be obtained from the wind generator, whereas S_t represents the forecasted maximum solar power that can be obtained from the solar generator. $P_{i,t}$ represents the power generated by conventional generator i at time t, and Pw_t represents the power generated by the wind generator at time t. The variable Ps_t represents the power created by the solar generator at a specific moment, denoted as t. The variable C_i represents the fuel cost associated with conventional generator t. D_t denotes the entire system demand at a given time t. $P_{i,min}$ and $P_{i,max}$

represent the minimum and maximum capacity of generator i, respectively. The variable Pr_{max} represents the maximum power that can be exchanged between the main grid and MG. The variables DR_i and UR_i denote the maximum rates at which conventional generator i can decrease and increase its power output, respectively. The variables a_i and b_i represent the fuel cost coefficients associated with conventional generator i. The variables i and t correspond to the total number of conventional generators and the duration of the dispatch interval, respectively.

The subsequent passage provides a concise overview of the limitations.

The power balancing constraint (16.6) assures that, at any moment t, conventional, wind, and solar generators and the main grid match the total demand.

Conventional generator generation limitations are protected by restriction (16.7).

The third and fourth limitations are renewable generator generation limits (16.8) and (16.9). They guarantee that wind and solar generator ideal values are less than or equal to anticipated or maximum values.

Constraint (16.10) limits main grid–MG power transmission. The physical parameters of transmission infrastructure between the main grid and MG determine this.

The standard generator ramp rate limitations restriction (16.11) prevents violations.

The traditional generator fuel cost in Equation (16.12) is considered to be a quadratic function of active power output [24] for simplicity:

$$C_i\left(P_{i,t}\right) = a_i P_{i,t}^2 + b_i P_{i,t} \tag{16.12}$$

Other types of conventional generating units can be used as long as they have similar fuel cost functions and ramp rate constraints.

16.2.2 Mathematical Modeling of Demand Response

Let $c(\theta, x)$ be defined as the cost incurred by a customer of type θ who decreases power consumption by x MW. The benefit function of the customer is given as follows [25]:

$$V_1(\theta, x, y) = y - c(\theta, x) \tag{16.13}$$

where y represents the monetary reward received by the consumer. It may be deduced logically that the client will only engage in the activity if V_1 is greater than or equal to zero. The utility's benefit function is expressed as follows [25]:

$$V_2(\theta, y) = \lambda x - y \tag{16.14}$$

The variable lambda represents the economic cost associated with the absence of electricity supply to a certain place within the power system. In some circumstances, there may be a significant financial burden for the power company when providing electric power to certain load buses within the grid [18]. The cost of power outage may be readily computed by the electric company. The aforementioned determined value has been previously designated as the "value of power interruptibility" (λ) [26], and it is commonly derived from the optimum power flow (OPF) analysis. The primary goal of the utility is to optimize its benefit function [25]:

$$\max_{x,y} [\lambda x - y] \tag{16.15}$$

where θ is the "customer type," normalized in [0,1].

The variable x represents the amount of electricity that is decreased by a client who is actively involved in the process.

The variable $c(\theta, x)$ represents the cost incurred by a client of type θ when lowering their electricity consumption by x kilowatts.

The variable lambda, referred to as the "value of power interruptibility," may be determined by utilizing OPF with LMPs.

16.2.2.1 Customer Cost Function

As stated before, $c(\theta, x)$ is the cost incurred by a customer of type θ who decreases power consumption by x MW. In this work, it is assumed that the mathematical function is given as in ref. [25]:

$$c(\theta, x) = K_1 x^2 + K_2 x - K_2 x\theta \tag{16.16}$$

where K_1 and K_2 are cost coefficients. θ is the customer type [27, 28] and is used to categorize the different kinds of customers based on their desire/readiness to curb electric power. θ is normalized in the interval thus $\theta = 1$

for the most willing customer and $\theta = 0$ for the least willing. We provide a summary of all the conditions that the cost function must satisfy.

The term $K_2 x\theta$ can be used to categorize customers depending the value of θ.

As θ increases, marginal cost decreases: The most willing customer $\theta = 1$ has the least marginal cost and thus has the highest marginal benefit, whereas the least willing customer $\theta = 0$ has the highest marginal cost and thus the lowest marginal benefit.

$$\partial C / \partial x = 2K_1 x + K_2 - K_2\theta \tag{16.17}$$

The concept of contract formulations to more than one customer is given as in ref. [27]: Thus, if y_j is the amount of payment paid to customer j, the customer benefit is obtained from [24]:

$$u_j = y_j - (K_1 x^2 + K_2 x - K_2 x\theta),\ for\ j = 1,\ldots,J \tag{16.18}$$

The utility benefit is determined from [24]:

$$u_0 = \sum_{j=1}^{J} \lambda_j x_j - y_j \tag{16.19}$$

The objective is thus to maximize the expected utility benefit [24]:

$$max_{x,y} = \sum_{j=1}^{J} [\lambda_j x_j - y_j] \tag{16.20}$$

Such that

$$y_j = \left(K_1 x_j^2 + K_2 x_j - K_2 x_j \theta_j\right) \geq 0,\ for\ j = 1,\ldots,J \tag{16.21}$$

$$y_j - \left(K_1 x_j^2 + K_2 x_j - K_2 x_j \theta_j\right) \geq y_{j-1} - \left(K_1 x_{j-1}^2 + K_2 x_{j-1} - K_2 x_{j-1} \theta_{j-1}\right) \quad for\ j = 2,\ldots,J \tag{16.22}$$

The mathematical formulation described previously consists of two variables: the electricity curtailed (x MW) and the incentive provided (y \$).

In addition, the two limitations are delineated and explicated as follows [24].

The "individual rationality constraint" [constraint (16.20)] guarantees that the benefit of each consumer is higher than or equal to zero. The incentive compatibility constraint, as represented by constraint (16.21), guarantees that clients get suitable compensation corresponding to the extent of their load curtailment. The formulas (16.19) to (16.21) for demand management contracts are expanded to encompass several time intervals. Additionally, we propose a modification to the constraint of individual rationality and the constraint of incentive compatibility, extending their enforcement across the whole optimization horizon, which spans a day, as opposed to a single time period, such as every hour. We contend that this approach is more pragmatic and cost-effective. Incorporation of maximum power objectives and total budget as practical restrictions is included in the model. The ultimate mathematical model is presented as [24].

$$max_{x,y} \sum_{t=1}^{T} \sum_{j=1}^{J} [\lambda_{j,t} x_{j,t} - y_{j,t}] \tag{16.23}$$

Such that

$$\sum_{t=1}^{T} \left(K_{1,j} x_{j,t}^2 + K_{2,j} x_{j,t} - K_{2,t} x_{j,t} \theta_j \right)] \geq 0, \; for \; j = 1, \ldots, J \tag{16.24}$$

$$\sum_{t=1}^{T} \left[y_{j,t} - \left(K_{1,j} x_{j,t}^2 + K_{2,j} x_{j,t} - K_{2,t} x_{j,t} \theta_j \right) \right] \geq \left[y_{j-1,t} - \left(K_{1,j-1} x_{j-1,t}^2 + K_{2,j-1} x_{j-1,t} - K_{2,t} x_{j-1,t} \theta_{j-1} \right) \right] for \; j = 2, \ldots, J \tag{16.25}$$

$$\sum_{t=1}^{T} \sum_{j=1}^{J} y_{j,t} \leq UB \tag{16.26}$$

$$\sum_{t=1}^{T} x_{j,t} = CM_j \tag{16.27}$$

The variable UB represents the entire budget of the utility, whereas CM_j denotes the daily limit of interruptible electricity for customer j. Constraint (16.24) guarantees that the aggregate daily incentive earned by a client is more than or equal to their daily cost of interruption. The constraint (16.25) guarantees that there is a positive correlation between the amount of customer power cut and the corresponding benefit experienced by the customer. Constraint (16.26) is designed to guarantee that the aggregate incentive disbursed by the utility remains within the limits of the utility's allocated budget. Constraint (16.27) guarantees that the aggregate amount of power curtailed by each individual customer on a daily basis remains below their designated limit of interruptible power.

16.2.3 Combined Grid-Connected Microgrid with Demand Response Model

The grid-connected MG with a DR model has two distinct objective functions. The primary goal of a specific objective function is to reduce both the fuel cost associated with conventional generators and the transaction cost related to the trading of transferrable power. The primary aim of the second objective function is to optimize the grid operator's DR benefit, with the intention of maximizing its overall effectiveness.

The mathematical formulation is presented below [24]:

$$\min w[\textstyle\sum_{t=1}^{T}\sum_{i=1}^{I} C_i(P_{i,t}) + \sum_{t=1}^{T} C_r(Pr_t)] + (1-w)[\sum_{t=1}^{T}\sum_{j=1}^{J}[y_{j,t} - \lambda_{j,t}x_{j,t}]] \tag{16.28}$$

The objective function weights, denoted as w and $1-w$, are subject to the network restrictions specified in Equations (16.6) to (16.11) and Equations (16.24) to (16.27). Additionally, it is necessary to satisfy the condition outlined in equation [16.24] while selecting these weights.

$$w + (1-w) = 1 \tag{16.29}$$

The variables to be determined by the optimization model are x, $y_{j,t}$, Pwt, Pst, Prt, and $Pi;t$.

16.3 Fuzzy Logic-Based Assortment of Finest Compromise Solution

The process of selecting the most optimum compromise solution from a group of accessible options is a crucial aspect of decision-making. The researchers utilized a fuzzy membership technique [29] in order to determine the most optimal compromise option. The objective function f_j of individual k is characterized by a membership function μ_j^k, which accounts for the uncertain nature of the decision maker's conclusion.

$$\mu_j^k = \begin{cases} 1 & f_j \leq f_j^{min} \\ \dfrac{f_j^{max} - f_j}{f_j^{max} - f_j^{min}} & f_j^{min} < f_j < f_j^{max} \\ 0 & f_j \geq f_j^{max} \end{cases} \tag{16.30}$$

The symbols f_j^{max} and f_j^{min} represent the upper and lower bounds of the jth fitness function inside the set of non-dominated solutions. Additionally, the normalized membership function μ^k is computed in a similar manner for each non-dominated result k, as described in ref. [30].

$$\mu^k = \frac{\sum_{j=1}^{N} \mu_j^k}{\sum_{k=1}^{r} \sum_{j=1}^{N} \mu_j^k} \tag{16.31}$$

where r denotes the whole non-dominated solutions. The finest compromise solution consists of the maximum value, μ^k.

16.4 Enhanced Pelican Optimization Algorithm

The MG's optimal allocation problem is a multi-objective, multi-constraint, severely nonlinear complex integer optimization problem. The swarm intelligence algorithm is the most often utilized approach to address this issue because the conventional mathematical optimization technique is frequently inefficient. Numerous swarm intelligence algorithms have been used in recent years to optimize MG configuration, but the majority of studies have used the continuous optimization algorithm without any integration, which causes the algorithm to fall into the local optimal and leaves no feasible configuration scheme. As a result, this is a critical issue that must be resolved in order to identify the optimization technique that can resolve the MG's optimization allocation issue.

A population-based method, the pelican optimization algorithm (POA) includes pelicans as population members [31]. Each population in a population-based algorithm is member stands for a potential answer. Each group participant suggests a price for the dependent on a variable's position in the search space and the optimization issue. At first, an equation is utilized to initialize population members at random using the lower and the problem's top boundaries.

Step 1:
Initialize a population of potential solutions (individuals or "pelicans") randomly or using some heuristic method [31].

$$x_{i,j} = l_j + rand.\left(u_j - l_j\right), i = 1,2,\ldots,N, j = 1,2,\ldots,m \tag{16.32}$$

where $x_{i,j}$ is the value of the jth variable specified by the ith candidate solution; N is the number of population members; m is the number of problem variables; $rand$ is the random number in the interval [0, 1]; l_j is the jth lower bound; and u_j is the jth upper bound of the problem variable.

The pelican population members in the proposed POA were identified using a matrix known as the population matrix in Equation (16.29). The problem variable's recommended values are represented by the matrix's columns, with each row representing a potential solution.

$$X = \begin{bmatrix} X_1 \\ \vdots \\ X_i \\ \vdots \\ X_N \end{bmatrix}_{N\times m} = \begin{bmatrix} x_{1,1}\cdots x_{1,j}\cdots x_{1,m} \\ \vdots \quad \ddots \quad \vdots \quad \iddots \quad \vdots \\ x_{i,1}\cdots x_{i,j}\cdots x_{i,m} \\ \vdots \quad \iddots \quad \vdots \quad \ddots \quad \vdots \\ x_{N,1}\cdots x_{N,j}\cdots x_{N,m} \end{bmatrix}_{N\times m} \tag{16.33}$$

where X is the pelican population matrix and X_i is the ith pelican.

Step 2:
Define the fitness function that quantifies the quality of each solution [31]. Each group member in the proposed OPOA represents a potential solution to a certain problem as a pelican. As a result, it is possible to assess the objective function of a given problem using the solutions of each contender. The value obtained for the objective function is calculated using the vector known as the objective function vector in Equation (16.30).

$$F = \begin{bmatrix} F_1 \\ \vdots \\ F_i \\ \vdots \\ F_N \end{bmatrix}_{N\times 1} = \begin{bmatrix} F(X_1) \\ \vdots \\ F(X_i) \\ \vdots \\ F(X_N) \end{bmatrix}_{N\times 1} \tag{16.34}$$

where F is the objective function vector, and F_i is the objective function value of the ith candidate solution.

Step 3:

Population Initialization Based on Stochastic Opposition Learning

The conventional POA still has certain issues that need to be fixed, like premature convergence and subpar local search capabilities. To tackle the aforementioned optimization model, a better pelican optimization technique was therefore provided. The POA population was initialized using a random approach, and the produced population was not uniform, which had an impact on the accuracy and speed of convergence. This paper used a stochastic contrastive learning technique to determine the most recent location in order to obtain a better initial population [32]. The latest position $X_{i,new}$ can be calculated as follows:

$$X_{i,new} = (l + u) - kX_i \tag{16.35}$$

where X_i is the current individual's location information, and k is a random number between (0, 1).

The fitness function value for the relevant person is determined using the random opposition learning optimization. The superior fitness value is chosen by comparing the fitness function values of the current individual and the optimized individual.

$$X_i = \begin{cases} X_{i,new}, & f(X_{i,new}) < f(X_i) \\ X_i, & else \end{cases} \tag{16.36}$$

Step 4:

In order to improve potential solutions, the proposed OPOA simulates the actions and tactics used by pelicans when they are attacking and hunting prey. Two stages of this hunting technique are simulated [32]:

(1) Move toward prey (exploration stage).
(2) Wingspan above water (development stage).

16.4.1 Phase 1: Move to Prey (Exploration Phase)

The pelican locates where its prey is in the first stage and then flies toward that location. The suggested OPOA's ability to explore and find new areas of the search space is enhanced by modeling this pelican technique, which permits scanning of the search space. The OPOA's key feature is that the prey's placement in the search space is determined at random. This improves the OPOA's capacity for exact search problem solving through exploration of the environment. The equation simulates the idea and the pelican's movement to the place of prey mathematically [32].

$$x_{i,j}^{P_1} = \begin{cases} x_{i,j} + rand.\left(p_j - I.x_{i,j}\right), F_p < F_i \\ x_{i,j} + rand.\left(x_{i,j} - p_j\right), \; else \end{cases} \tag{16.37}$$

where p_j is the position of the prey in the jth dimension, I is a random number equal to 1 or 2, F_p is its objective function value, and $x_{i,j}^{P_1}$ is the new state of the ith pelican in the jth dimension based on stage 1. The value of the parameter I, a number, could at random be either 1 or 2. For each iteration and each member, this parameter is chosen at random. When this parameter's value is 2, it causes the member to be more displaced, which could lead to the member entering a newer region of the search space.

As a result, parameter I influences how precisely the EPOA probe scans the search space.

In the proposed EPOA, the pelican's new location is allowed if it increases the value of the objective function. The algorithm is stopped from going to a non-optimal location in this type of update, known as a legitimate update. Equation (16.32) is used to model the procedure.

$$X_i = \begin{cases} X_i^{P_1}, F_i^{P_1} < F_i \\ X_i, \; else \end{cases} \tag{16.38}$$

Where $F_i^{P_1}$ is the objective function value based on stage 1, and $X_i^{P_1}$ is the new state of the ith pelican.

16.4.2 Phase 2: Flying Over Water (Development Phase)

The pelicans lift the fish upward in the second step after they reach the surface by spreading their wings above the water, and, then, they gather their meal in neck bags. Pelicans in target regions catch more fish as a result of this tactic. By simulating this behavior in pelicans, it is possible to improve the planned POA's convergence in the hunting area [32].

The POA's local search and development skills are enhanced by this procedure. To arrive to a better solution, the program must mathematically explore points close to the pelican position. Equation (16.6) simulates this pelican behavior while hunting mathematically [32].

$$x_{i,j}^{P_2} = x_{i,j} + R.\left(1 - \frac{t}{T}\right).(2.rand - 1).x_{i,j} \tag{16.39}$$

where R is constant and equal to 0.2, $R(1-(t/T))$ is the radius of the neighborhood surrounding $x_{i,j}^{P_2}, t$ is the iteration counter, and T is the greatest number of iterations. $x_{i,j}^{P_2}$ is the ith point based on phase 2 of the new state of the jth dimension. Each group member in the close-by local search converges to a better solution; the neighborhood radius of the group members is represented by the coefficient $R(1-(t/T))$. The EPOA development capability is successfully brought closer to the ideal global solution by this coefficient. The area surrounding each component is more significant in the initial iteration since the value of this coefficient is higher. The $R(1-(t/T))$ coefficient and each member's neighborhood radius decrease as the algorithm of repetition is increased. This enables us to scan the region surrounding each person in the population with a smaller and more exact step size, allowing the EPOA to get closer to an overall (or perhaps entirely overall) ideal solution based on usage.

Valid updates are also used at this point to approve or disapprove the new pelican location predicted by Equation (16.34).

$$X_i = \begin{cases} X_i^{P_2}, F_i^{P_2} < F_i \\ X_i, \quad else \end{cases} \tag{16.40}$$

where $X_i^{P_2}$ is the new state of the ith pelican, and $F_i^{P_2}$ is its objective function value based on the second stage.

16.4.2.1 Disturbance Suppressor

The global and local search capabilities of the algorithm can be successfully balanced by perturbation inhibitors [32]. The disturbance inhibition factor θ is a new feature of the pelican optimization method, and its formula is as follows:

$$\theta = \sin\left(\frac{\pi.t}{2.T_{max}} + \pi\right) + 1 \tag{16.41}$$

where t is the number of current iterations, and T_{max} is the maximum number of iterations.

The POA's optimization models for the prey moving stage and the flying stage are expressed as follows after the addition of disturbance inhibitors:

$$x_{i,j}^{P_1} = \begin{cases} x_{i,j} + \theta.rand.(p_j - I.x_{i,j}), F_p < F_i \\ x_{i,j} + \theta.rand.(x_{i,j} - p_j), \quad else \end{cases} \tag{16.42}$$

$$x_{i,j}^{P_2} = x_{i,j} + \theta.R.\left(1 - \frac{t}{T}\right) \cdot (2.rand - 1).x_{i,j} \tag{16.43}$$

To address the issue of losing EPOA variety in the later stage and effortlessly settling into the local optimal, the Levy flight approach was developed. A common random technique in optimization algorithms is called Levy flight. Its high step probability in a random walk can significantly increase the algorithm's randomness [29, 30]. Phase 2 of the EPOA formula included Levy flights to further improve the method's exploratory capabilities.

$$X_{i,new} = X_{best} + \omega.R.\left(1 - \frac{t}{T}\right) \cdot (2.rand - 1).X_{best}.Levy(D) \tag{16.44}$$

$$Levy(D) = 0.01 \times \frac{c \times \sigma}{|d|^{\frac{1}{\beta}}} \tag{16.45}$$

$$\sigma = [\Gamma(1+\beta) * \sin(\pi * \frac{\beta}{2}) / (\Gamma\left(\frac{1+\beta}{2}\right) * \beta * 2^{\frac{(\beta-1)}{2}})]^{1/\beta} \tag{16.46}$$

where $Levy$ represents Levy flight function, and D is the dimension of the problem. c and d are random values between 0 and 1, and b is a constant equal to 1.5.

The random number in Equation (16.44) can be determined using sine map. The initial value C_t and a are chosen as 0.36 and 2.8, respectively [33].

$$C_{t+1} = \frac{a}{4}\sin(\pi C_t), 0 < a < 4 \tag{16.47}$$

where t is the current iteration number.

16.5 Results and Discussion

The system being examined consists of a grid-connected MG and a DR model. The system consists of three conventional diesel generators, one wind generator, one solar generator, and three remote customers. The scheduling period under consideration is 24 h; however for the solar generator, a scheduling interval of 8 h (from 8 a.m. to 6 p.m.) is considered. $x_{j,t}, y_{j,t}, P_{wt}, P_{st}, P_{rt}$, and $P_{i,t}$ are the decision variables.

Table 16.1 displays the attributes associated with typical generators, including fuel cost coefficients, output power limitations, and ramp rate restrictions. Table 16.2 displays the initial hourly MG demand and the values of hourly power interruptibility ($\lambda_{j,t}$). The wind generator has a maximum power rating of 11 kW, whereas the solar generator has a maximum power rating of 15 kW. The maximum power that can be transferred between the main grid and MG is 4 kW. Table 16.3 presents W_t and S_t values that have been taken from those in ref. [24]. Data about solar radiation and wind power are obtained from ref. [24]. The cost coefficients for the generator are taken from ref. [24]. It is initially assumed for this MG that the power interruptibility of each of the three consumers is equal. The daily customer power limit, customer category, and cost function coefficients are all listed in Table 16.4. To establish a hierarchy of customers based on their inclination to decrease electricity consumption, it is presupposed that the MG operator have knowledge on the daily interruptible energy limit (CM_j) for each individual customer. In alternative terms, the variable CM_j assists the MG operator in ascertaining the value of parameter θ_j. Furthermore, it is worth noting that the MG operator possesses a daily budget denoted as UB, which amounts to $500. Additionally, the MG operator possesses knowledge on the coefficients of the outage cost function for the customers involved, namely, $K_{1,j}$ and $K_{2,j}$. The ideal power provided by conventional generating units and the power transferred between the utility and MG

Table 16.1 Input data of the three-unit system [24].

i	a_i	b_i	$P_{i,min}$	$P_{i,max}$	Dr_i	Ur_i
1	0.06	0.5	0	4	3	3
2	0.03	0.25	0	6	5	5
3	0.04	0.3	0	9	8	8

Table 16.2 Total initial hourly microgrid demand and hourly values of power interruptibility [24].

Time(h)	$D_t(kW)$	$\lambda_{j,t}$
1	31.83	1.57
2	31.4	1.4
3	31.17	2.2
4	31	3.76
5	31.17	4.5
6	32.1	4.7
7	32.97	5.04
8	34.1	5.35
9	37.53	6.7
10	38.33	6.16
11	40.03	6.38
12	41.17	6.82
13	39.67	7.3
14	41.7	7.8
15	42.1	8.5
16	41.67	7.1
17	40.7	6.8
18	40.07	6.3
19	38.63	5.8
20	36.4	4.2
21	34.1	3.8
22	32.8	3.01
23	32.5	2.53
24	32	1.42

Table 16.3 Forecast power from the wind and solar powered generators [24].

Time(h)	W_t(kW)	S_t(kW)
1	7.56	0
2	7.5	0
3	8.25	0
4	8.48	0
5	8.48	0
6	9.42	0
7	9.82	0
8	10.35	7.99
9	10.88	10.56
10	11.01	13.61
11	10.94	14.97
12	10.68	15
13	10.42	14.78
14	10.15	14.59
15	9.67	13.56
16	8.98	11.83
17	8.37	10.17
18	7.61	7.66
19	6.7	0
20	5.72	0
21	7.21	0
22	7.75	0
23	7.88	0
24	7.69	0

Table 16.4 Customer cost function coefficients, customer type and daily customer curtailable energy limit [24].

j	$K_{1,j}$	$K_{2,j}$	θ_j	CM_j(kWh)
1	1.079	1.32	0	30
2	1.378	1.63	0.45	35
3	1.847	1.64	0.9	40

are presented in Table 16.5. The convergence characteristic curve pertaining to the reduction of operational costs in the suggested EPOA algorithm exhibits a smoother and more-quicker progression compared to that of the POA algorithm, as seen in Figure 16.1. The convergence characteristic of the EPOA algorithm indicates that the ideal value of operation cost is achieved at the completion of the 43rd iteration. A weighting factor of 0.5 has been used for the grid-connected MG. The purpose of this action is to ensure that both objective functions are given equal consideration. Figure 16.2 illustrates the optimal output power generated by the three conventional generating units. Figure 16.3 illustrates the ideal customer power curtailed by each MG consumer. A comprehensive investigation of the simulation findings reveals intriguing underlying viewpoints on the MG's operational mode. For case study 1, Figure 16.4 demonstrates how the RES supports the conventional generators' operation during the 24-h scheduling window. Because of this, when the solar generator starts operating, conventional generators 1 and 3 start producing less power. It has been noted that the MG's traditional generators cannot meet demand on their own. It is now crucial that the MG implement the DR program and engage in business with the main grid. According to Figure 16.4, power is being purchased from the main grid when P_{rt} is positive and is being sold to the main grid when P_{rt} is negative. Thus, it can be seen from the figure that power is purchased early in the morning and late at night, when RESs are not producing at their highest levels. Power is available to sell to the main grid when RESs are fully operational, particularly when the solar generator starts up. The traditional generators must operate near to their maximum output since their energy costs less to produce than that obtained from the main grid. The statistical findings obtained using the suggested enhanced particle optimization algorithm (EPOA) and other meta-heuristic optimization strategies are shown in Table 16.6. Therefore, after careful analysis, it can be concluded that the EPOA algorithm, which

Table 16.5. Optimal power produced by conventional generators and transfer power between the microgrid and main grid.

$P_{j,t}$(kW)			
$i = 1$	$i = 2$	$i = 3$	Pr_t(kW)
3.9364	5.9995	8.8346	−3.9821
4	5.8631	8.9958	3.9252
3.9543	5.8806	8.9269	3.3743
3.9994	5.9478	8.9812	0.8682
3.9626	5.9995	9	0.7687
3.9389	5.9783	8.9918	−0.1651
4	5.998	9	−1.0619
2.7513	5.7659	8.4119	−3.9329
3.2309	4.0203	8.1291	−3.997
3.3386	2.944	7.4892	−4
2.9298	4.2162	6.4468	−3.9998
2.0651	3.7637	7.8465	−3.9964
1.8462	3.5253	7.4546	−4
2.8552	4.3556	8.5581	−4
3.6028	5.5629	7.4275	−3.9994
3.7595	5.9844	8.9974	−3.9827
3.9996	5.9247	8.9993	−3.8876
4	5.9939	9	−1.0416
4	6	9	2.9584
3.9471	5.9946	8.9992	4
3.9999	6	8.9928	3.9736
3.8573	6	8.9674	3.2672
3.9997	5.9955	8.9921	3.89
4	5.9666	8.9877	3.9502

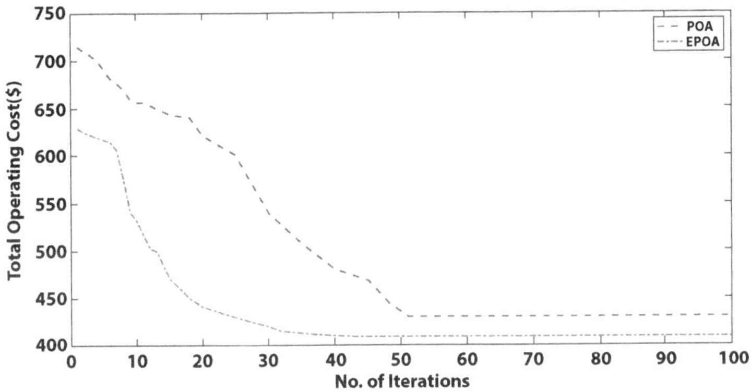

Figure 16.1 Convergence characteristic of test system.

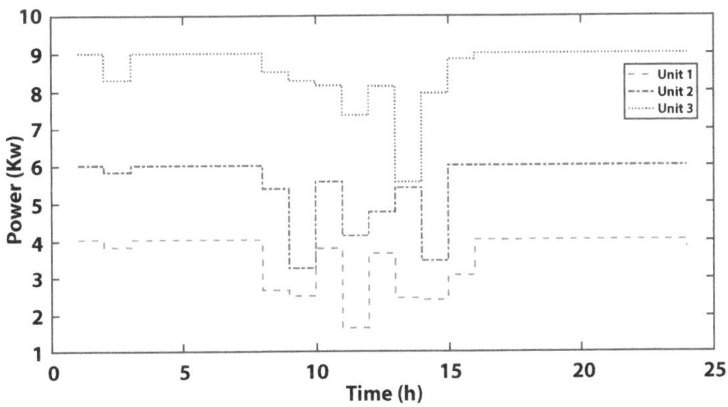

Figure 16.2 Optimal output power from conventional generators.

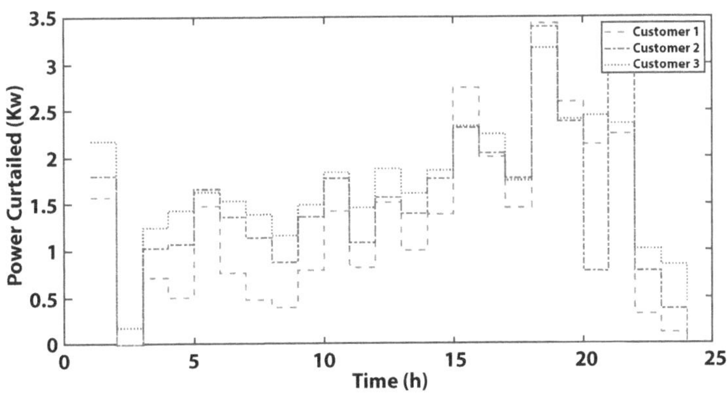

Figure 16.3 Optimal customer power curtailed.

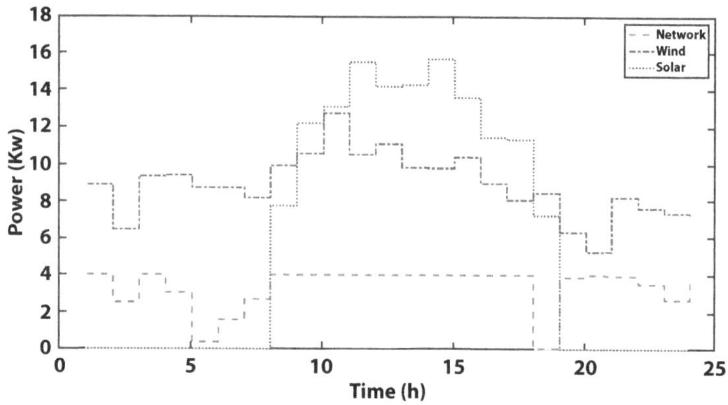

Figure 16.4 Output power from transferred power between the main grid and microgrid, wind generator, and solar powered generator.

Table 16.6 Statistical comparison of operating cost.

	Total operating cost ($)			
Algorithm	**Max**	**Min**	**Mean**	**Std. dev.**
PSO [34]	807.35	477.4	658.49	90.47
JAYA [34]	790.67	647.49	711.32	41.57
HBA [34]	672.81	460.39	545.31	40.11
INFO [34]	747.1355	509.374	615.9763	66.6
LINFO [34]	356.18	532.38	441.24	38.85
POA	714.58	430.32	450.92	35.75
EPOA	629.35	409.58	412.15	28.38

incorporates the chaotic levy mechanism, demonstrated a somewhat limited performance compared to the POA, Jaya [34], HBA [34], INFO [34], and LINFO [34] algorithms. Table 16.7 presents the detailed comparison of cost breakdown using EPOA and other optimization algorithms. The findings obtained from the optimization table demonstrate that the suggested EPOA approach yields the most cost-effective outcome in the examined case study. This statement affirms that EPOA have the necessary competence to effectively build a worldwide optimum solution. Moreover, the findings of this study indicate that the EPOA algorithm, which was

Table 16.7 Comparison of simulation results of cost breakdown for the microgrid test system.

Parameters	EPOA	POA	AIMMS [24]
Total Conventional Power Cost ($)	248.5691	250.5224	250
Total Transferred Power Transaction Cost ($)	387.271	398.6959	427
Total Customer Incentive ($)	373.5545	371.9369	371
Total Customer Energy Curtailed (kWh)	104.8264	104.9857	105
Total Conventional Energy Generated (kWh)	421.0849	424.8958	428
Total Transferred Energy (kWh)	77.0223	78.7953	84.5
Total Cost ($)	1009.395	1021.155	1048

suggested herein, has the capability to provide high-quality solutions when compared to other algorithms.

16.6 Conclusion

This research looked into the DR program's impact on a MG's energy management problem. This study examines the operating mode of a MG in relation to a DR program that is based on game theory principles. The objective is to optimize the benefits of grid operator DR while simultaneously lowering the fuel costs incurred by traditional producers and the transaction costs associated with the exchange of transferable power. The determination of the optimal customer power curtailment, the optimal customer incentive, the ideal power generation schedule for the conventional generators, the ideal power to be transferred between the main grid and MG, and the optimal customer incentive is accomplished through the utilization of an optimization model. This model operates within a scheduling interval of 24 h. The findings indicate that the integration of DR programs into the energy management of MG systems has advantageous outcomes, enhancing efficiency in both the supply and demand aspects of the MG. The generated model is solved utilizing the suggested EPOA technique.

The results indicate that cost reductions in the MG may be achieved by prioritizing the maximization of DR benefits for the grid operator, even if it means reducing fuel and transaction costs. The results of the study also indicated that the prerequisites of individual rationality and incentive compatibility, as outlined in game theory, were satisfied. The proposed EPOA can be applied to implement multi-objective optimal MG operation by scheduling DGs economically and segmenting the current distribution system into various self-sufficient MGs. Furthermore, the proposed EPOA can be employed to solve economic emission dispatch problem with the integration of DR resources in a large-scale power system.

References

1. Karthik, N., Parvathy, A. K., & Arul, R. (2020). A review of optimal operation of microgrids. *Int. J. Electr. Comput. Eng. (2088-8708)*, *10*(3).
2. Karthik N, A K Parvathy, Arul Rajagopalan, S Baskar. (2019). A review of optimization techniques applied to solve unit commitment problem in microgrid. *Indonesian J. Electr. Eng. Comput. Sci., 15(3),* 1161-1169.
3. Karthik, N., Rajagopalan, A., Prakash, V. R., Montoya, O. D., Sowmmiya, U., & Kanimozhi, R. (2023). Environmental Economic Load Dispatch Considering Demand Response Using a New Heuristic Optimization Algorithm. *AI Techniques for Renewable Source Integration and Battery Charging Methods in Electric Vehicle Applications*, 220-242.
4. Karthik, N., Parvathy, A. K., Arul, R., & Padmanathan, K. (2021). Multi-objective optimal power flow using a new heuristic optimization algorithm with the incorporation of renewable energy sources. *Int. J. Energy Environ. Eng.*, *12*, 641-678.
5. Rajagopalan, A., Nagarajan, K., Montoya, O. D., Dhanasekaran, S., Kareem, I. A., Perumal, A. S., ... & Paramasivam, P. (2022). Multi-Objective Optimal Scheduling of a Microgrid Using Oppositional Gradient-Based Grey Wolf Optimizer. *Energies*, *15*(23), 9024.
6. Karthik, N., Parvathy, A. K., Arul, R., Jayapragash, R., & Narayanan, S. (2019, March). Economic load dispatch in a microgrid using Interior Search Algorithm. In *2019 Innovations in Power and Advanced Computing Technologies (i-PACT)* (Vol. 1, pp. 1-6). IEEE.
7. Karthik, N., Parvathy, A. K., & Arul, R. (2019). Multi-objective economic emission dispatch using interior search algorithm. *Int. Trans. Electr. Energy Syst.*, *29*(1), e2683.
8. Rajagopalan, A., Kasinathan, P., Nagarajan, K., Ramachandaramurthy, V. K., Sengoden, V., & Alavandar, S. (2019). Chaotic self-adaptive interior search algorithm to solve combined economic emission dispatch problems with security constraints. *Int. Trans. Electr. Energy Syst.*, *29*(8), e12026.

9. Nagarajan, K., Parvathy, A. K., & Rajagopalan, A. (2020). Multi-objective optimal reactive power dispatch using levy interior search algorithm. *Int. J. Electr. Eng. Inform, 12*, 547-570.
10. Jin, M., Feng, W., Liu, P., Marnay, C., & Spanos, C. (2017). MOD-DR: Microgrid optimal dispatch with demand response. *Appl. Energy, 187*, 758-776.
11. Xu, H., Meng, Z., & Wang, Y. (2020). Economic dispatching of microgrid considering renewable energy uncertainty and demand side response. *Energy Rep., 6*, 196-204.
12. Ishraque, M. F., Shezan, S. A., Ali, M. M., & Rashid, M. M. (2021). Optimization of load dispatch strategies for an islanded microgrid connected with renewable energy sources. *Appl. Energy, 292*, 116879.
13. Zhao, Y., Song, X., Wang, F., & Cui, D. (2020). Multiobjective optimal dispatch of microgrid based on analytic hierarchy process and quantum particle swarm optimization. *Global Energy Interconnect., 3*(6), 562-570.
14. Tan, B., Chen, H., Zheng, X., & Huang, J. (2022). Two-stage robust optimization dispatch for multiple microgrids with electric vehicle loads based on a novel data-driven uncertainty set. *Int. J. Electr. Power Energy Syst., 134*, 107359.
15. Jordehi, A. R. (2021). Economic dispatch in grid-connected and heat network-connected CHP microgrids with storage systems and responsive loads considering reliability and uncertainties. *Sustain. Cities Soc., 73*, 103101.
16. Zaery, M., Wang, P., Wang, W., & Xu, D. (2021). A novel fully distributed fixed-time optimal dispatch of DC multi-microgrids. *Int. J. Electr. Power Energy Syst., 129*, 106792.
17. Moazeni, F., & Khazaei, J. (2020). Dynamic economic dispatch of islanded water-energy microgrids with smart building thermal energy management system. *Appl. Energy, 276*, 115422.
18. España, N., Barco-Jiménez, J., Pantoja, A., & Quijano, N. (2021). Distributed population dynamics for active and reactive power dispatch in islanded microgrids. *Int. J. Electr. Power Energy Syst., 125*, 106407.
19. Wen, L., Zhou, K., Yang, S., & Lu, X. (2019). Optimal load dispatch of community microgrid with deep learning based solar power and load forecasting. *Energy, 171*, 1053-1065.
20. Hou, H., Xue, M., Xu, Y., Xiao, Z., Deng, X., Xu, T., ... & Cui, R. (2020). Multi-objective economic dispatch of a microgrid considering electric vehicle and transferable load. *Appl. Energy, 262*, 114489.
21. Wu, K., Li, Q., Chen, Z., Lin, J., Yi, Y., & Chen, M. (2021). Distributed optimization method with weighted gradients for economic dispatch problem of multi-microgrid systems. *Energy, 222*, 119898.
22. Nayak, A., Maulik, A., & Das, D. (2021). An integrated optimal operating strategy for a grid-connected AC microgrid under load and renewable generation uncertainty considering demand response. *Sustain. Energy Technol. Assess., 45*, 101169.

23. MansourLakouraj, M., Shahabi, M., Shafie-khah, M., & Catalão, J. P. (2022). Optimal market-based operation of microgrid with the integration of wind turbines, energy storage system and demand response resources. *Energy, 239*, 122156.
24. Nwulu, N. I., & Xia, X. (2017). Optimal dispatch for a microgrid incorporating renewables and demand response. *Renew. Energy, 101*, 16-28.
25. Nwulu, N. I., & Fahrioglu, M. (2013). A soft computing approach to projecting locational marginal price. *Neural Comput. Appl., 22*, 1115-1124.
26. Nwulu, N. I., & Xia, X. (2015). Multi-objective dynamic economic emission dispatch of electric power generation integrated with game theory based demand response programs. *Energy Convers. Manage., 89*, 963-974.
27. Fahrioglu, M., & Alvarado, F. L. (2000). Designing incentive compatible contracts for effective demand management. *IEEE Trans. Power Syst., 15*(4), 1255-1260.
28. Fahrioglu, M., & Alvarado, F. L. (2001). Using utility information to calibrate customer demand management behavior models. *IEEE Trans. Power Syst., 16*(2), 317-322.
29. Karthik, N., Parvathy, A. K., Arul, R., & Padmanathan, K. (2022). A New Heuristic Algorithm for Economic Load Dispatch Incorporating Wind Power. In *Artificial Intelligence and Evolutionary Computations in Engineering Systems: Computational Algorithm for AI Technology, Proceedings of ICAIECES 2020* (pp. 47-65). Springer Singapore.
30. Karthik, N., Parvathy, A. K., Arul, R., & Padmanathan, K. (2021). Levy interior search algorithm-based multi-objective optimal reactive power dispatch for voltage stability enhancement. In *Advances in Smart Grid Technology: Select Proceedings of PECCON 2019—Volume II* (pp. 221-244). Springer Singapore.
31. Trojovsky, P., & Dehghani, M. (2022). Pelican optimization algorithm: A novel nature-inspired algorithm for engineering applications. *Sensors, 22*(3), 855.
32. Li, J., Chen, R., Liu, C., Xu, X., & Wang, Y. (2023). Capacity Optimization of Independent Microgrid with Electric Vehicles Based on Improved Pelican Optimization Algorithm. *Energies, 16*(6), 2539.
33. Song, H. M., Xing, C., Wang, J. S., Wang, Y. C., Liu, Y., Zhu, J. H., & Hou, J. N. (2023). Improved pelican optimization algorithm with chaotic interference factor and elementary mathematical function. *Soft Comput., 27*(15), 10607-10646.
34. Alamir, N., Kamel, S., Hassan, M. H., & Abdelkader, S. M. (2023). An improved weighted mean of vectors algorithm for microgrid energy management considering demand response. *Neural Comput. Appl.*, 1-22.

17

Dual Battery Management System for Hybridization of Photovoltaic Systems with Batteries

P. Devadharshini[1], S. Arun Kumar[1], S. Rishwanth[1] and G. Kanimozhi[2,*]

[1]School of Electrical Engineering, Vellore Institute of Technology, Chennai, India
[2]Center for Smart Grid Technologies, SELECT, Vellore Institute of Technology, Chennai, India

Abstract
This paper provides a survey of dual battery management systems (DBMSs) that work with a solar cell array to manage two separate battery banks in a smart way. The system is designed to maintain the maximum power point of the PV module while delivering an uninterrupted power supply to the load via two lithium-ion batteries. The DBMS works on a unique principle where battery bank 1 and battery bank 2 switch between charging and discharging. This creates a steady flow of energy and makes the most of the energy that has been stored. The batteries' state of charge (SOC) is kept within the threshold by monitoring the voltage and temperature levels. The intelligent control algorithm of the DBMS optimizes the charging and discharging cycles of the batteries based on the load demand, the amount of sunlight, and the SOC of the batteries. Through real-time monitoring and data analysis, the system balances the flow of energy, extends the life of the batteries, and cuts down on energy waste as a whole. A hardware prototype is designed and operated in three modes based on load requirement and solar supply availability. Data collected from the system, such as battery voltages, charging rates, and energy outputs, are systematically analyzed to quantify the system's efficiency and reliability. The result proves that the DBMS effectively manages the battery to charge, adapts to various energy scenarios, and maximizes energy utilization.

**Corresponding author*: kanimozhi.g@vit.ac.in

O.V. Gnana Swathika, K. Karthikeyan, Milind Shrinivas Dangate and Nicoletta Ravasio (eds.) Biomass and Solar-Powered Sustainable Digital Cities, (285–304) © 2024 Scrivener Publishing LLC

***Keywords*:** Super capacitor, battery management system (BMS), maximum power point tracking (MPPT), energy management, rechargeable battery

17.1 Introduction

The incorporation of photovoltaic (PV) systems alongside battery storage has significantly transformed the methods capture and regulate renewable energy. This convergence not only tackles the issue of the intermittent nature of solar power generation but also facilitates the achievement of greater energy self-sufficiency and improved system reliability. The fundamental component of this system is the dual battery management system (DBMS), which serves as an advanced control mechanism that coordinates the smooth interaction between the PV array and the battery storage units. Its primary objective is to enhance the efficiency of the PV system while also optimizing the charging process for electric vehicles (EVs) and other electrical devices.

In order to comprehend the importance of this technology, it is imperative to delve into pertinent terminology and fundamental concepts. Battery charging, in its most basic sense, refers to the act of restoring electrical energy in energy storage devices, such as lithium-ion batteries, in order to make them ready for subsequent utilization. Within the realm of hybrid PV systems, the crucial aspect of battery charging plays a pivotal role in guaranteeing the efficient storage of excess energy generated by the PV array for subsequent utilization. One of the key advantages offered by DBMS is its capacity to deliver a consistent and dependable power source, particularly in critical contexts such as EVs, renewable energy systems, and uninterruptible power supplies [1].

EV charging [2] represents a highly transformative application of this technology. With the global push toward sustainable transportation, EVs have experienced a significant surge in popularity. The utilization of a DBMS facilitates a dynamic interaction between PV generation and EV charging. The surplus solar energy can be harnessed for the purpose of charging EV batteries, thereby mitigating the environmental impact of transportation and alleviating the burden on traditional power grids.

Moreover, this system further enables the process of charging electrical appliances in residential and commercial settings. This technology allows individuals who own residential properties to effectively utilize solar energy by capturing it during daylight hours, storing it in battery systems, and subsequently employing the stored energy to operate household appliances during periods of limited or no sunlight. This practice leads to a notable decrease in dependence on grid electricity, resulting in reduced energy expenses and diminished environmental consequences [3].

The incorporation of PV systems with battery storage via a DBMS signifies a noteworthy progression in sustainable energy technology. The implementation of this comprehensive approach not only improves the effectiveness and dependability of energy systems but also expedites the shift toward a more environmentally sustainable and robust energy landscape. As ongoing research and development efforts persist in this domain, it is anticipated that additional enhancements will result in the advancement of management systems, thereby promoting the extensive utilization of hybrid PV-battery solutions for the purpose of efficient EV charging, powering household appliances, and other applications.

Among the six methods of battery charging algorithm that seems promising is the hysteric control algorithm that coordinates solar panels, batteries, and supercapacitors to charge the battery and track the solar panel's maximum power output during maximum power point tracking (MPPT) mode [4].

To maximize the use of energy produced by hybrid PV systems with batteries, a DBMS was used instead of a battery management system (BMS). The DBMS controls the charging and discharge of the primary and secondary batteries effectively. This offers better energy management, improved system performance, and increased efficiency when compared to a traditional BMS [4, 5].

17.2 Battery Charging Algorithms

MPPT algorithms are normally used across the solar charging techniques in order to extract the maximum power from the solar panels. Some methodologies are also used for charging the battery that helps in efficiency. These techniques prove helpful in charging an EV battery. As EVs are supposed to be in the open air during their running times, the collection of PV arrays seems frequent, and charging the battery with MPPT techniques strengthens the battery conditions. Some of the MPPT algorithms that have been used in recent times are quoted below with a brief explanation.

17.2.1 Battery Switching Technique

The charging strategy employed by PV-based battery switch stations (BSSs) considers both the availability of service and the self-consumption of PV energy. The company employs a service model that facilitates the exchange of batteries, as well as a power distribution model. In order to ensure the availability of service, the battery-swapping service model establishes the minimum charging power by utilizing short-term predictions of the anticipated requirements of EVs. The power distribution model is responsible for transmitting the charging power from both the PV system and the power grid. The algorithm employed for forecasting the demand of EV battery swapping utilizes a wavelet neural network, yielding a mean absolute percentage error of approximately 23%. By utilizing the short-term forecasting engine, it becomes feasible to ascertain the frequency at which battery replacements will be required within the upcoming hours. The algorithm determines the optimal number of EVs that should be charged during each time slot in order to maintain the continuous operation of the battery swapping service, as illustrated in Figure 17.1. The suggested charging strategy has the potential to enhance the self-consumption of PV energy and optimize the performance of PV-based BSS holistically. The determination of the duration required for a battery replacement in this specific system involves the consideration of various factors.

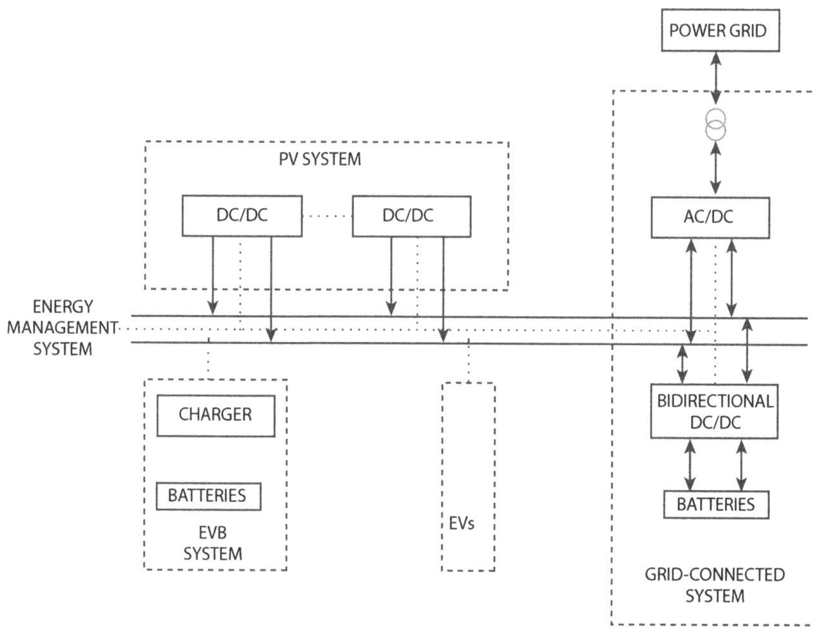

Figure 17.1 Battery switching technique.

Battery capacity, denoted as C, refers to the overall energy storage capacity of a battery that is being replaced. The charging/discharging rate (I) refers to the speed at which the battery can be discharged or charged. The parameter denotes the rate at which the battery is capable of transferring energy. The variable denoted as T represents the duration required to physically extract the discharged battery and substitute it with a fully charged one. This encompasses the duration required for mechanical operations, such as the act of opening and closing battery compartments, as well as ensuring secure connections. The equation used to determine the duration needed for a battery replacement is as follows:

$$T = \frac{C}{I} \tag{17.1}$$

where T represents the duration, measured in hours, needed for the battery replacement process. The variable C represents the battery capacity, which is measured in units of ampere-hours (Ah) or watt-hours (Wh). The charging/discharging rate is denoted by the symbol "I" and is measured in amperes (A) [7].

17.2.2 Schiffer Model

A novel configuration for dual-battery banks is proposed for the purpose of enhancing the performance of isolated microgrids that rely on renewable energy sources. This would potentially enhance the longevity of the largest battery bank and reduce the expenses associated with its operation and maintenance. This strategy has the potential to yield economic advantages for systems characterized by varying levels of installed PV capacity, particularly those with lower PV capacity. The system is composed of two separate banks. The initial financial institution accumulates energy during periods when renewable sources are capable of satisfying the demand for power consumption.

The second bank of the power system is comparatively smaller in size and is primarily utilized for the purpose of daily snapshot balancing and network voltage regulation. The objective of this study is to forecast the lifespan of lead-acid batteries and determine the optimal size for hybrid energy systems and microgrids that utilize renewable energy sources, as depicted in Figure 17.2. This study also examines the optimal control strategies for autonomous PV systems equipped with battery-supercapacitor

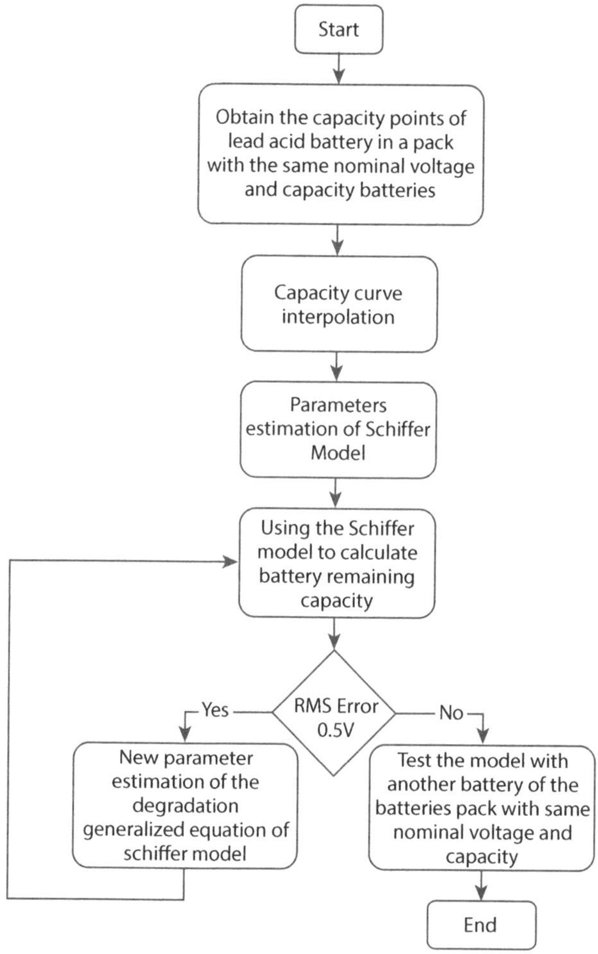

Figure 17.2 Schiffer model.

hybrid energy storage systems (HESSs). Additionally, it explores the potential application of the novel HESS [8] control strategy in regulating the voltage levels within the direct current (DC) grid.

17.2.3 Hysteric Control Algorithm

This algorithm is to manage a solar car's hybrid energy system made up of solar panels, batteries, and supercapacitors as shown in Figure 17.3. The energy management control is the most important part of the algorithm.

Figure 17.3 Supercapacitors.

This is done by coordinating super capacitors, batteries, and solar panels, which are the three energy sources and storage devices.

A hysteric control algorithm is used for the solar panel to charge the battery. During MPPT mode, this algorithm uses a hill climbing method to track the solar panel's maximum output power. To get MPPT, the boost converter changes the input voltage based on voltage and current coming out of the solar panel. As battery voltage reaches threshold, constant charge mode is turned on. This keeps the battery from being overcharged. This mode stays on until the voltage goes below the specified voltage and reverts MPPT back [6].

The bidirectional DC/DC converter is a key part of the integrated supercapacitor and battery. It works in three ways: battery current discharging mode, super capacitor current charging mode, and super capacitor voltage charging mode. The switch between these modes depends on how fast the battery is being drained and how much voltage is in the supercapacitor. The control strategy makes sure that the battery output current stays within a certain range and that the supercapacitor is charged or drained as needed to meet the load demands.

17.2.4 Fuzzy Logic Technique

The construction of an isolated PV system with a unique BMS and fuzzy logic-based MPPT controller is described in the study [11]. The MPPT controller processes error and change in error inputs through fuzzification, inference, and defuzzification steps in order to effectively track the

PV array's maximum power output. The suggested BMS method controls the battery's state of charge (SOC) and load connection using fuzzy logic as well. For the best performance and security, three SOC-based scenarios decide whether the battery should be charged, discharged, or kept in its current condition while managing the load connection. The usefulness of the fuzzy logic-based technology is shown by simulation results and compared to traditional approaches, showcasing its potential for improved battery management and PV system performance [9].

17.2.5 Coulomb Counting Method

The SOC of a rechargeable battery can be determined using the Coulomb counting method, an energy estimating method. The idea behind it is that by integrating the current going into or out of the battery over time, the quantity of energy stored or released in a battery can be estimated. Coulomb counting, to put it simply, is the process of detecting the electric current entering or leaving the battery and then integrating these readings over time to determine the total charge added to or taken from the battery. The SOC, which represents how full the battery is, may be calculated by comparing this projected charge with the battery's known capacity. Although the Coulomb counting approach is conceptually simple, it has certain drawbacks. Errors in the SOC estimate can be caused by things like battery efficiency, temperature changes, and self-discharge. As a result, it is frequently used in combination with other techniques, such as voltage-based techniques or algorithms that take these affecting variables into consideration, to produce SOC values that are more accurate [10].

17.2.6 PWM Control Method (Perturb and Observe Method)

The present approach employs a control strategy in order to facilitate rapid battery charging within a PV energy system. The system comprises two converter circuits, a resistive load, and battery storage that are interconnected to a shared DC bus. The boost converter is employed to achieve MPPT of the PV array, whereas the bidirectional converter is responsible for regulating the charging and discharging of the battery. The bidirectional controller is equipped with two control schemes, namely, a proportional-integral controller and a Takagi–Sugeno fuzzy controller [11]. This methodology offers a comparative evaluation of battery management with respect to charging-discharging dynamics, rise time, overshoot and undershoot phenomena, peak amplitude, and peak timing [12].

17.3 DBMS Algorithms

A BMS is an important part of keeping track of and managing the performance, safety, and overall health of battery packs in a variety of applications, such as EVs, renewable energy storage systems, and consumer electronics. A DBMS has redundant systems that make the system more reliable and safer by providing backup functions in case the primary BMS fails. The various methods used in DBMSs are the electro static discharge algorithm (ESDA), dual-balancing mechanism, percentage of discharge (POD), deep-discharge method, cell balancing, unitized charging and discharging (UCD) BMS for rechargeable batteries, SOC estimation, SOH estimation, battery balancing, and thermal management.

17.3.1 Dual-Balancing Mechanism

For passive methods of balancing, resistors are used to use up extra energy and balance the charge or voltage of battery cells while they are being charged or discharged. This method is easy to use, but it may waste energy. Active balancing methods, on the other hand, use active parts like switches and inductors to move energy between battery cells or packs and achieve balance. Active balancing may be more effective, but it may also be harder to set up and cost more.

The proposed dual-balancing mechanism combines the best parts of both active and passive methods of balancing. It uses a passive balancing circuit to make sure that each battery cell in a pack has the same amount of charge or voltage. This reduces the cost of design and the size of the circuit. It also has an active pack-balancing circuit that can work in either buck or boost mode to move energy between battery packs and make them equal as shown in Figure 17.4. The goal of this combination is to waste as little energy as possible while keeping the capacity of each cell and pack the same. Overall, the algorithm used by the BMS probably involves measuring the current, voltage, and temperature of the battery module and controlling the balancing circuits based on these measurements [13].

17.3.2 Unitized Charging and Discharging (UCD) BMS for Rechargeable Batteries

This system is made to handle charging and discharging battery units in a way that lets some units be discharged while others are being charged. The UCD BMS is made up of different parts that work together to make it work.

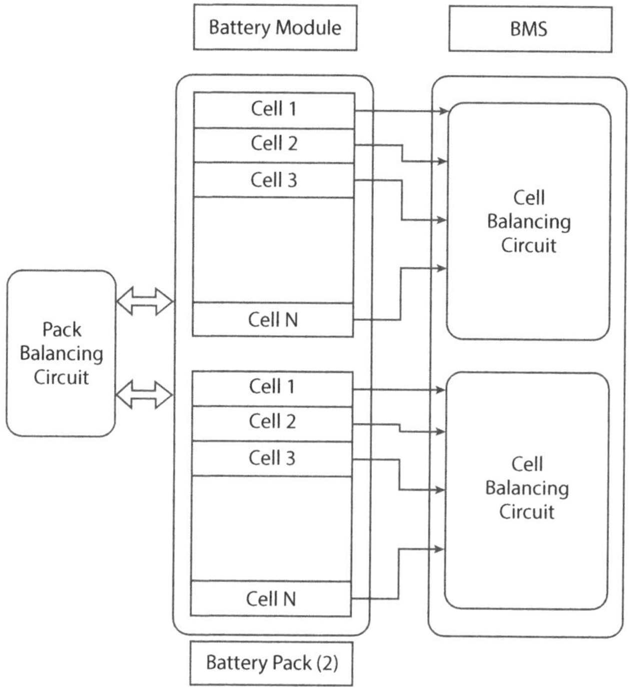

Figure 17.4 Dual-balancing mechanism.

First, there is a smart battery module, which is made up of individual battery units. These units have sensors that track their charge and discharge cycles and stop them from being overcharged or over-discharged.

Another important part of the UCD BMS is the programmable battery management module. It has a control circuitry and a programmable path array (PPA) module. The switches in the PPA module are controlled by the control circuitry, which makes two paths: the battery path and the charging/discharging path. This lets battery units be charged or drained in different ways. The charging module and the discharging module oversee connecting the battery path to the smart battery charger and the load, respectively. This lets the battery units be charged or drained depending on the configuration.

The UCD BMS has different configurations for different uses, such as charging, discharging, charging and discharging, and balancing the battery. It gives more options for output voltages and makes better use of the energy that is discharged, which improves the overall power efficiency. In conclusion, the unitized charging and UCD BMS is an innovative way to manage rechargeable batteries. It lets battery units be charged and drained in different places. This improves efficiency and flexibility in many applications [14].

17.3.3 Percentage of Discharge (POD)

The used algorithm is a way to model how the discharge of the battery cells in a battery pack works. The POD parameter is the basis for the method. POD is a way to measure how fast a battery cell drains at a certain current. It is a direct measurement that does not need any correction factors and can be used in non-standard situations. It looks at how different cells discharge and looks for patterns that are related to the POD. The model is based on these patterns, which show how the electrochemistry works inside the cells.

With this method, a polynomial equation is used to fit the residual curves, which are found by subtracting the reference curve from the other discharge curves. A linear relation between the discharge current and the coefficients of the polynomial equation can be used to estimate the coefficients. By estimating the coefficients, it is possible to model the discharge curve of various cells. Experiments with lithium-ion batteries and lead-acid batteries proved that the method worked. The results showed that the estimated discharge curves were between 1.7% and 5% off from the original curves. This shows that the method can be used to accurately model how battery cells discharge. Overall, the algorithm models the discharge curves of battery cells by using the POD parameter and polynomial fitting. This is a simple and effective way for BMSs to work.

$$POD = \left(\frac{Remaining\ Capacity}{Full\ capacity} \right) \times 100 \tag{17.2}$$

where POD shows how much of the battery's total capacity has been used up. The amount of energy left in a battery after it has been used is called its remaining capacity. As the battery is fully charged, full capacity defines the total amount of energy it can hold [15].

17.3.4 Deep-Discharge Method

The deep-discharge method is a way to test a battery's capacity by draining it until its voltage falls to a certain level. In this experiment, the lead-acid batteries were drained until their voltage in a closed circuit reached 9.5 volts. The deep-discharge method is often used to test how well and how much a battery can do. By draining the battery to a low voltage, battery's capacity can be analyzed.

In this study, researchers tested lead-acid batteries at different discharge rates and ambient temperatures. Multimeter logger is used to measure the voltage and current per minute, then analyzed the data to figure out the voltage drop and battery capacity in different situations. Using the deep-discharge method, the researchers were able to figure out how the rate of discharging and the temperature around the battery affect its capacity. The results showed that the battery capacity went down when the rate of discharge is higher and the temperature outside is lower. This information is helpful for understanding how lead-acid batteries work in renewable energy storage systems, where things like the rate at which they are discharged and the temperature can affect how well they work overall and how long they last. Overall, the deep-discharge method is a good way to test a battery's performance and capacity. It lets researchers gather important information that can be used to improve how batteries are used in different situations [16].

17.3.5 SOC Estimation

Estimating the SOC is done with an algorithm that uses measurements of voltage, current, and temperature to figure out how much power the battery still has. It is a key part of figuring out how much energy is in the battery and stopping it from being overcharged or deeply discharged. In solar power generation systems, it is important to get an accurate SOC estimate to optimize the charging and discharging process, make sure that the battery's capacity is used efficiently, and extend its life. By keeping an eye on the SOC all the time, the BMS can decide when to charge or drain the battery in a way that will not hurt its performance.

$$SOC = SOC_{initial} + \int((I)t/C)dt \tag{17.3}$$

where SOC stands for state of charge, which is shown as a percentage from 0% (empty) to 100% (fully charged). $SOC_{initial}$ is the initial state of charge at the start of the estimation cycle. I(t) is the current from a battery over time in A, and C is the capacity of the battery [measured in ampere-hours (Ah)] [17].

17.3.6 State of Health (SOH) Estimation

Estimating the battery's SOH is done with an algorithm that looks at the battery's overall health and how it has changed over time. To figure out how long the battery has left, it looks at things like the number of cycles, the internal resistance, and the loss of capacity. In solar power generation systems, SOH estimation helps figure out how well and how reliable the batteries will be in the long run. By constantly checking the SOH, the BMS can see if the battery's performance is getting worse and take the right steps to keep it working at its best. This includes taking care of the battery by doing things like making sure the charge levels are the same or replacing worn-out cells.

$$\text{SOH} = (\text{Current Capacity}/\text{Initial Capacity}) \times 100 \qquad (17.4)$$

This is shown as a percentage, from 0% for a completely dead battery to 100% for a brand-new battery that is in perfect health. Current capacity is the actual capacity of the battery at the time of estimation [measured in ampere-hours (Ah)]. Initial capacity is the battery's capacity when it was new or at an earlier point in time, as measured in ampere-hours (Ah) [17].

17.3.7 Thermal Management

Thermal management means keeping an eye on and controlling the battery's temperature to keep it from getting too hot or too cold. It is essential for keeping batteries in solar power generation systems in good shape and making sure they work safely and reliably. Too much heat can hurt the battery's performance and shorten its life, and too much cold can make it less powerful. Thermal management helps prevent thermal runaway and makes the batteries last longer by constantly checking the temperatures and putting in place cooling or heating systems as needed. This can be done with active cooling methods like fans or liquid cooling systems or passive cooling methods like heat sinks or insulation as shown in Figure 17.5. Thermal management is the process of keeping an eye on and controlling the battery's temperature to keep it from getting too hot or too cold, which can hurt its performance and life [17].

17.3.8 Cell Balancing

This report outlines the discharge test performed on a LiFePO4 battery pack in order to assess the impact of discharge current on cell balancing.

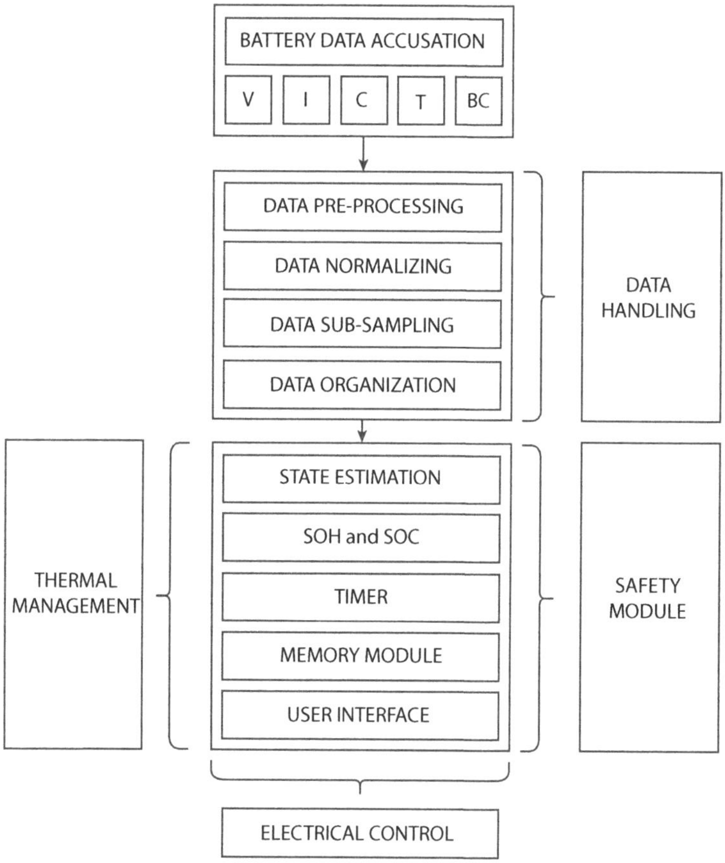

Figure 17.5 Thermal management.

The experiment consisted of quantifying the voltage disparity between the cell with the highest voltage and the cell with the lowest voltage within the battery pack while subjecting it to varying discharge currents. The objective of the experiment was to ascertain the optimal amount of balancing current required among cells in order to achieve efficient balancing time and maximize the battery's lifespan and performance, as depicted in Figure 17.6. Through the examination of the voltage disparity, the researchers were able to evaluate the degree of asymmetry within the battery pack and comprehend its fluctuations in response to varying discharge currents.

The experimental protocol consisted of an initial step wherein the battery pack was charged, followed by a subsequent stage aimed at balancing each individual cell. Subsequently, the discharge process was conducted

Figure 17.6 Cell balancing.

at various constant currents, specifically 50 A, 100 A, and 150 A, while simultaneously monitoring the voltage of each individual cell. The empirical findings demonstrated a negative correlation between the discharge current and the voltage differential observed between the highest and lowest cells [18].

17.3.9 Electrostatic Discharge Algorithm (ESDA)

ESDA is the name of the algorithm being talked about. It is made to estimate the unknown polynomial coefficients of the electric circuit model of lithium-ion batteries used in EVs. The goal of the algorithm is to get a good estimate of the values of the battery circuit elements, which are needed to figure out the battery's SOC.

ESDA starts by putting objects (solutions) in the search space in a random order. At each iteration, ESD equations are used to update three objects. The number of objects involved in the ESD interaction affects how the update works. If there are two objects, then the position of the one with the lowest fitness value is changed based on the position of the one with the highest fitness value. If there are three objects, then the position of the one with the lowest fitness value is changed based on where the other two objects are.

The algorithm also has steps for resetting to the beginning and exploring in a random way to improve accuracy and the ability to explore. Objects that have been changed more than a certain number of times are reset within the search space. Also, objects with fewer than three updates are given a random new start so that the search space can be explored more.

The study's results show that ESDA is more accurate at estimating battery parameters than other recently proposed metaheuristic algorithms like the artificial hummingbird optimization technique, the chameleon swarm algorithm, and the tuna swarm optimization. ESDA is at least 10% more accurate than these other algorithms.

Overall, ESDA is a promising algorithm for accurately estimating the values of circuit elements. This is important for figuring out the SOC and improving the BMS in EVs [19].

17.4 DBMS Algorithm Comparison

The ESDA is made to estimate the unknown polynomial coefficients of lithium-ion battery circuit models in EVs. It uses ESD equations to change the objects in the search space and has steps for resetting and exploring randomly [18]. The passive and active ways of balancing battery cells are both used by the dual-balancing mechanism. It uses an active pack-balancing circuit to move energy between packs and a passive balancing circuit to make sure that each cell has the same amount of charge or voltage [19].

The POD algorithm patterns related to POD to model how a battery cell discharges. Polynomial fitting is used to estimate coefficients and model discharge curves accurately. By draining a battery to a certain voltage level, the deep-discharge method checks how much power it can hold. It helps people understand how the rate of discharge and the temperature affect the capacity of a battery. The voltage difference between the highest and lowest cells in a battery pack is measured by cell balancing. This is done at different discharge currents to find the best time to balance the cells and get the most life out of the battery.

The UCD BMS let the battery units be charged and drained at the same time. It makes different applications more efficient and flexible. SOC estimation uses measurements of voltage, current, and temperature to figure out how much power is left in the battery. It improves how charging and discharging work and makes the battery last longer. SOH estimation looks at things like cycles, internal resistance, and capacity loss to figure out how healthy a battery is overall. It helps predict how a battery will work and how reliable it will be. Battery balancing makes sure that each cell in a pack charge and drains at the same rate. This improves the pack's overall performance and lifespan. It can be done with either active or passive ways to balance.

17.5 Future Scope

The DBMS is a technology that is currently undergoing continuous development, with significant potential for future advancements and promising applications. There are several domains that exhibit potential for additional advancement and integration: Future iterations of DBMS can prioritize the enhancement of energy density and capacity, thereby enabling longer battery runtimes and greater energy storage capabilities. This characteristic would enhance their suitability for deployment in remote regions, where ensuring a consistent power source is frequently problematic.

The integration of DBMS into smart grid systems has the potential to enhance the management of energy consumption and mitigate peak load demands for both utilities and consumers. This phenomenon has the potential to enhance the stability of the grid and mitigate energy expenditures. The utilization of artificial intelligence and machine learning algorithms can effectively augment the intelligence of DBMS. This augmentation involves optimizing battery consumption by leveraging historical data, weather conditions, and user preferences. This would lead to a further enhancement in energy management efficiency.

The EV sector is anticipated to experience substantial expansion within the automotive industry. The utilization of a DBMS has the potential to significantly impact the expansion and durability of EV batteries, thereby enhancing their feasibility and accessibility for consumers. The integration of Internet of Things (IoT) devices is becoming increasingly widespread in both residential and commercial settings. The integration of a DBMS with IoT devices has the potential to establish a cohesive and automated ecosystem for managing energy. This integration enables users to remotely regulate and monitor their energy consumption. The topic of environmental sustainability has garnered significant attention globally. In this context, DBMS can play a crucial role in promoting this cause by prioritizing the optimization of energy storage systems to enhance efficiency and minimize adverse environmental effects. In summary, the efficacy of the DBMS has been substantiated across diverse applications, thereby indicating its promising prospects for future utilization. The ongoing advancement of technology and increasing dependence on energy storage solutions necessitate the pivotal role of DBMS in shaping a more dependable, efficient, and environmentally sustainable energy infrastructure. The domain exhibits great potential for innovation and investment, holding the capacity to fundamentally transform the methods by which energy is stored and utilized in the foreseeable future.

17.6 Conclusion

In conclusion, DBMS signifies a notable progression in the field of energy storage and management technology. This system employs intelligent control mechanisms to optimize energy utilization, improve reliability, and prolong the lifespan of two batteries: a primary high-capacity battery and a secondary auxiliary battery. Upon conducting an analysis of the present condition of DBMS technology and its practical implementations, it becomes evident that this advancement exhibits significant potential within diverse industries and sectors. This technological advancement guarantees a continuous and reliable power supply, thereby minimizing periods of inactivity and augmenting user contentment. Furthermore, the utilization of DBMS enables efficient regulation of energy transfer between the two batteries, thereby mitigating the risks associated with overcharging and over-discharging. These factors are known to have a substantial impact on battery degradation. One notable benefit of a DBMS is its capacity to enhance the longevity of batteries. The system achieves an equitable distribution of the workload between the primary and auxiliary batteries, thereby mitigating the detrimental effects of wear and tear on both components. Consequently, this approach leads to energy storage solutions that exhibit enhanced longevity and cost-effectiveness. This holds significant importance in sectors where the cost of battery replacement constitutes a substantial financial burden, as exemplified by the EV industry.

In addition, it is important to note that DBMS has a significant impact on the advancement and implementation of sustainable energy practices. With the increasing prevalence of renewable energy sources such as solar and wind, it is possible for DBMS to store surplus energy produced during peak production hours and subsequently release it during periods of high demand or when renewable energy generation is low. This phenomenon contributes to the mitigation of dependence on fossil fuels and improves the overall efficacy of energy systems. In conclusion, by effectively controlling solar panels, batteries, and supercapacitors, the hysteric control algorithm excels at optimizing hybrid energy systems. Its dynamic strategy that combines MPPT and continuous charge modes, guarantees optimum energy efficiency and battery safety. It is significant to switch from conventional BMS to DBMS because DBMS provides a complete solution in line with the rising need for effective and sustainable energy management across a range of applications.

References

1. Barca, G., *et al.* "Optimal energy management of a photovoltaic stand-alone dual battery system." *MELECON 2008-The 14th IEEE Mediterranean Electrotechnical Conference*. IEEE, 2008.
2. C. Duan *et al.*, "A Solar Power-Assisted Battery Balancing System for Electric Vehicles," in *IEEE Trans. Transp. Electrif.*, vol. 4, no. 2, pp. 432-443, June 2018, doi: 10.1109/TTE.2018.2817123.
3. M. A. Fatullah, A. Rahardjo and F. Husnayain, "Analysis of Discharge Rate and Ambient Temperature Effects on Lead Acid Battery Capacity," *2019 IEEE International Conference on Innovative Research and Development (ICIRD)*, Jakarta, Indonesia, 2019, pp. 1-5, doi: 10.1109/ICIRD47319.2019.9074667.
4. F. Rayya and N. Karami, "A Smart Lithium Battery Management System for Photovoltaic Plants (Review and Strategy design)," *2021 International Conference on Microelectronics (ICM)*, New Cairo City, Egypt, 2021, pp. 203-206, doi: 10.1109/ICM52667.2021.9664908.
5. M. Nirmala, R. Sanjay Kumar, V. Vijaya Varshini and S. Siva, "Iot Based Battery Monitoring System For Solar PV Fed DC-DC Converter," *2023 9th International Conference on Electrical Energy Systems (ICEES)*, Chennai, India, 2023, pp. 322-327, doi: 10.1109/ICEES57979.2023.10110230.
6. B. Wu, F. Zhuo, F. Long, W. Gu, Y. Qing and Y. Liu, "A management strategy for solar panel — battery — supercapacitor hybrid energy system in solar car," *8th International Conference on Power Electronics - ECCE Asia*, Jeju, Korea (South), 2011, pp. 1682-1687, doi: 10.1109/ICPE.2011.5944438.
7. N. Liu, Q. Chen, X. Lu, J. Liu and J. Zhang, "A Charging Strategy for PV-Based Battery Switch Stations Considering Service Availability and Self-Consumption of PV Energy," in *IEEE Trans. Ind. Electron.*, vol. 62, no. 8, pp. 4878-4889, Aug. 2015, doi: 10.1109/TIE.2015.2404316.
8. P. B. L. Neto, O. R. Saavedra and L. A. de Souza Ribeiro, "A Dual-Battery Storage Bank Configuration for Isolated Microgrids Based on Renewable Sources," in *IEEE Trans. Sustainable Energy*, vol. 9, no. 4, pp. 1618-1626, Oct. 2018, doi: 10.1109/TSTE.2018.2800689.
9. J. John, A. Yoonus, F. Shijad, M. Aslam Mm, A. Thasneem and L. Arun, "Isolated PV System with Fuzzy Logic based MPPT Controller and Battery Management System," *2021 5th International Conference on Electrical, Electronics, Communication, Computer Technologies and Optimization Techniques (ICEECCOT)*, Mysuru, India, 2021, pp. 194-199, doi: 10.1109/ICEECCOT52851.2021.9707930.
10. K. Simeonov, N. Mihailov, N. Valov and K. Gabrovska-Evstatieva, "Analysis of a PV Installation with a Battery Storage and BMS at a Residential Building," *2022 8th International Conference on Energy Efficiency and Agricultural Engineering (EE&AE)*, Ruse, Bulgaria, 2022, pp. 1-5, doi: 10.1109/EEAE53789.2022.9831207.

11. N. B. Halima, H. Kolsi, N. B. Halima and A. Oualha, "Energy Management of Electrical Vehicle Charging Station Based on Photovoltaic System," *2022 IEEE 21st international Conference on Sciences and Techniques of Automatic Control and Computer Engineering (STA)*, Sousse, Tunisia, 2022, pp. 623-628, doi: 10.1109/STA56120.2022.10019200.
12. A. Basu and M. Singh, "A Fast Battery Charging Scheme for a Solar Energy System Using Intelligent Bi-directional Converter," *2022 International Conference on Intelligent Controller and Computing for Smart Power (ICICCSP)*, Hyderabad, India, 2022, pp. 1-6, doi: 10.1109/ICICCSP53532.2022.9862393.
13. Kanimozhi.G and Sreedevi.VT, "Semibridgeless Interleaved PFC boost rectifier for PHEV battery chargers", *IETE J. Res.*, Taylor and Francis, Vol 65, issue:1 2019, pp-128-138.
14. C. -L. Wey and P. -C. Jui, "A unitized charging and discharging smart battery management system," *2013 International Conference on Connected Vehicles and Expo (ICCVE)*, Las Vegas, NV, USA, 2013, pp. 903-909, doi: 10.1109/ICCVE.2013.6799924.
15. S. Bhagyashri, G.Kanimozhi, "Soft Switching Bridgeless Rectifier for NEV Battery Charger" , *Int. J. Renewable Energy Res.*, Vol.10, no-2,2020, pp-988-997.
16. Amin, A. C. Budiman, A. Hapid, S. Sudirja and S. Kaleg, "Discharge Test to Detect Unbalancing in Electric Vehicle LiFePO4 Battery Pack," *2019 International Conference on Sustainable Energy Engineering and Application (ICSEEA)*, Tangerang, Indonesia, 2019, pp. 1-5, doi: 10.1109/ICSEEA47812.2019.8938628.
17. S. R.G, H. C and N. K. Marati, "Remote Electric Vehicle Battery Monitoring & Life Cycle Management System," *2022 IEEE 2nd International Conference on Sustainable Energy and Future Electric Transportation (SeFeT)*, Hyderabad, India, 2022, pp. 1-5, doi: 10.1109/SeFeT55524.2022.9909384.
18. M. A. Bandora, V. C. Ifrim and M. Artiom, "Battery Balancing System for Electric Vehicles Solar Power Assisted," *2023 10th International Conference on Modern Power Systems (MPS)*, Cluj-Napoca, Romania, 2023, pp. 1-6, doi: 10.1109/MPS58874.2023.10187487.
19. I. Pervez, C. Antoniadis, H. Ghazzai and Y. Massoud, "An ElectroStatic Discharge Algorithm for Electric Vehicle Li Ion Battery Parameters Estimation," *2023 IEEE International Symposium on Circuits and Systems (ISCAS)*, Monterey, CA, USA, 2023, pp. 1-5, doi: 10.1109/ISCAS46773.2023.10181565.

18

Advancing Smart City Energy Management: Very Short-Term Photovoltaic Power Generation Forecasting Using Multi-Scale Long Short-Term Memory Deep Learning

Battula Srinivasa Rao[1*], Mudiyala Aparna[1] and Mallela Siva Naga Raju[2]

[1]*School of Computer Science and Engineering, VIT-AP University, Andhra Pradesh, India*
[2]*Department of Computer Science and Engineering, V R Siddhartha Engineering College, Andhra Pradesh, India*

Abstract
The rapid adoption of renewable energy sources, particularly photovoltaic (PV) systems, in smart city energy management has prompted the need for accurate and forecasting PV power output in the short term. Accurate predictions are essential for efficient grid integration, demand response optimization, and enhanced energy management strategies. Traditional forecasting methods often struggle to handle the complexities of PV power generation due to its intermittent and highly variable nature. In order to accurately predict very short-term PV power generation, we investigate the use of multiscale long short-term memory, a deep learning algorithm. This development has a great deal of potential to improve energy management strategies for smart cities.

***Keywords*:** Photovoltaic (PV) power generation forecasting, very-short-term forecasting, smart city energy management, multi-scale long short-term memory (LSTM), deep learning, renewable energy integration, power grid stability, machine learning, time series forecasting, smart grids

**Corresponding author*: srinivas.battula@vitap.ac.in

O.V. Gnana Swathika, K. Karthikeyan, Milind Shrinivas Dangate and Nicoletta Ravasio (eds.) Biomass and Solar-Powered Sustainable Digital Cities, (305–316) © 2024 Scrivener Publishing LLC

18.1 Introduction

The rapid urbanization and escalating demands for energy have compelled cities worldwide to seek innovative solutions for sustainable and efficient energy management. As the concept of smart cities gains traction, the integration of advanced technologies, data-driven strategies, and renewable energy sources emerges as a critical pathway toward achieving energy resilience, reduced environmental impact, and improved quality of life for urban populations [1]. Among the various renewable energy sources, photovoltaic (PV) systems have garnered significant attention due to their potential to harness solar energy and contribute to a cleaner energy future. However, the dynamic and inherently unpredictable nature of solar energy, influenced by varying weather conditions, introduces unique challenges in the effective integration of PV systems into smart city energy grids [2]. For solar energy and traditional energy sources to coexist peacefully, accurate predictions of PV power generation are required. Grid stability, appropriate energy distribution, and effective demand response tactics have all made short-term forecasting (within minutes to hours) an absolute necessity. Traditional forecasting methods have struggled to cope with the complexities of PV power generation due to its intermittent and highly variable nature [3–5]. These methods often fall short in capturing the intricate temporal dependencies, nonlinear relationships, and sudden fluctuations that characterize solar energy production. Consequently, a paradigm shift is underway, where advanced artificial intelligence (AI) techniques are being harnessed to overcome these challenges and enable accurate predictions in real time.

18.1.1 The Role of Deep Learning in Smart City Energy Management

When compared to other AI methods, deep learning has been revolutionary, completely altering the terrain of forecasting and analysis. When it comes to capturing temporal dynamics, recognizing subtle patterns, and adapting to quickly changing inputs, deep learning models, especially long short-term memory (LSTM) networks, shine. Because of this, they are well-suited for predicting short-term PV power generation. This chapter delves into the intersection of deep learning, smart city energy management, and PV power generation forecasting. Specifically, it explores the application of a multiscale LSTM-based deep learning technique as a pioneering solution for addressing the challenges posed by short-term predictions of PV electricity output [6–8]. By integrating the strengths of LSTM networks with multiscale decomposition, this approach aims to unravel the complexities of PV power generation and provide accurate predictions that empower urban energy systems to operate efficiently.

18.1.2 Objectives and Structure of the Chapter

The principal aim of this chapter is to conduct a thorough assessment of the multiscale LSTM-based deep learning methodology in the context of forecasting extremely short-term PV power generation. It delves into the intricacies of the methodology, the benefits that it offers to smart city energy management, and the potential implications for shaping more sustainable and resilient urban environments. The subsequent sections of this chapter will elucidate the methodology behind the multiscale LSTM-based approach, provide insights into its evaluation and performance through comparative analyses with traditional methods, discuss its benefits and implications for advancing smart city energy management, highlight potential challenges and future directions, and, finally, conclude by emphasizing its role as a pivotal component in the evolution of modern urban energy systems. Through this exploration, this chapter endeavors to contribute to the ongoing discourse on smart city energy management, deep learning, and the integration of renewable energy sources, ultimately fostering a greener, more sustainable, and forward-looking future for cities worldwide.

18.2 Methodology: Multiscale LSTM-Based Deep Learning Approach

This chapter presents an innovative methodology for predicting PV power generation, which capitalizes on the multiscale LSTM—a deep learning framework. This approach leverages the strengths of LSTM networks, renowned for their adeptness in capturing sequential relationships. Furthermore, it incorporates multiscale decomposition to effectively handle the intricacies inherent in PV power generation time series data. The architecture of the multiscale LSTM-based deep learning model is illustrated in Figure 18.1. This diagram visually represents the configuration of the multiscale LSTM-based deep learning architecture, showcasing how the LSTM technique is employed for multiscale analysis within the framework of deep learning methodologies.

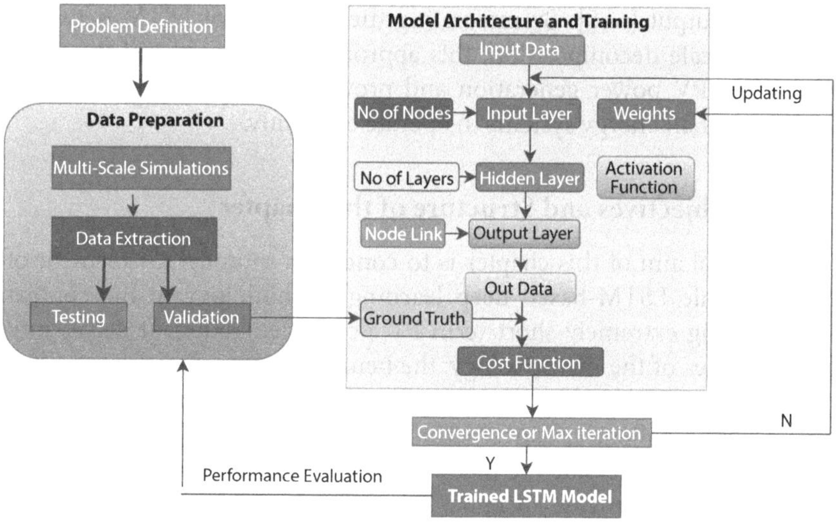

Figure 18.1 Architecture for multiscale LSTM-based deep learning approach.

18.2.1 Long Short-Term Memory (LSTM) Networks

The method relies heavily on recurrent neural networks (RNNs), specifically the LSTM architecture. LSTMs are able to overcome the limitations of traditional RNNs by including memory cells and gating mechanisms, which allow them to record both short-term patterns and long-term associations within sequences [9]. In the context of very-short-term PV power generation forecasting, LSTMs excel at learning the complex relationships between solar energy production and various environmental factors. In Figure 18.2 depicts the workflow of LSTM network.

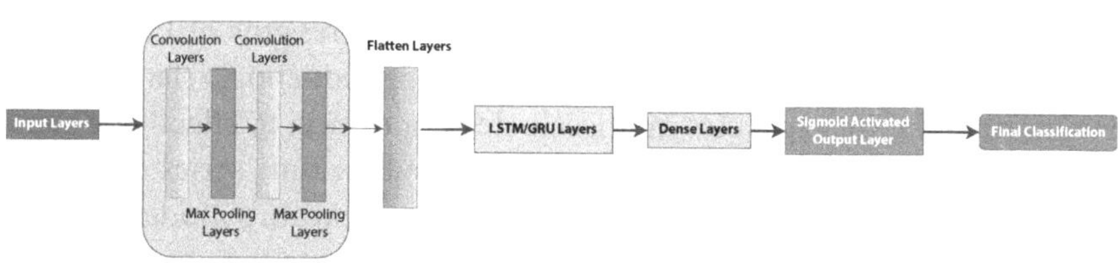

Figure 18.2 Architecture for LSTM networks.

18.2.2 Incorporating Multiscale Decomposition

The suggested method incorporates multiscale decomposition, which is a novel step forward. This technique allows the model to analyze data at multiple levels of granularity. By disentangling the time series into various scales, from fine to coarse, the model gains the ability to capture both local variations and global trends present in the PV power generation data. This multiscale perspective enhances the model's adaptability and accuracy, particularly in situations where rapid fluctuations and overarching patterns coexist.

18.2.3 Data Collection and Preprocessing

Historical PV power generation data are collected alongside key meteorological and environmental factors including solar irradiance, temperature, and cloud cover to execute the multiscale LSTM-based deep learning technique. The data are pre-processed to ensure consistency and compatibility, and suitable features are engineered to provide meaningful inputs to the model.

18.2.4 Training and Validation Strategies

The multiscale LSTM model uses a sliding window method for training. Predictions of future PV power generation values are made using series of previous power generation data and accompanying weather variables. Model hyperparameters like LSTM layer depth, hidden unit size, and learning rate are optimized using methods like grid search and Bayesian analysis. Regularization techniques like dropout and batch normalization are employed to prevent overfitting and enhance generalization.

18.2.5 Evaluation and Testing

The evaluation of the multiscale LSTM-based approach is performed using real-world PV power generation datasets sourced from smart city energy management systems. To guarantee accurate evaluation, the dataset is meticulously partitioned into training, validation, and testing sets [10, 11]. Using comparisons to other common forecasting methods, such as the autoregressive integrated moving average (ARIMA) and the support vector regression (SVR), provides a solid assessment of the method's efficacy.

Quantitative Performance Metrics: Mean absolute error (MAE), root mean squared error (RMSE), and mean absolute percentage error (MAPE) are generated as standard performance measures to objectively examine the precision of the projections. These measurements show how effectively the multiscale LSTM-based method compares to, or outperforms, conventional forecasting techniques.

Evaluation and Performance Analysis: It is important to conduct a thorough evaluation of the multiscale LSTM-based deep learning approach in order to comprehend its efficacy for very short-term forecasting of PV power generation. This section outlines the methodologies used to rigorously evaluate the approach, including dataset splitting, comparative analysis with traditional forecasting methods, the utilization of quantitative metrics, and the interpretation of comparative results.

Dataset Splitting for Rigorous Evaluation: To ensure the precision of the assessment, the dataset is partitioned into three distinct segments: a training set designed for model training, a validation set utilized for refining hyperparameters, and a testing set employed for the ultimate evaluation. This segmentation serves to mitigate overfitting and facilitates an unbiased validation of the model's effectiveness.

18.3 Comparative Analysis with Traditional Forecasting Methods

Comparisons are made between the multiscale LSTM-based method and other more established forecasting techniques, such as the ARIMA and the SVR. These techniques have been around for a long, they may not be able to fully capture the dynamic complexity of PV power generation. The comparative analysis aims to highlight the superior performance of the Multiscale LSTM-based approach.

18.3.1 Quantitative Metrics: Mean Absolute Error and Root Mean Squared Error

Forecasting models can be evaluated objectively using quantitative performance metrics. Commonly employed metrics include the following:

Mean Absolute Error (MAE): The mean of all discrepancies between forecasted and measured quantities.

Root Mean Squared Error (RMSE): The average squared deviation from the forecasted value to the actual value, expressed as a square root.
Mean Absolute Percentage Error (MAPE): The mean of the projected and actual value percentage discrepancies.

These measures assist evaluate the accuracy of the multiscale LSTM method by revealing the magnitude of the forecasting errors.

18.3.2 Interpretation of Comparative Results

The interpretation of the comparative results plays a pivotal role in assessing the practical implications of the multiscale LSTM-based approach. By analyzing the performance metrics and comparing them against those of traditional methods, the chapter gains a nuanced understanding of the multiscale LSTM's superiority. The analysis also helps in identifying scenarios where the proposed approach excels, such as during rapid weather changes or extreme variations in solar irradiance.

As a result of this comparison, the advantages of the multiscale LSTM-based method to extremely short-term PV power generation forecasts are made clear. These benefits extend to various aspects of smart city energy management, from grid stability and demand response optimization to efficient energy distribution and a reduced carbon footprint. The evaluation and performance analysis of the multiscale LSTM-based approach contribute a data-driven validation of its effectiveness. By utilizing rigorous dataset splitting, conducting comprehensive comparative analyses, employing quantitative metrics, and interpreting results, this section offers a comprehensive understanding of how the proposed approach transforms the landscape of PV power generation forecasting in the context of smart city energy management.

18.4 Benefits and Implications for Smart City Energy Management

Short-term forecasting of PV power generation utilizing a multiscale LSTM-based deep learning approach has significant implications for the growth of smart city energy management. The following key points highlight these benefits:

Grid Integration and Stability Enhancement: Integrating solar energy into the grid is facilitated by accurate predictions of PV power generation.

By predicting fluctuations in advance, the grid can be prepared to balance supply and demand, reducing the risk of voltage fluctuations and instability.
Supporting Demand Response and Load Optimization: The ability to anticipate PV power generation variations empowers utilities and consumers to strategically manage electricity demand. This facilitates the implementation of demand response strategies, where energy consumption can be adjusted based on forecasted solar generation, resulting in cost savings and reduced strain on the grid.
Facilitating Effective Energy Storage and Distribution: Accurate forecasts enhance the efficiency of energy storage systems. By predicting when excess solar energy will be generated, energy storage systems can be charged optimally, ensuring that stored energy is available during periods of high demand or low sunlight.
Reducing Dependence on Non-Renewable Energy Sources: Improved forecasting mitigates the need for immediate activation of backup non-renewable energy sources during cloudy periods or unexpected changes in solar irradiance. This reduction in reliance on conventional sources contributes to cleaner energy production and reduced greenhouse gas emissions.
Contributing to Sustainable and Resilient Urban Energy Systems: The integration of accurate PV power generation forecasting strengthens the foundation of smart city energy management. The alignment of renewable

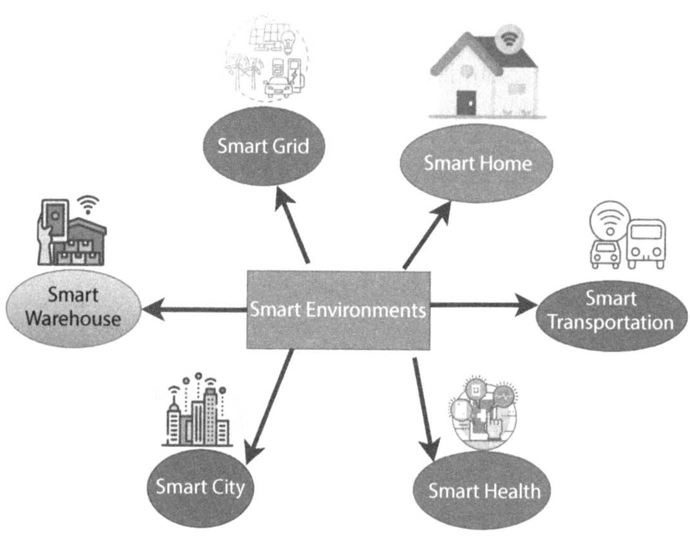

Figure 18.3 Various applications for smart energy systems.

energy sources, demand response mechanisms, and advanced grid infrastructure fosters energy systems that are both environmentally conscious and resilient in the face of energy supply variability.

In Figure 18.3 represents the various applications for smart environment.

18.4.1 Challenges and Future Directions

Despite the fact that multiscale LSTM-based deep learning approach shows significant promise for very-short-term PV power generation forecasting in smart city energy management, it is important to investigate a number of hurdles and potential areas for further improvement.

Addressing Data Availability and Quality: Quality and availability of data are crucial to the success of any forecasting model. Data gaps, discrepancies, and inaccurate information may provide problems. The accuracy and dependability of the multiscale LSTM-based method can be improved by ensuring a thorough data collection process and adopting quality control measures.
Enhancing Model Interpretability: The complexity of deep learning models, including LSTMs, can hinder their interpretability. As stakeholders increasingly demand transparency in decision-making processes, efforts to develop methods for explaining the rationale behind model predictions are crucial. To further understand how the model makes decisions, methods like attention mechanisms and feature visualization can be used.
Scalability for Large-Scale Urban Energy Systems: The scalability of the multiscale LSTM-based approach is paramount, especially as smart cities grow and encompass diverse energy systems. Careful consideration of computational resources and model architecture is required to ensure the model can handle the growing volume of data from diverse sources without compromising its ability to produce reliable forecasts.
Emerging Research Frontiers: The field of deep learning is dynamic, with continuous advancements. Exploring emerging techniques like transformer architectures, graph neural networks, and hybrid models that combine multiple forecasting methodologies can potentially enhance the multiscale LSTM-based approach. Incorporating such innovations can lead to more accurate and adaptable predictions.
Integration of Multi-Domain Data: Future developments could involve incorporating data from various domains, such as urban infrastructure, social dynamics, and economic indicators. A more complete understanding of the elements affecting PV power generation and improved forecasts could be achieved by including these varied datasets into the forecasting model.

Real-Time Adaptability: Urban energy systems are subject to rapid changes and unexpected events. Enhancing the multiscale LSTM-based approach's real-time adaptability is essential to ensure that it can quickly respond to sudden fluctuations, weather anomalies, or unforeseen energy demand patterns, thus maintaining accurate forecasts.
Collaboration and Standardization: Promoting cooperation between academics, business leaders, and politicians is crucial for the widespread use of cutting-edge forecasting methods. Establishing standardized evaluation protocols, benchmark datasets, and best practices can facilitate cross-comparisons of different forecasting models and foster innovation in the field.

As smart city technologies evolve, ethical and societal considerations must be addressed. These considerations span data privacy, equitable access to energy resources, and ensuring that advanced forecasting methods do not inadvertently exacerbate existing urban inequalities. In navigating these challenges and capitalizing on the potential future directions, the multiscale LSTM-based deep learning approach can contribute significantly to the evolution of smart city energy management. By addressing these obstacles and leveraging emerging research opportunities, the approach can evolve into a robust, adaptable, and indispensable tool for realizing sustainable and resilient urban energy systems.

18.5 Conclusion

In the pursuit of enhanced smart city energy management, the rising adoption of renewable energy, especially PV systems, mandates precise and very-short-term power generation forecasts. This chapter delves into a groundbreaking solution: An LSTM-based multi-scale deep learning approach. Amidst the complexities posed by PV's intermittent nature, deep learning's prowess to capture temporal dependencies stands out. Leveraging LSTM networks and multiscale decomposition, the proposed approach masters both short-term patterns and long-term trends. The promising results showcased in comparative analyses against traditional methods demonstrate its superiority. As this methodology spearheads accurate forecasting, it fuels efficient grid integration, demand response, and sustainable energy strategies, charting a course toward resilient smart cities. While challenges persist, the potential of multiscale LSTM-based deep learning in shaping a greener urban future remains resolute.

References

1. Char, D. S., Shah, N. H., & Magnus, D. (2018). Implementing machine learning in health care—addressing ethical challenges. *N. Engl. J. Med.*, 378(11), 981-983.
2. Paszke, A., Gross, S., Massa, F., Lerer, A., Bradbury, J., Chanan, G., ... & Chintala, S. (2019). PyTorch: An Imperative Style, High-Performance Deep Learning Library. In *Adv. Neural Inf. Process. Syst.* (Vol. 32).
3. Shao, Z., Gao, J., Hu, Z., Fan, S., & Yu, X. (2020). A Comprehensive Review of Photovoltaic Power Short-Term Forecasting Methods. *Renewable Sustainable Energy Rev.*, 134, 110347.
4. Zhang, H., Li, W., & Hong, T. (2020). A Review of Short-Term Load Forecasting Methods in the Era of Smart Grids. *Energies*, 13(8), 2014.
5. Brown, D. G., Gaddis, E. J., & Rivers, L. (2015). Evolution of the Smart City: A Systems Perspective. *Environ. Innovation Societal Transitions*, 17, 23-40.
6. Chen, T., & Guestrin, C. (2016). XGBoost: A Scalable Tree Boosting System. In Proceedings of the 22nd ACM SIGKDD International Conference on Knowledge Discovery and Data Mining (pp. 785-794).
7. Hochreiter, S., & Schmidhuber, J. (1997). Long Short-Term Memory. *Neural Comput.*, 9(8), 1735-1780.
8. Hodge, B. M., & Austin, P. (2004). A survey of outlier detection methodologies. *Artif. Intell. Rev.*, 22(2), 85-126.
9. Lipton, Z. C., Kale, D. C., & Elkan, C. (2015). Learning to Diagnose with LSTM Recurrent Neural Networks. *arXiv preprint arXiv*:1511.03677.
10. Miranskyy, A., Zhou, X. S., & Hossain, M. J. (2012). A Novel ARIMA Model for Short-Term Forecasting of Photovoltaic Array Power. *IEEE Trans. Sustainable Energy*, 3(4), 834-844.
11. Srivastava, N., Hinton, G., Krizhevsky, A., Sutskever, I., & Salakhutdinov, R. (2014). Dropout: A Simple Way to Prevent Neural Networks from Overfitting. *J. Mach. Learn. Res.*, 15(1), 1929-1958.

19

Artificial Intelligence Applied to the Management and Operation of Solar Systems

Neel Ghoshal[1] and B.K. Tripathy[2]*

[1]School of Computer Science and Engineering, VIT, Vellore, TN, India
[2]School of Information Technology and Engineering, VIT, Vellore, TN, India

Abstract

Solar systems have come into limelight due to the current trends in acquiring clean and sustainable energy sources and have become the crux of many discussions pertaining to energy. Recently, the development of integrated systems has emerged as a revolutionary approach toward optimization and enhancement of solar systems structurally and, productionally. Artificial intelligence (AI) has entered deep into such systems in the aspects of management and operations. In this chapter, it is aimed to cover the nuanced and intricate details covering the critical aspects of AI with respect to the governance and performance of integrated solar systems. The topics covered here provide a comprehensive exploration of case studies and real-world examples, aimed at explaining the tangible advantages and efficiencies achieved through AI-driven optimization. Use of AI has facilitated the early detection of potential faults and inefficiencies in solar panels, invertors etc., which has consequently allowed for the optimal performance and minimum downtime. Modern solar systems are embedded with Integrated monitoring systems in order to use AI for analysis of data. Functional components within the solar systems like anomaly detection and malfunctions are to be presented. Adaptive control strategies for adjustment of parameters for dynamic systems that allow and ensure a stable and reliable solar energy source while varying unstable environmental patterns are to be analyzed. It may be noted that AI is advantageous in providing grid integration and energy distribution optimizations. Energy and demand management techniques, basically relating to the application of AI techniques in developing and sustaining solar energy demand fluctuations, framework enhancements and

**Corresponding author*: tripathybk@vit.ac.in

O.V. Gnana Swathika, K. Karthikeyan, Milind Shrinivas Dangate and Nicoletta Ravasio (eds.)
Biomass and Solar-Powered Sustainable Digital Cities, (317–338) © 2024 Scrivener Publishing LLC

conservation requirements of solar energy from the perspective of enhancement of reliability and longevity of abstract frameworks and physical necessities will be looked upon. Simultaneously, we shall shed light on the basis for strategic operations for solar systems which includes the nuances of increasing productivity in relation to performance and cost factors. Finally, some challenges faced in these systems are presented, which culminates in getting insights for future scopes and possible innovations in this field.

Keywords: Machine learning, solar systems, renewable energy, AI, management

19.1 Introduction

In the 21st century, there has been a significant gain of importance concerning awareness factors pertaining to environmental pollution and drastic fossil-fuel depletion [1]. Due to these causes and factors, there has been an inevitable rise of research and interest toward the domains of renewable and sustainable energy sources along with their mediums. These newer and recent domains have generated a simultaneous need of creation of technology and functionalities that correspond with the capture, transportation, conversion, and storage of energy from their respective sources. The use, generation, accumulation, and transportation of solar power currently are done under two different methodologies, which are photovoltaic (PV) solar power and concentrated solar power (CSP). The former refers to the procedures involving the use of semiconductor materials to make PV cells, which, in turn, absorb sunlight and generate and electric current [2]. The latter refers to a utility-scale solution in an energy driven demand structure allowing for dispatchability, cost-effectiveness, and versatility, following through with a set of procedures for energy conversions [3]. Figures 19.1 and 19.2 provide a visual depiction of both these terminologies.

There is a plethora of corresponding requirements that are efficiently met by AI technologies with regard to solar systems; a few of these requirements would be algorithm development for solar panel placement needs

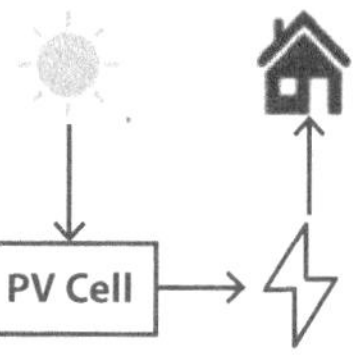

Figure 19.1 PV cell energy framework.

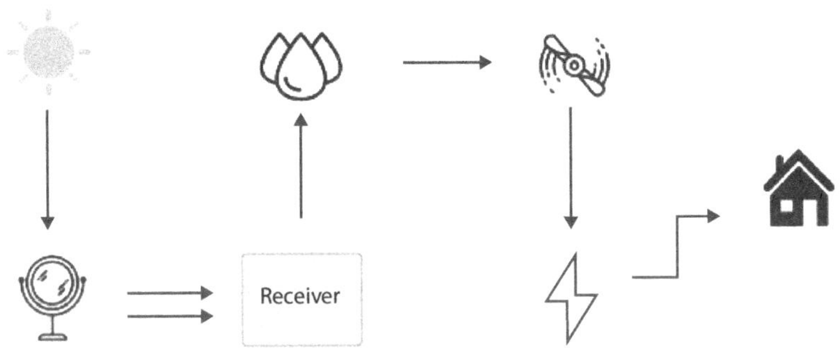

Figure 19.2 CSP energy framework.

that is identifying the best location for solar panels according to given satellite and weather data for maximizing solar irradiance; there also is the use of AI in energy production forecasting, which, in essence, allows for the prediction of the amount of energy produced via a solar system; detection of future problems and diagnosis is another issue dealt with by AI; and other corresponding issues would be maintenance scheduling, battery storage management, and residential demand response needs of solar systems. All of these metrics and functionalities and others will be discussed in detail in the upcoming subsections.

19.2 Background

There has been a significant amount of work in relation to building and researching AI systems for solar systems and renewable energy. For instance, there have been efforts put into the design needs of PV systems, where AI is used to create optimal sizing and efficient designs for the same [4]. These corresponding efforts have aimed at a multitude of different needs concerning renewable energy sources including their design, production, and functioning. For instance, Justin *et al.* have used deep learning in order to optimize solar-operated charging stations for transport passenger vehicles [5]. Work has also been carried out concerning demand-side management for smart grids driven by renewable energy [6]. Advancements toward smart cities have also been carried out via using deep learning utilization involving the creation of predictive models allowing for the prediction of power-plant cycles resulting in the development of an intelligent energy management solution [7]. Work related to energy demand response systems using artificial intelligence (AI) methods has been carried out via Esnaola-Gonzalez *et al.* [8]. Recent demonstrable work has also been carried out with regard to usage of the deep learning platform, namely, TensorFlow, pertaining to the functionality of simulation needs for intelligent energy management in smart cities [9].

As discussed above, solar systems and research efforts surrounding it have become a widespread phenomenon with which other simultaneous and corresponding factors also come into play. This chapter is further formatted and divided into two major subtopics relating to the management and operations of the use of AI used for solar systems, and each of these subtopics are further subdivided into five subtopics each, relating to the specific and nuanced details and functionalities of the practical necessities of these systems.

19.3 AI for Solar System Management

An important factor concerning the use of modern solar-based systems would be the management of such technologies, which is a crucial and essential factor to be considered due to the need of ensuring reliability, availability, maintainability, and safety [10]. Almost every aspect regarding the overall management of solar energy generation has tended to be transformed by the use of AI in solar energy management systems [11]. The specific use-cases of AI in this regard are further elaborated upon in the following sub-sections.

19.3.1 Maintenance

The use of solar systems, which are inherently composed of panels, associated mechanical systems, and other components, has proven to be integral in the current scenario of the global energy landscape. It is widely essential to ensure optimum capacities when dealing with the issue of maintenance scheduling. Conventionally, there has been reliance on pre-determined schedules, impromptu reactive responses to system failures, each of which contribute to the increase of financial and manpower related inefficiencies, increased downtime, and calibration costs. There are many steps that are taken by individuals and organizations when scheduling or reactively planning the maintenance of solar systems, including general checks like waterlogging, wire checks, and infestation checks, and other specific checks such as component-related checks including modules, inverters, cabling connectors, transformers, external element protection, and remote monitoring requirements [12]. All of these subcategories require specific and detailed maintenance-based tasks, which require a plethora of time, effort, money, personnel labor, and constant monitoring. Historically, everything from leaks, to preventative maintenance, to pump maintenance, control unit checks, and module maintenance is done periodically and meticulously, concerning itself with a wide array of checklists, routines, procedures, and other requirements [13]. Figure 19.3 depicts a visual list of substructures, which generally require maintenance in solar systems.

Artificial intelligence possesses the ability to create a paradigm shift with regard to solar system maintenance. Its ability to analyze vast datasets and identify patterns have the potential to do a vast majority of tasks previously assigned and expected to be done manually and via the use of personnel. Using machine learning algorithms and predictive analysis, AI can predict potential system failures, performance degradation, and other maintenance needs with remarkable accuracies [14]. Typically, AI-powered predictive models possess the ability to localize abnormalities in modules, panels, and other respective substructures via the continuous monitoring of real-time data and its comparison to established performance model depictions [15]. Figure 19.4 is a visual portrayal of the common reasons why faults occur in solar systems.

Tripathy *et al.* have demonstrated the use of Fuzzy C Means for classification [16] whereas there also has been advancements using fuzzy based clustering algorithm to basically analyze the various faults and anomalies present in PV arrays by cluster grouping [17]. Work has also been carried out using artificial neural network (ANN)-based [18] frameworks for the purpose of detection and classification of faults in the corresponding PV systems [19]. Quintero *et al.* have successfully implemented two separate

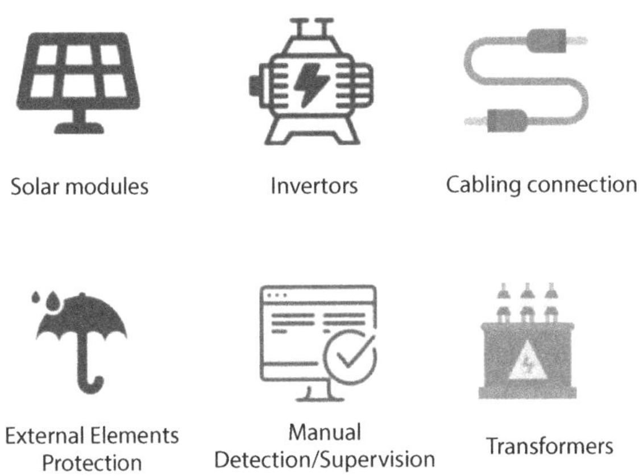

Figure 19.3 Generalized domains requiring maintenance routines in solar systems.

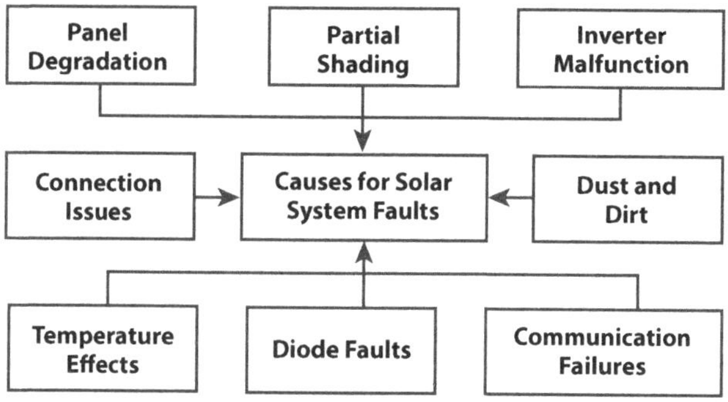

Figure 19.4 Segmented depiction of common causes concerned with faults and system abnormalities in solar systems.

models toward this endeavor, which are based on neural networks, a binary classifier for fault occurrence classification and a multiclass classifier to detect the type of the specific fault [20]. Work has also been carried out toward the creation of visual fault detection methodologies, in which unmanned aerial vehicles were used for the purpose of data acquisition, after which CNNs (convolutional neural networks) [21] were used in unison with decision tree algorithms for providing visual classification results primarily focused on features such as burn marks, glass breakage, discoloration, and snail trails, as pertaining to the category of maintenance faults [22]. Consequently, Panda *et al.* have used soft granular computing based classification using KNNs and SVMs as a core mechanism [23]. There has also been advancements in performance improvements using fault identification in solar photovoltaic systems [24]. For the task of classification of faults and the corresponding detection of them, CNNs have been used to classify [25] the presence of faults, and a simultaneous neuro-fuzzy inference-system has also been proposed to detect faults according to the signals provided via sensors [26]. Tanimura *et al.* have developed an AI-based diagnostic system, in which they follow a particular framework, that is, they use string data collected by the concerned solar system PLC system, then run diagnostic tests on this string data, and finally apply algorithmic methods to analyze certain criterion such as power generation output-based detections, short circuit fault detection, etc. [27]. Appiah *et al.* have worked on the basis of four major faults as their groundwork, including line-line, ground, arc, and hotspot, according to which they use statistical, ML-based (MLBTs), signal, and comparison-based systems [28].

19.3.2 Market-Based Strategy Optimization

The integrated requirement and the need for a market-based forefront for energy markets have always prevailed as a crucial step toward an efficient economy. Artificial intelligence in this kind of economy can be considered as a key enabler of a complex, new, and data-related energy industry, providing a key magic tool to increase operational performance and efficiency in an increasingly cut-throat environment [29]. Due to its vast prevalence in modern economies, there exists a vast repository of databases containing information pertaining to many aspects of this domain including supply and demand trends, real-time variation patterns, forecast-related data, and many more. All of these factors enable AI systems to come into play and provide necessary inferences and predictions, which may prove pivotal to create a robust and coherent modern solar power market. In Figure 19.5, a visual conceptual view is provided concerning the factors associated with the solar energy market.

Gao *et al.* have worked toward utilizing AI techniques, particularly fog-based AI along its implementation on blockchain technology, essentially creating a novel solar power trading system, namely, "FogChain," whose main ideology revolves around peer-to-peer trading of solar power along with factors such as transparency, security, and trust [30]. Work has also been carried out concerning the randomized and unintended loss of energy particularly via "Energy Theft," referring to electricity or power theft carried out without proper metering or payment generally done via tampering with the concerned systems. Pereira *et al.* have worked toward

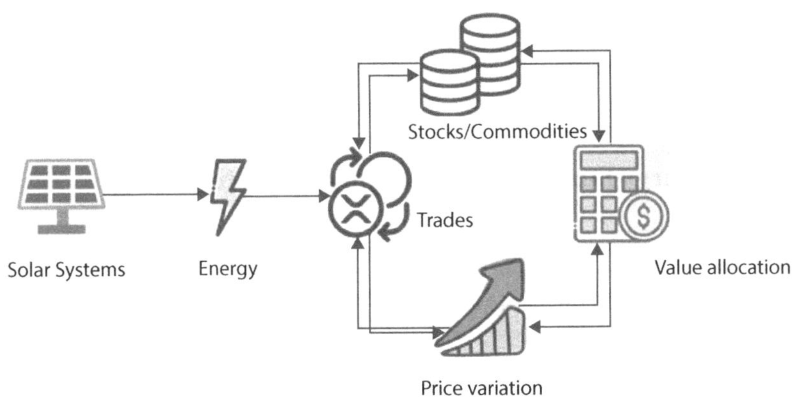

Figure 19.5 Conceptual view of factors related to the energy market environment.

developing a social spider optimization–based SVM-based application for energy-based detection, which is employed to identify instances of energy theft within a given dataset [31]. Research has also been done toward the development of vulnerability assessment systems for protection against cyber-physical attacks for electricity distribution systems, using a particular method, namely, "Stochastic Counterfactual Risk Analysis," which is a method that quantifies the risk associated with such attacks by considering counter factual scenarios [32]. Wang *et al.* have researched and explored the possibilities concerning artificial societies as a means to achieve integrated and sustainable development within metropolitan systems, wherein the approach aims to inform urban planning, policy-making, and decision-making processes to achieve sustainable and harmonious development within metropolitan environments [33]. There has been much work done in the domain of researching different economic aspects of electricity-based power markets. Newly developed solar power markets can easily develop functionalities and aspects of its own system which resemble these already researched and proven ecosystems while allowing for an expedited but bolstered framework along with the simultaneous development of newer novel ideas. Work in this regard has been carried out by Wang *et al.*, for instance, who have developed a novel model, namely, "Dynamic Choice Artificial Neural Network," which they have employed for developing an electricity price forecasting system, which essentially allows them to approach this problem through a hybrid solution via combining the use cases of ANNs and dynamic choice modeling techniques, finally enabling more informed decision-making in energy markets [34]. Similar work has been also done on developing an electricity price forecasting model, which is online-based and robust; essentially, the researchers have incorporated outlier detection and handling mechanisms, through which the hybrid model aims to provide more reliable and robust electricity price forecasts, which are crucial for effective energy market management [35].

19.3.3 Energy Storage Management and Optimization

Artificial intelligence methods have proven to be a much-needed stepping stone toward the overall creation of optimized solar systems. The comprehensive nature of the factors related to the increase in efficiency of a solar energy system is cohesively supplemented by energy management methodologies. This subsection will primarily deal with the AI-driven management techniques employed in increasing the adaptabilities, capabilities, and dexterity of the concerned mechanisms.

One of the primary storage mechanisms employed by solar energy systems is batteries. The generated solar energy, once converted into appropriate energy modalities, is then stored in these batteries for later usage needs. Although the use of these storage mechanisms is widespread, advantageous, and track-proven, there are a number of issues that must be dealt with while employing these systems, for example, discharge issues, leakages, and temperature-based issues. In this specific subdomain, AI has demonstrated revolutionary techniques for the management of these batteries, including abilities ranging from optimization of battery charging, discharging, and maintenance, along with other fulfilled requirements such as monitoring of battery health, prevention of overcharging, and extension of battery life.

Bendary *et al.* have implemented different AI algorithms for optimal selection of parameters concerning battery charger controllers employed in the solar energy system field [36]. Work has also been carried out toward the creation of strategic applications for the utilization of renewable energy, that is, by employing fuzzy logic, researchers have been able to enhance efficiency of energy utilization from solar energy sources [37]. Ismail *et al.* have developed software with the ability to estimate the state of charge (SoC) for rechargeable batteries using ANN techniques using input data like voltage and temperature. This enhances battery performance and utilization for efficient energy storage [38]. Research has also gone into the use of fuzzy logic and other machine learning techniques into developing an intelligent battery charge controller tailored for PV panels, which aims to create a sophisticated charge controller utilizing intelligent algorithms to manage battery charging in PV systems, basically by incorporating smart strategies; the controller optimizes battery charging based on real-time solar generation, battery state, and load demand; it contributes to efficient energy storage and improved PV system performance [39]. A hybrid fuzzy–proportional-integral-derivative (PID) controller framework has also been experimented on aimed toward combining fuzzy logic control with PID control techniques to optimize the operation of the converter within solar energy-battery systems, allowing for the adjustment of the converter's duty cycle based on inputs like solar irradiance and battery state to efficiently regulate voltage levels and manage energy flow between the solar panels and batteries [40]. Similar work has also been done toward the development of an advanced battery management system including optimizing SoC, prioritizing load demand, utilizing surplus renewable energy for charging, regulating discharge rates, and maintaining optimal voltage levels; the framework enables the controller adapts to energy fluctuations, integrates system components, and employs adaptive learning for improved performance; and, overall, it enhances energy efficiency, system stability, and battery longevity within the hybrid setup [41]. Rubavathy *et al.* have introduced a machine learning strategy for optimizing solar energy utilization within distributed systems, essentially by leveraging machine learning techniques; the strategy optimizes energy generation, distribution, and consumption across distributed solar energy sources allowing for proper optimization of storage and distribution in solar energy-based systems [42]. Other than these utilizations, ANNs are also currently utilized toward energy prediction, allowing for avoidance of over or under charging, load imbalances, and, ultimately, battery health preservation [43].

19.3.4 Remote Monitoring and Control

Currently, concerning the management of solar systems, there exists enormous amounts of tasks that either require manual and laborious supervision or the need of physical interference of corresponding personnel. The main problem with this methodology is the unnecessary use of physical intervention and the need of hiring specialized personnel for a vast spectrum of related work. All of these factors correspondingly point toward a counterproductive and outdated framework. Here arises the need of automation toward the tasks of remote monitoring and control, in which AI systems have proven to produce valuable and essential results conceding efficiency, robustness, and overall increase in managerial standards. Figure 19.6 is a depiction of the transitory phase from manual to autonomous labor in solar system monitoring and control.

Recently, there have been many advances in the subdomain of AI with respect to remote monitoring and control of solar systems. For instance, work has been carried out toward a novel approach utilizing deep reinforcement learning for voltage control in hybrid pumped storage

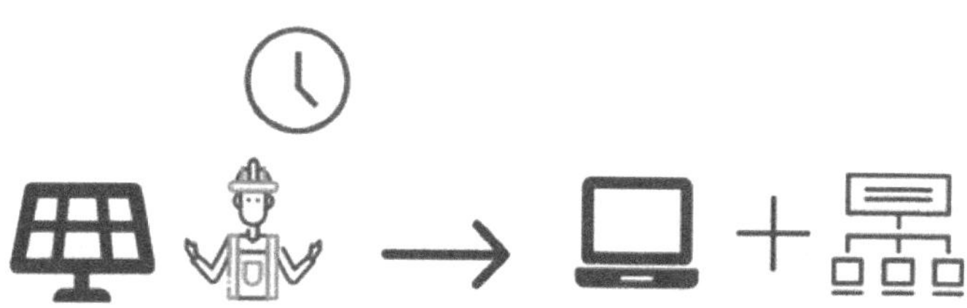

Figure 19.6 Transition from manual investigations/schedules to automated AI-based environments.

hydro-wind-solar systems, which employs an agent that learns and adapts voltage control strategies through reinforcement learning, ultimately by optimizing voltage levels in the context of dynamic renewable energy inputs, the method contributes to efficient system operation and stability within complex hybrid energy systems [44]. Yerramsetti *et al.* have implemented an autonomous robot for a floating solar panel cleaning system; utilizing AI techniques, their system employs AI to enable the robot to efficiently navigate, detect dirt and debris, and clean solar panels on floating platforms [45]. Rosyadah *et al.* have developed a smart optimization of PV panel output using a solar tracker controlled by a fuzzy logic controller (FLC) that focuses on enhancing the energy yield of PV panels by intelligently adjusting their orientation based on real-time solar tracking enabling dynamic tracking decisions by considering factors like solar position, intensity, and weather conditions [46]. Kumar *et al.* concisely explain in their articles about the technologies that empower drones for intelligent monitoring of solar PV power plants and how drones equipped with advanced sensors, cameras, and data analytics capabilities are employed to efficiently inspect and monitor the performance of PV panels and associated infrastructure basically highlighting how these technologies enable real-time data collection, fault detection, performance analysis, and maintenance planning, contributing to enhanced operational efficiency and optimized energy output in solar PV power plants [47]. Energy extraction is also targeted by researchers for proper optimization of storage systems, researching into hybrid AI-based techniques for maximum power point tracking (MPPT) in solar energy systems [48].

19.4 AI for Solar System Operation

One major aspect that takes vast importance in the overall life-cycle of a solar system is its operation and measures taken to allow and efficient and smooth functioning of the concerned system. Hence, it becomes pivotal to adopt techniques which allow for leveraging every available aspect and consideration that goes into the working of these systems, ultimately resulting in a system that utilizes all its resources in a coherent and lucrative manner. The use of AI methods as tools for the use of solar system operation optimization has seen an exponential amount of growth, in terms of interest toward the topic and the recent increase in the amount of research being done toward this domain [49].

19.4.1 Solar Energy Optimization

During and after the initial PV steps of the conversion of solar energy into electrical or usable energy, there arises a need for a mechanism to allow for optimization of the energy thus generated. This step simultaneously and pragmatically allows for a better, more cohesive and highly efficient solar system. When carrying out these essential functionalities, there is a prevalence of large amounts of data pertaining to various aspects and facets of the system. The availability of the given amounts of data allows the use of AI and machine learning algorithms to further optimize and quantify the necessary steps and functionalities to be encoded and taken into consideration while operating the corresponding solar system. Maximum power point tracking (MPPT) is one of the crucial mechanisms that is taken into consideration when developing modern solar systems. It basically refers to the ensuring of optimal PV panel performance via adjusting panel parameters such as voltage and current where the highest cohesive amount of energy output is generated. This dynamic adjustment enables the panels to extract the most energy from changing environmental conditions, increasing overall energy efficiency and maximizing electricity production from solar sources. MPPT and its integrated products with AI systems play a vital role in enhancing the efficiency and effectiveness of solar power systems. Figure 19.7 provides a visual understanding of how the MPPT algorithm works within the overall solar system framework.

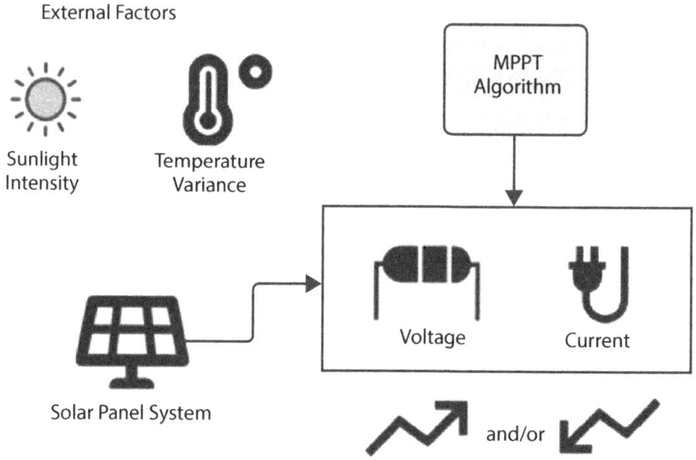

Figure 19.7 Conceptual description of MPPT algorithms in practice.

Zahraee *et al.* have focused on the application of AI methods for hybrid energy system optimization; AI is employed to intelligently manage and enhance the performance of hybrid energy systems that combine different energy sources and technologies including a focus toward optimal energy generation essentially improving the overall sustainability and effectiveness of the system [50]. Ajiboye *et al.* concisely cover many AI techniques that aim to achieve optimal system performance and energy generation [51]. Work has also been carried out toward the goal of integrating machine learning methods to renewable power generation, Sharifzadeh *et al.* use ANNs, support vector regression, and Gaussian process–based algorithms to test and compare effectiveness metrics toward this endeavor [52].

19.4.2 Demand Response

An important and crucial factor to consider when a solar energy system has been deployed is the need of the intrinsic ability of the system to respond appropriately to fluctuations in energy demand, grid conditions, and other related requirements. There needs to be a proper and streamlined strategy built in the concerned system to be able to handle these spikes, falls, and variations to allow for a stable environment. Overall, currently, research has been done toward the problem of having a specialized framework for the task of supplying necessary units of energy appropriately at various periods of time, which is done via effectively incorporating the use of machine learning and statistical algorithms.

The use of AI techniques in the subdomain of solar system demand response has been historically and recently well researched into. Azunatalam *et al.*, for instance, employ the use of reinforcement learning for enhancing HVAC-based energetically control and demand response in entire buildings, concepts of similar nature can be configured and optimized toward the aim of solar generated energy [53]. Pallonetto *et al.*, in their paper, focuses on using machine learning models to forecast electricity demand in commercial buildings, specifically to support the implementation of demand response programs, particularly utilizing historical data, weather conditions, and other relevant factors, machine learning algorithms create predictive models that can anticipate future electricity consumption patterns, which is an essential parameter to be considered when optimizing solar systems from an operational point of view [54]. Similar work has been carried out, taking into consideration the use of simulations for refinement and testing purposes while aiming toward developing layout-based reinforcement learning techniques for demand response-related algorithms [55]. Mohammad *et al.* have worked toward exploring integration of AI techniques into demand response strategies, primarily discussing how AI methods are employed to enhance the effectiveness of demand response programs by optimizing energy consumption patterns in response to changing grid conditions, pricing signals, and supply-demand dynamics [56].

19.4.3 Solar Power Forecasting

The term "solar power forecasting" generally refers to the use of computational and logical methods in order to ascertain and estimate standardized values pertaining to the amount of solar energy that is expected to input or reach a solar power harvesting system. The use of AI methods provide several modalities to work with the computational data related to various subdomains including the weather, environmental changes, previous trends, and location. These vast amounts of data are used to create predictive forecasting models that ultimately assist in providing efficient solar systems. Figure 19.8 provides the various categories of features that are taken into consideration when developing a predictive model for solar power forecasting needs.

Wan *et al.*, in their paper, highlight the importance of solar power forecasting within smart grid energy management and explore various forecasting techniques, discuss challenges posed by factors like weather uncertainty, and showcase applications in optimizing energy scheduling, grid stability, and renewable integration [57]. Chen *et al.* delve into the creation of an advanced statistical method for solar power forecasting using a self-organized map that is trained to classify the local weather type

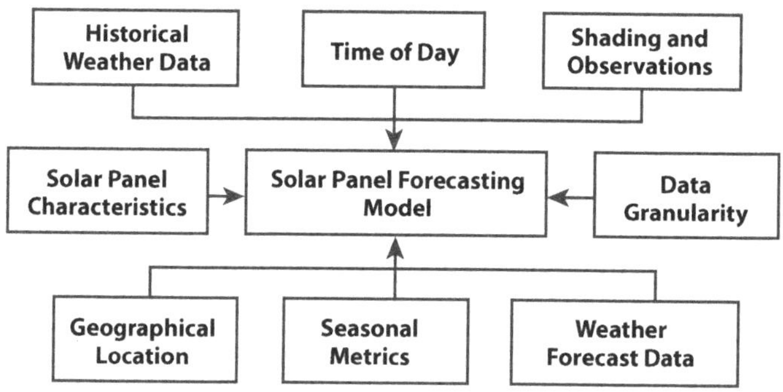

Figure 19.8 Features taken into consideration when developing a solar power forecasting model.

of 24 h ahead provided by the online meteorological services as its basis [58]. Pattanaik *et al.* have developed a novel approach to solar power forecasting by combining genetic algorithms and ANNs, basically using this hybrid technique allows for the optimization of neural network parameters through genetic algorithms to enhance accuracy in capturing complex solar energy patterns [59]. Liu *et al.* have introduced an enhanced PV power forecasting model that incorporates aerosol index data to improve accuracy, which, by integrating historical solar radiation data, weather parameters, and aerosol index information, the model aims to better predict PV power generation, accounting for atmospheric conditions [60].

19.4.4 Grid Integration

Grid integration in the context of solar energy refers to the process of a seamless incorporation of solar energy systems into existing electrical grid infrastructure. Abstractly, it involves the connection of solar power generative sources, like solar panels, to the broader electricity grid in a way that allows for an efficient, reliable, and safe energy exchange between the two architectures. The technologies surrounding the domain of solar grid integration has the involvement of advanced developments concerning advanced inverters, anti-islanding, grid-plant protection, solar-grid forecasting, and smart grids [61]. The advent and rise of AI has also seen various applications in the solar grid integration domain. The inherent features of AI including automation, intelligent identification, management, and monitoring have allowed for its use as an assistive technology in providing these systems with rational techniques to handle situational problems [62]. Figure 19.9 is a visual depiction of the need and use of algorithmic substructures in solar power grid integration.

Boretti, in his research, identifies and facilitates AI-based solutions to the issues surrounding solar power grid integration, ultimately allowing for a sustainable, synergy-oriented, and stable grid-based environment within

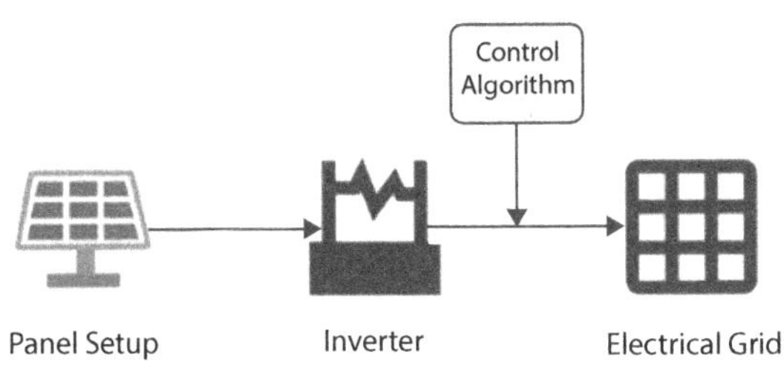

Figure 19.9 General overview of the solar grid integration procedure.

their corresponding city [63]. Work has also been carried out toward the development of robust energy forecasting–based solutions for solving the solar energy power and grid integration problem, focusing on deep learning models and autoencoders [64]. The uses of fuzzy algorithms are also used in the practice of tailored integration of renewable energy and optimizing its flow [65]. Boza *et al.*, in their paper, demonstrate the facilitation of renewable energy sources for a flexible grid integration mechanism covering forecasting, optimization, demand response, grid flexibility, energy storage, and the broader goal of achieving a sustainable energy transition [66]. Yap *et al.* focus on exploring the application of machine learning for synthesizing virtual inertia in a synchronverter, addressing challenges in grid integration of solar PV systems, essentially aiming to enhance grid stability, particularly by mitigating solar power fluctuations and contributing to efficient solar energy integration [67].

19.5 Challenges and Future Scope

Some inherent characteristic faults that may lie in solar power–based electricity include intermittency, uncertainty, and non-synchronous power generation, which may lead to some technical challenges to large-scale power grid integration [68]. Other common issues would be data quality and availability, due to limited data sources. Developing scalable predictive models that are viable and integration friendly is another important factor to consider. Apart from these, there arise challenges concerning model interpretability, which is allowing stakeholders and personnel to understand the reasoning behind decisions made by the model. Regulations, policy frameworks, adoption, and acceptance are a few more socio-economic challenges that may hinder the development of AI systems in the management and operation of solar power systems. Other significant challenges may include current deployment and infrastructural issues, along with hindrances concerning computational needs and awareness.

There is a plethora of subdomains related to solar system management and operation related endeavors, in which research work is viable and yet to be properly explored and executed. For instance, there lies a gap in the development of hybrid incorporated models for advanced forecasting techniques, along with the use of un-streamlined data. Autonomous AI systems are yet to be implemented on a wide scale in solar power systems, which would increase reliability and reduce downtime. Another worthwhile topic to be explored is the integration of edge computing and AI for decentralized decision-making and control in solar systems, which would improve the efficiency and reduce latency. Resilience support, including safety against cyber-threats and attacks is yet to be properly explored. All of these scenarios and pointers are consolidated viewpoints into exploring the challenges and future scopes of the use of AI in the management and operation of solar systems, and, simultaneously, as time goes by, many other newer problem statements and innovative solutions will arise.

References

1. Wo Qazi, Atika, *et al.* "The artificial neural network for solar radiation prediction and designing solar systems: a systematic literature review." *J. Clean. Prod.* 104 (2015): 1-12. World Health Organization. The Global Burden of Disease: 2004 Update. Geneva: World Health Organization; 2008.
2. Fahrenbruch, Alan, and Richard Bube. *Fundamentals of solar cells: photovoltaic solar energy conversion.* Elsevier, 2012.
3. Dunham, Marc T., and Brian D. Iverson. "High-efficiency thermodynamic power cycles for concentrated solar power systems." *Renew. Sustain. Energy Rev.,* 30 (2014): 758-770.
4. Mellit, Adel, *et al.* "Artificial intelligence techniques for sizing photovoltaic systems: A review." *Renew. Sustain. Energy Rev.* 13.2 (2009): 406-419.
5. Justin, S.; Saleh, W.; Lashin, M.M.A.; Albalawi, H.M. Design of Metaheuristic Optimization with Deep-Learning-Assisted Solar-Operated On-Board Smart Charging Station for Mass Transport Passenger Vehicle. *Sustainability* 2023, 15, 7845. https://doi.org/10.3390/su15107845
6. Chaurasia, Kiran, and H. Ravishankar Kamath. "New Approach using Artificial Intelligence-Machine Learning in Demand Side Management of Renewable Energy integrated Smart Grid for Smart City." *Proceedings of the International Conference on Innovative Computing & Communication (ICICC).* 2021.
7. S. Abbas *et al.*, "Modeling, Simulation and Optimization of Power Plant Energy Sustainability for IoT Enabled Smart Cities Empowered With Deep Extreme Learning Machine," in *IEEE Access*, vol. 8, pp. 39982-39997, 2020, doi: 10.1109/ACCESS.2020.2976452.
8. Esnaola-Gonzalez, I.; Jelić, M.; Pujić, D.; Diez, F.J.; Tomašević, N. An AI-Powered System for Residential Demand Response. *Electronics* 2021, 10, 693. https://doi.org/10.3390/electronics10060693
9. Vázquez-Canteli, José R., *et al.* "Fusing TensorFlow with building energy simulation for intelligent energy management in smart cities." *Sustain. Cities Soc.* 45 (2019): 243-257.
10. Gonzalo, Alfredo Peinado, Alberto Pliego Marugán, and Fausto Pedro García Márquez. "A review of the application performances of concentrated solar power systems." *Appl. Energy* 255 (2019): 113893.

11. Mohammad, Ashif, and Farhana Mahjabeen. "Revolutionizing Solar Energy: The Impact of Artificial Intelligence on Photovoltaic Systems." *International Journal of Multidisciplinary Sciences and Arts 2.1* (2023).
12. https://www.cleanmax.com
13. Supply, CCDS Power, and Pastoral Areas. "*Operations & Maintenance Manual*." (1998).
14. B. K. Tripathy, Raju, H. and Kaul, D.: *Comparative Analysis of Pure and Hybrid Machine Learning Algorithms for Risk Prediction of Diabetes Mellitus, Helix: The Scientific Explorer*, vol. 7 (5), (2017), pp. 2029-2033.
15. Mohammad, Ashif, and Farhana Mahjabeen. "Revolutionizing Solar Energy with AI-Driven Enhancements in Photovoltaic Technology." *BULLET: Jurnal Multidisiplin Ilmu* 2.4 (2023): 1031-1041.
16. B.K. Tripathy and P. Swarnalatha: A Novel Fuzzy C-Means Approach with Bit Plane Algorithm for Classification of Medical Images, *IEEE International Conference on Emerging Trends in Computing, Communication and Nanotechnology (ICECCN 2013)*, March (2013), pp.360 – 365.
17. Liu, Shengyang, *et al.* "Photovoltaic array fault diagnosis based on Gaussian kernel fuzzy C-means clustering algorithm." *Sensors* 19.7 (2019): 1520.
18. B.K. Tripathy and J. Anuradha: *Soft Computing- Advances and Applications*, Cengage Learning publishers, New Delhi, (2015). ASIN : 8131526194, ISBN-10 : 9788131526194
19. W. Chine, A. Mellit, V. Lughi, A. Malek, G. Sulligoi, A. Massi Pavan, A novel fault diagnosis technique for photovoltaic systems based on artificial neural networks, *Renew. Energy*, Volume 90, 2016.
20. Ramón Fernando Colmenares-Quintero, Eyberth R. Rojas-Martinez, Fernando Macho-Hernantes, Kim E. Stansfield & Juan Carlos Colmenares-Quintero | (2021) Methodology for automatic fault detection in photovoltaic arrays from artificial neural networks, *Cogent Eng.*, 8:1, 1981520, DOI: 10.1080/23311916.2021.1981520
21. S. Bhattacharyya, V. Snasel, A. E. Hassanian, S. Saha and B. K. Tripathy: *Deep Learning Research with Engineering Applications*, De Gruyter Publications, (2020). ISBN: 3110670909, 9783110670905. DOI: 10.1515/9783110670905
22. Naveen Venkatesh Sridharan & Vaithiyanathan Sugumaran (2021) Visual fault detection in photovoltaic modules using decision tree algorithms with deep learning features, *Energy Sources Part A*, DOI: 10.1080/15567036.2021.2020379
23. M. Panda, A. Abraham and B.K. Tripathy: Soft granular computing based classification using hybrid fuzzy-KNN-SVM, *Intell. Decis. Technol.*, Volume 10, Issue 2, 1 March 2016, pp. 115-128
24. Sakthivel Ganesan, Prince Winston David, Pravin Murugesan & Praveen Kumar Balachandran (2023) Solar Photovoltaic System Performance Improvement Using a New Fault Identification Technique, *Electr. Power Compon. Syst.*, DOI: 10.1080/15325008.2023.2237013

25. Karan Maheswari, Aditya Shaha, Dhruv Arya, B. K. Tripathy and R. Rajkumar: Convolutional Neural Networks: A Bottom-Up Approach, (Ed: S. Bhattacharyya, A. E. Hassanian, S. Saha and B.K. Tripathy, *Deep Learning Research with Engineering* Applications), De Gruyter Publications, (2020), pp.21-50. DOI: 10.1515/9783110670905-002
26. Pa, Mary. Application of Artificial Intelligence in fault detection and classification of solar power plants and prediction of power generation of combined cycled power plants. *Diss.* 2022.
27. Tanimura, Kotaro, *et al.* "AI-based Diagnostic System for Utility-Scale Solar Power Plants." *SEI Tech. Rev.* 91 (2020): 53.
28. Appiah, A.Y.; Zhang, X.; Ayawli, B.B.K.; Kyeremeh, F. Review and performance evaluation of photovoltaic array fault detection and diagnosis techniques. *Int. J. Photoenergy* 2019, 2019.
29. Ahmad, Tanveer, *et al.* "Artificial intelligence in sustainable energy industry: Status Quo, challenges and opportunities." *J. Clean. Prod.* 289 (2021): 125834.
30. Gao, Guanyu, *et al.* "FogChain: A blockchain-based peer-to-peer solar power trading system powered by fog AI." *IEEE Internet Things J.* 9.7 (2021): 5200-5215.
31. Pereira, D.R., Pazoti, M.A., Pereira, L.A.M., Rodrigues, D., Ramos, C.O., Souza, A.N., Papa, J.P., 2016. Social-Spider Optimization-based Support Vector Machines applied for energy theft detection. *Comput. Electr. Eng.* 49, 25–38. https://doi.org/10.1016/j.compeleceng.2015.11.001
32. Oughton, E.J., Ralph, D., Pant, R., Leverett, E., Copic, J., Thacker, S., Dada, R., Ruffle, S., Tuveson, M., Hall, J.W., 2019. Stochastic Counterfactual Risk Analysis for the Vulnerability Assessment of Cyber-Physical Attacks on Electricity Distribution Infrastructure Networks. *Risk Anal.* 39, 2012–2031. https://doi.org/10.1111/risa.13291
33. Wang, F.Y., Tang, S., 2004. Artificial societies for integrated and sustainable development of metropolitan systems. *IEEE Intell. Syst.* 19, 82–87. https://doi.org/10.1109/MIS.2004.22
34. Wang, J., Liu, F., Song, Y., Zhao, J., 2016. A novel model: Dynamic choice artificial neural network (DCANN) for an electricity price forecasting system. *Appl. Soft Comput. J.* 48, 281–297. https://doi.org/10.1016/j.asoc.2016.07.011
35. Wang, J., Yang, W., Du, P., Niu, T., 2020. Outlier-robust hybrid electricity price forecasting model for electricity market management. *J. Clean. Prod.* 249, 119318. https://doi.org/10.1016/j.jclepro.2019.119318
36. Bendary, Ahmed F., and Mohamed M. Ismail. "Battery charge management for hybrid PV/wind/fuel cell with storage battery." *Energy Procedia* 162 (2019): 107-116.
37. R.Karthick, S.Manoharan, "FUZZY BASED OPTIMIZED ENERGY MANAGEMENT STRATEGY OF RENEWABLE ENERGY SOURCES FOR STAND ALONE APPLICATIONS", *J. Electr. Eng.*

38. M. M. Ismail and M. A. M Hassan. "The state of charge estimation for rechargeable batteries based on artificial neural network techniques", *International Conference on Control, Decision and Information Technologies (CoDIT)*. Page(s):733 – 739, May 2013.
39. S. M. Çınar and E. Akarslan, "On the Design of an Intelligent Battery Charge Controller for PV Panels", *J. Eng. Sci. Technol. Rev.* 5 (4) (2012) 30-34.
40. Karime Farhood Hussein, "*Hybrid Fuzzy-PID Controller for Buck-Boost Converter in Solar Energy-Battery Systems*", for the degree of Master of Science in Engineering, Electrical and Computer Engineering Western Michigan University, May 2015
41. M.Venkateshkumar, R.Raghavan, "Hybrid Photovoltaic and Wind Power System with Battery Management System using Fuzzy Logic Controller", *International Journal of Applied Power Engineering (IJAPE)* Vol. 5, No. 2, August 2016, pp. 72~78.
42. S. Jaanaa Rubavathy, Nithiyananthan Kannan, D. Dhanya, Santaji Krishna Shinde, N.B. Soni, Abhishek Madduri, V. Mohanavel, M. Sudhakar, Ravishankar Sathyamurthy, Machine Learning Strategy for Solar Energy optimisation in Distributed systems, *Energy Reports*, Volume 8, Supplement 8, 2022.
43. Mirjalili, M.A., Aslani, A., Zahedi, R. *et al.* A comparative study of machine learning and deep learning methods for energy balance prediction in a hybrid building-renewable energy system. *Sustain. Energy Res.* 10, 8 (2023). https://doi.org/10.1186/s40807-023-00078-9
44. Huang, Qin, *et al.* "A novel deep reinforcement learning enabled agent for pumped storage hydro-wind-solar systems voltage control." *IET Renew. Power Gener.* 15.16 (2021): 3941-3956.
45. Yerramsetti, Jaswanth, Dhatri Sri Paritala, and Ramesh Jayaraman. "Design and implementation of automatic robot for floating solar panel cleaning system using AI technique." *2021 International Conference on Computer Communication and Informatics (ICCCI)*. IEEE, 2021.
46. Rosyadah, Marhani, R. Kusumanto, and Tresna Dewi. "Smart optimization of PV panel output using Fuzzy Logic Controller based solar tracker." *Sinergi* 26.1 (2022): 73-80.
47. Kumar, Nallapaneni Manoj, *et al.* "On the technologies empowering drones for intelligent monitoring of solar photovoltaic power plants." *Procedia Comput. Sci.* 133 (2018): 585-593.
48. Taylor, Adeyemi R. "*Performance Analysis Of Hybrid Ai-Based Technique For Maximum Power Point Tracking In Solar Energy System Applications.*" (2023).
49. Milidonis, Kypros, *et al.* "Review of application of AI techniques to Solar Tower Systems." *Solar Energy* 224 (2021): 500-515.
50. Zahraee, S. M., M. Khalaji Assadi, and R. Saidur. "Application of artificial intelligence methods for hybrid energy system optimization." *Renew. Sust. Energy Rev.*, 66 (2016): 617-630.

51. Ajiboye, Olalekan Kunle, *et al.* "A review of hybrid renewable energies optimisation: design, methodologies, and criteria." *Int. J. Sustain. Energy* 42.1 (2023): 648-684.
52. Sharifzadeh, Mahdi, Alexandra Sikinioti-Lock, and Nilay Shah. "Machine-learning methods for integrated renewable power generation: A comparative study of artificial neural networks, support vector regression, and Gaussian Process Regression." *Renew. Sust. Energy Rev.*, 108 (2019): 513-538.
53. Azuatalam, Donald, *et al.* "Reinforcement learning for whole-building HVAC control and demand response." *Energy and AI* 2 (2020): 100020.
54. Pallonetto, Fabiano, Changhong Jin, and Eleni Mangina. "Forecast electricity demand in commercial building with machine learning models to enable demand response programs." *Energy and AI* 7 (2022): 100121.
55. Liu, Huafeng, *et al.* "An effective energy management Layout-Based reinforcement learning for household demand response in digital twin simulation." *Solar Energy* 258 (2023): 95-105.
56. Mohammad, Ashif, and Farhana Mahjabeen. "Revolutionizing Solar Energy with AI-Driven Enhancements in Photovoltaic Technology." *BULLET: Jurnal Multidisiplin Ilmu* 2.4 (2023): 1031-1041.
57. Wan, Can, *et al.* "Photovoltaic and solar power forecasting for smart grid energy management." *CSEE J. Power Energy Syst.*, 1.4 (2015): 38-46.
58. Chen, Changsong, *et al.* "Online 24-h solar power forecasting based on weather type classification using artificial neural network." *Solar Energy* 85.11 (2011): 2856-2870.
59. Pattanaik, Debasish, *et al.* "An innovative learning approach for solar power forecasting using genetic algorithm and artificial neural network." *Open Eng.*, 10.1 (2020): 630-641.
60. Liu, Jun, *et al.* "An improved photovoltaic power forecasting model with the assistance of aerosol index data." *IEEE Trans. Sustain. Energy* 6.2 (2015): 434-442.
61. Nwaigwe, K. N., Philemon Mutabilwa, and Edward Dintwa. "An overview of solar power (PV systems) integration into electricity grids." *Mater. Sci. Energy Technol.*, 2.3 (2019): 629-633.
62. Liu, Zhengxuan, *et al.* "Artificial intelligence powered large-scale renewable integrations in multi-energy systems for carbon neutrality transition: Challenges and future perspectives." *Energy and AI* (2022): 100195.
63. Boretti, Alberto. "Integration of solar thermal and photovoltaic, wind, and battery energy storage through AI in NEOM city." *Energy and AI* 3 (2021): 100038.
64. Said, Yahia, and Abdulaziz Alanazi. "AI-based solar energy forecasting for smart grid integration." *Neural Comput. Appl.* 35.11 (2023): 8625-8634.
65. Aurangzeb, Muhammad, *et al.* "Single algorithm mpso depend solar and wind mppt control and integrated with fuzzy controller for grid integration." *2019 IEEE 3rd Conference on Energy Internet and Energy System Integration (EI2).* IEEE, 2019.

66. Boza, Pal, and Theodoros Evgeniou. "Artificial intelligence to support the integration of variable renewable energy sources to the power system." *Appl. Energy* 290 (2021): 116754.
67. Yap, Kah Yung, Charles R. Sarimuthu, and Joanne Mun-Yee Lim. "Grid integration of solar photovoltaic system using machine learning-based virtual inertia synthetization in synchronverter." *IEEE Access* 8 (2020): 49961-49976.
68. Mohammadi, Fazel, and Mircea Neagoe. "Emerging issues and challenges with the integration of solar power plants into power systems." *Solar Energy Conversion in Communities: Proceedings of the Conference for Sustainable Energy (CSE)* 2020. Springer International Publishing, 2020.

20

Robust Rearrangement of Interconnected Microgrids to Reduce the Effects of Physical Cyberattacks on Intelligent Distribution Networks

Sanjeevikumar Padmanaban[1], Mostafa Azimi Nasab[1], Omid Halimi Milani[2], Mohammad Zand[1], Mohsen Hatami[3], Morteza Azimi Nasab[1]* and Mohsen Khalili[1]

[1]*Department of Electrical Engineering, IT and Cybernetic, University of South-Eastern Norway, Porsgrunn, Norway*
[2]*Department of Electrical and Computer Engineering, University of Illinois at Chicago, IL, USA*
[3]*Department of Electrical and Computer Engineering, Binghamton University-SUNY, Binghamton, NY, USA*

Abstract
Smart grids have grown more vulnerable to cyber threats and assaults as communication and control technologies have advanced, owing to the combination of comprehensive control and monitoring technologies. Power grids are extremely susceptible due to poor design of power grid telecommunication systems and the usage of insecure IP protocols. These modifications may enable hackers to gain access to hierarchical control systems and exploit weaknesses to take control of the system. Denial of service (DOS) assaults, database alteration attacks, and efforts to introduce fake information into the system are all common types of cyberattacks. The physical domain of the system is prone to sabotage and terrorist operations, while hundreds of cyber assaults on smart grids occur every day across the world. Many of these attacks go unnoticed or are successfully countered. Many of these attacks go unnoticed or are successfully countered. Cyber assaults have gotten more sophisticated and destructive as communication and communications technology have improved. The smart zone's growth has increased the number

**Corresponding author*: Morteza1368morteza@gmail.com

O.V. Gnana Swathika, K. Karthikeyan, Milind Shrinivas Dangate and Nicoletta Ravasio (eds.) Biomass and Solar-Powered Sustainable Digital Cities, (339–362) © 2024 Scrivener Publishing LLC

of cyber assault penetration sites. Three cyberattack tactics for the network were studied in this chapter. Of course, cyber assaults vary. They exclusively launch assaults against the system's benefits by deploying LR and FEDI cyberattacks in cyberattack mode. Untraceable and undetected attacks were carried out. The operator finds a problem with his control system. However, the reason for the failure remains unknown. If the attack was successful, then the attacker had a budget to halt it. The intensity and incidence of these assaults can be reduced by undertaking the essential tests and investments.

Keywords: Microgrids, rearrangement, cyber-physical, assault, two-stage optimization, defense

20.1 Smart Grid

Smart grids (SGs) are one of the world's most recent technologies, the product of specialists' attempts to upgrade distribution networks and usher in the digital era. A SG is defined as a grid that combines two-way and integrated information and communication technologies to provide a safe, clean, flexible, efficient, and highly reliable grid that incorporates all voltage levels from production to consumption. It is entirely up to the subscribers. The primary purpose is to supply reliable electricity while also meeting the demands of the world's first SG, which was launched in March 2002. Balder, Colorado, has been designated as the first city to have a SG for distribution, including subscribers, equipment, and communications. Smart technology has the potential to alter generation, transmission, and distribution [1]. Distribution and usage of power with economic and environmental benefits that eventually lead to consumer satisfaction and the availability of dependable and sustainable electricity. The system may gather data in need to make decisions in crucial situations to prevent unwanted outages.

20.2 The Benefits of Smart Grids

Smart grids are a new type of electricity grid that uses digital technology to improve efficiency, reliability, and sustainability. Here are some of the benefits of SGs:

- Improved efficiency: Smart grids can help to reduce peak and energy losses, as well as by reducing the drop of distribution lines. This can lead to a reduction in the amount of fossil fuels that are needed to generate electricity.

- Increased reliability: Smart grids can help to increase the reliability of the electricity grid by making it easier to predict and prevent power outages. They can also help to reduce the duration of power outages when they do occur.
- Reduced costs: Smart grids can help to reduce the costs of operating and maintaining the electricity grid. This can be achieved by reducing the need for manual meter reading and by making it easier to manage demand.
- Improved customer service: Smart grids can help to improve customer service by providing customers with real-time information about their energy usage and by making it easier for them to manage their energy consumption.

20.3 Reduce Subscriber Shutdown Time

An intelligent distribution network (IDN) can make outage prevention a key to customer satisfaction by quickly identifying and shutting down devices that are likely to cause distribution network failure. Smart grids can also quickly detect leakage currents and areas that need repair. Smart networks quickly identify out-of-service subscribers using advanced metering software. Providing such information to incident personnel is very important and increases performance. Smart networks reduce subscriber downtime by using the following solutions:

- Reset the system with the help of smart automatic switches that are compatible with smart posts.
- Remote detection.
- Determine the size and location of the load out of the circuit remotely and instantly.
- Remote control of distributed energy products and their aggregation for use.
- Remote detection of network disconnection and reconnection.

20.4 The Benefits of the Smart Distribution Network for Subscribers

Due to the changing nature of renewable energy, namely, wind power, subscribers must select the quantity of product information and its availability in order to make the best decision based on their financial and environmental concerns. This necessitates subscriber education and empowerment, as well as investment in this area. Instant pricing software assists in demand management by increasing the retail price when power is expensive and decreasing the retail price when electricity is cheap. People, however, are unfamiliar with this form of software. A new system has been devised to regulate shared energy use depending on the subscriber's selections for price and time. This method enables subscribers to regulate the source of energy production and utilize energy that has a lower environmental impact. The following are predictable and attainable goals:

Reduce costs associated with meter readiness: Smart grids can reduce the costs associated with meter readiness by automatically reading meters with high accuracy. This can eliminate the need for manual meter reading, which can save time and money.

- Reduction in human resources that required the accuracy of meter cycles: Smart grids can also reduce the number of human resources required to ensure the accuracy of meter cycles. This is because SGs can automatically detect and correct meter errors.
- Reduce calls to the response center: Smart grids can reduce the number of calls to the response center by providing subscribers with self-service tools to manage their energy usage. This can free up response center staff to focus on more complex issues.
- Annual benefits from preventing energy theft: Smart grids can help to prevent energy theft by providing real-time data on energy usage. These data can be used to identify and investigate potential cases of energy theft.
- Remote connection and the benefits of not visiting a technician: Smart grids can enable remote connection to meters, which can eliminate the need for technicians to visit homes or businesses to read meters. This can save time and money for both customers and utility companies.
- An 8.2% reduction in energy consumption reduces emissions of CO_2 by one million tons: Smart grids can help to reduce energy consumption by providing customers with real-time data on their energy usage. These data can be used to help customers make informed decisions about how to reduce their energy consumption.

- Reduce energy sales costs and service costs to subscribers: Smart grids can help to reduce energy sales costs and service costs to subscribers by automating many of the tasks that are currently performed manually. This can save money for both customers and utility companies.

20.5 Capital Optimization

The smart distribution network enables us to be aware of the health and safety of the network. Data collection and transmission create a system that can make automatic decisions. The result is the optimal utilization of capital, the reasons for which are as follows:

- Avoid breakdowns by timely replacement of cables, equipment, substations, and distribution transformers
- Dynamic adjustment of transformers to help delay investment in this area
- It increases the life of production equipment that prevents reinvestment from producing the energy needed

20.6 Concepts of Security in Telecommunication Systems

Any organization's most valuable asset is its information. Government organizations, military bases, corporations, financial institutions, hospitals, and other private firms have access to sensitive information about personnel, customers, goods, research, and financial position. These data are now gathered and stored on computers before being sent to other computers over the network. Allowing this information to become public can result in financial, personal, and political losses, as well as societal implications. As a result, preserving secret information is both a commercial and, in most cases, a moral and legal need.

An IDN has an integrated information and monitoring system that can monitor and manage any equipment connected to the network. Therefore, it is necessary to consider security issues at all system levels. Smart metering systems and distribution networks, like other information systems, have security concerns. In smart metering systems, there are potential security threats to using smart meters, and the necessary security considerations must be considered in controlling and dealing with these hazards. The security of intelligent measurement systems is essential for the security of customer service. If the system's security is compromised, it could lead to global power outages. Therefore, it is important to invest in information security and to pay special attention to the discussion of this topic. By doing so, the quality of electricity supply services to subscribers can be improved and customer satisfaction can be guaranteed.

20.7 Information Security

Protection of information, regardless of its form (including electronic, physical, etc.), is called information security. In information security, unauthorized access, misuse, disclosure, disruption, alteration, reading, recording, and destruction can be discussed. Information security aims to ensure the continuity of current activities, minimize risks, and increase the return on investment and opportunities. The history of information security begins at the same time as computer security. External threats summarize it.

Today, the Internet continues to connect millions of insecure networks. The security of the information stored on each computer depends on the level of security of each computer to which it is connected. The rise of cyberattacks has required companies and governments to protect computer-controlled systems in public industry and other critical infrastructures, highlighting the importance of information security to national security. Each organization chooses its security policy according to the budget and the amount of security required. A security policy determines what is allowed and what is not allowed in terms of security.

20.8 Information Technology Security

It is sometimes called computer security. Information technology security is the information security applied to computer systems technology. Computer systems are not just home computers but can be any device with a processor and some memory. IT professionals are often hired in large companies due to information in large businesses. These professionals keep information and information systems safe from malicious cyberattacks, often seeking access to critically private information or gaining control of internal systems. Activity to ensure information protection is guaranteed when critical issues, such as natural disasters, server and computer failures, and physical thefts occur. At present, information security is usually the responsibility of security experts, and one of the most common methods of information security in some issues is data backup. Although these issues are interrelated, and all have a common goal of maintaining the confidentiality, accessibility, and accessibility of information, there are subtle differences between them. These differences are primarily related to the approach to information security, problem-solving methods, and centralized issues. Information security is concerned with the confidentiality, integrity, and availability of data without regard to the form of information, whether electronic, print, or other forms.

Easy and cheap access to the information highway through high-speed Internet lines and wireless-based technologies has gradually forced companies and organizations to comply with information security tips and install firewalls and intrusion detection systems. Having an effective security policy and creating the right procedures are inevitable.

20.9 Information Security Management System 1 (ISMS)

Like other assets, information is considered an important and valuable asset for any organization and therefore requires the provision of necessary protection solutions to maintain them. The information security management system (ISMS) establishes information security. It is a set of security policies and mechanisms for managing an organization's sensitive information, and its purpose is to minimize system risks and ensure the continuity of active business by limiting the impact of a security breach.

20.9.1 Information Security Management Standard

One of the existing standards and indicators in information security management, ISO/IEC 27000, sets out a collection of essential requirements for managing information security.

The ISMS set of standards includes information security standards. The International Organization for Standardization (ISO) and the International Electoral Commission (IEC) collaborated on the publication of this collection. Management of information security is stated. Table 20.1 lists a collection of essential requirements for managing information security.

Table 20.1 Information and communication security standards.

Title	Standard
Information Management Systems	ISO 27000
Information Security Management Systems	ISO 27001
Guide to Implementing Information Security Management Systems	ISO 27004
Information Security Management System Measurement	ISO 27005
Telecommunication Information Security Management	ISO 27006
Information Security Management Systems Evaluation Guide	ISO 27007
Guide for Assessors in ISMS Controls	ISO TR 27008
Information Technology	ISO 27010
Information Management Policy for Telecommunication Organizations	ISO 27011
Integrated Implementation Path	ISO 27013
Security of Government Information	ISO 27014
Information Security Management Guide for Financial Services	ISO TR 27015
Information Security Management Guide for the Electricity Field	ISO 27031
Cyber Security Management Guide	ISO 27032
Introduction and Concepts of Network Security	ISO 27033-1
Guide to Designing and Implementing Network Security	ISO 27033-2

20.9.2 Security in Smart Grid Telecommunication Systems

The security of telecommunication systems ensures the security of smart power networks. Security can be divided into two general parts: physical security and cybersecurity. Physical security refers to the security of equipment of telecommunication systems and power networks and the entire power network against threats such as theft of electricity and equipment to the disruption of the network to destabilize the power network. Due to the extent and dispersion of the power grid, it is impossible to provide complete physical security in the power grid. Cyber security means the disruption of the smart power grid due to the penetration of cyber threats. The existence of malware such as Stuxnet in 8010 has highlighted the need to pay attention to cybersecurity in older power networks. The malware targeted supervisory control and data acquisition (SCADA) and caused physical damage to the physical infrastructure. Currently, the SCADA system is isolated, and the virus enters the SCADA system via a USB drive.

The SG is increasingly using the telecommunication network to increase the efficiency of the power system. Therefore, the SG should be considered as two parallel graphs connected. An illustration of this graph is given in Figure 20.1.

20.9.3 Threats in the Smart Power Grid

Different cyber risks have been explored in computer networks, and different remedies to these concerns have been offered. However, in smart power networks, network security and physical security of the power system must be considered concurrently. Cyberattacks in these systems damage both cyberspace and the electricity grid. Stuxnet virus is an example of a cyberattack having widespread effects. The telecommunications system, like the electrical system, is divided into levels. The communications network is dispersed throughout the surface of the power network like a graph, increasing the likelihood of cyberattacks. The susceptibility of the system to cyberattacks is determined by the type of telecommunications infrastructure utilized. Inside a smart house, Zigbee technology is also utilized for communication infrastructure. A set of smart meters is also communicated through Wi-Fi communications infrastructure or 4G internet.

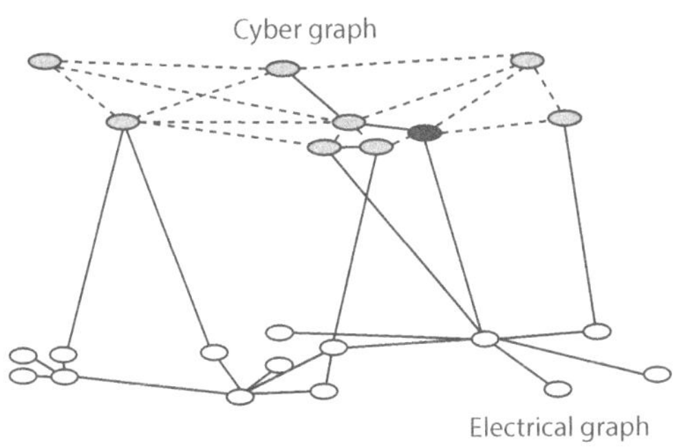

Figure 20.1 Smart grid graph supported by the cyber system.

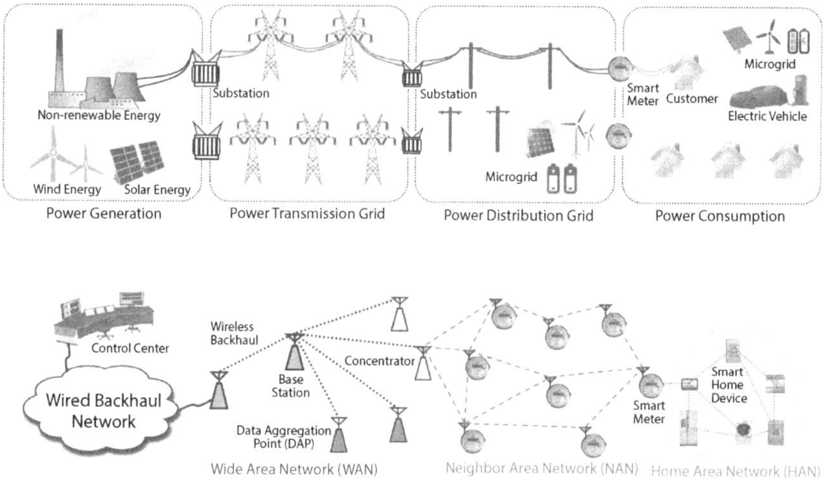

Figure 20.2 Transfer of power to customers and the existing communication infrastructure for network operation.

Figure 20.2 shows the communication infrastructure of smart networks. These communication networks are necessary for the operation of smart networks.

20.9.4 Comparison of Security of Telecommunication Systems

Because there are different options at the level of distribution networks as a candidate for the desired telecommunications system, it is essential to evaluate these options based on various criteria, including security, bandwidth, and latency, to determine the most suitable choice. In Table 20.2, the security of these systems is compared from different points of view.

Today, there are widely available public telecommunication networks with adequate bandwidth and latency. Some public cellular networks, such as 4G, have bandwidth and latency comparable to wired telecommunications technologies.

20.9.5 Attack on the Architecture of the Intelligent Measurement System

Advanced metering infrastructure (AMI) makes energy consumption clearer and more efficient in SGs. Smart meters are part of AMI, which measures, stores, and transmits energy from data and voltage data from residential units and commercial buildings used in a SG. Unlike traditional

Table 20.2 Cyber and physical security of important telecommunication systems in intelligence.

Wired telecommunication system				**Wireless telecommunication system**								
Fiber optic private	PLC	Coaxial cable	Twisted pair (TP)	Wi-Fi	Zigbee	WiMAX	Cellular (4G)	Cellular (3G)	Cellular (25G)	Microwave	Satellite	**Criterion**
3	3	3	3	2	2	2	2	2	2	2	3	**Physical security**
3	2	2	3	3	3	3	3	3	2	2	2	**Cybersecurity**

energy meters, smart meters are often used in AMI systems. A two-way communication mechanism is used in sending and receiving data for the operators of water and electricity organizations through wireless communications.

In this mechanism, the center will remotely interrupt or connect the power or water transmission process, but, in these systems, a hacker may threaten the control of the AMI server, thus providing access to electricity for the meter. Disable smart meters and ultimately target the building by cutting off or preventing energy from reaching it. If a hacker gains control of an AMI server, he will control all aspects of a local AMI system. Here, the hacker will block or manipulate energy data or other functional features remotely. Smart meters, for example, are designed to allow organizations to cut off power to a shared building if the customer does not pay the bill, but a hacker with AMI server control can remotely power the smart meter. Cut off or prevent energy from reaching a building or a neighborhood is a potential consequence of a cyberattack on an advanced metering infrastructure (AMI) server that controls smart meters. If a hacker gains control of the AMI server, they could remotely disconnect or manipulate the smart meters, cutting off power supply to buildings or entire neighborhoods. This highlights the importance of robust cybersecurity measures for critical infrastructure like smart grids and AMI systems to prevent such disruptive attacks. Of course, hackers' access to the AMI server may not be limited to power outages, damage to the meter's physical circuit, or psychologically sending hefty bills to the subscriber, causing them psychological stress.

20.9.6 Internet of Things

The Internet of Things (IoT) is an emerging Internet-based information architecture that develops the interaction between objects and services in a secure and secure environment. It aims to reduce the distance between objects in the physical world and information systems. In the IoT debate, intelligent objects are expected to become active members of the business, information, and social processes to interact between themselves and the external environment by exchanging data and sensed information. IoT technology has been developed with the same philosophy. Receive, record, and send up-to-date and instantaneous information for the use, analysis, and improvement of products and services, and, of course, for the use of consumers. An athlete who studies the physical changes of their body during a workout program, a chimpanzee who is the subject of a scientific study, a can of fish in a grocery store, or a smart gas tank of a passenger car that reports emptying—all of these examples illustrate how IoT technology can be used to instantly send necessary information to any connected database anywhere in the world. All of them will use IoT technology to instantly send the necessary information to any connected database anywhere in the world. Sending data in IoT technology does not require "human-to-human" or "human-to-computer" interaction, and data are sent automatically, according to settings, and at specific times (usually instantaneously and instantaneously). The IoT is one thousand results of the spread of the Internet and the development of wireless technologies and micro-electromechanical systems. Energy and gas have been widely used in industries (especially in various manufacturing plants). Other smart products that can communicate "machine to machine," such as smart labels and smart meters, also benefit from the IoT technology.

Examples of using the IoT in everyday life are as follows:

- If it is forecast to rain today, then the color of the umbrella handle will change [2].
- Caring for older adults remotely becomes much easier.
- Tracking and tracking of goods are done in a very efficient and effective warehousing process.
- The grid-connected refrigerator reminds us that the food is over.

The Internet of Energy is defined as a dynamic grid infrastructure that connects the energy grid via the Internet so that power units (locally created, stored, and sent) can transmit information, where needed. Related data/information follows energy flows, so essential information is exchanged with energy transfer. In smart cities, the energy Internet has many applications, examples of which are the intelligence of buildings and facilities. Internet of service refers to software components delivered over networks and the Internet [3]. Resources include skills, technical systems, information, land, and uses. In Internet services, service refers to web services and service networks. Internet services are formed when the dimensions of business and technology. Organizations must define their services, which must be automatically implemented through computers and linked to each other.

20.9.7 Cyber Security

OT systems transmit data and commands necessary to operate and protect the network. OT systems have a longer history than IT systems, and they may be vulnerable because they are older systems. Upgrading new technologies for OT systems is very challenging and costly. Cyber threats are becoming more advanced, and these systems cannot constantly update. Modern OT systems are often digital versions of analog systems, which are replaced to improve efficiency and functionality. Some drawbacks include complete failure to perform state estimation and risk management, which enters the system. OT systems have different operating times than IT systems. The data in OT systems may be much less than in IT systems.

Hence, the requirements of scheduling and data loss can be very difficult. The accuracy of the information is very important for both OT (and IT) systems, but the main difference is the high data transfer speed required for OT systems. Protection plans, especially large-scale protection, require careful scheduling. The result of delays caused by encryption systems or security entities is not acceptable to the system, although, as encryption systems develop, the use of older encryption systems (OT) in the environment may cause unwanted delays for the system that are not welcome [5].

New protection methods and new technologies require adaptation to the current system. A fundamental difference between the two systems is their maintenance. Unlike IT systems, whose components and software can be constantly updated and replaced, system updates (OTs) are performed by the system operator and developer after extensive testing and validation, usually weeks and months. An experiment is used to prove that these changes do not affect the system, as OT systems must be in standby mode for a long time, so restarting them after a fault is possible. IT systems are often restarted for maintenance. Although new technologies are advancing, OT systems can be updated soon as the need for attention to calibration and the interaction between the two systems increases.

The National Institute of Standards and Technology (NIST) has identified the basic parameters and important goals for cybersecurity research and development. Table 20.1 shows these parameters. NIST introduces cyber-physical systems as "intelligent computable systems."

These systems are integrated systems that can understand and change the real world. For the electrical sector, the term smart grids are more commonly used. But now, the term has expanded to include smart homes and connecting components and devices to the Internet, known as the IoT. These intelligent systems and equipment connected to the Internet increase the number of access points for cyberattacks. However, it is unclear whether intelligent distribution will make the system vulnerable. Distributing intelligence across the network may increase system resilience. With the increase of automation installed to control the system equipment, demand consumption and electrical energy production will decrease. Limited physical-cyber advances can give customers access to distribution and transmission systems, which is unwelcome. For example, wireless systems and applications installed on smartphones designed to control equipment and control demand consumption can be a source of influence. Assessing the current state of cybersecurity in the energy sector is complex for several reasons. Companies determine part of their level of risk acceptance [4]. They select equipment based on an acceptable level of risk and perform equipment updates based on them.

There are infrastructure protection standards for transmission system protection. Distribution system investors and governments set specific standards for cybersecurity. In assessing the resilience of smart grid technologies against various cybersecurity threats, it is crucial to understand the specific vulnerabilities associated with different components of the grid. Table 20.3 provides a comprehensive analysis of these vulnerabilities,

Table 20.3 Research and development parameters for smart grid telecommunication systems.

Research and development parameters	Items
Time delay	1 ms for protection relays It takes a few seconds to monitor and show the state of an extensive network; for transfer, a few seconds to monitor and control feeders and posts and obtain their information; a few minutes to monitor unnecessary information and power market information; and a few hours to read smart meter information and market information; for electricity, days, weeks, and months to collect long-term information such as power quality.
Ensure correct operation	Do not change the information without the approval of the system. Sources of information are clear. The saw time of the specified information should be obvious. The quality of the data is specified.
Reliability	Security of subscribers' information; electricity market information; confidential information such as payment statements and strategic plans.

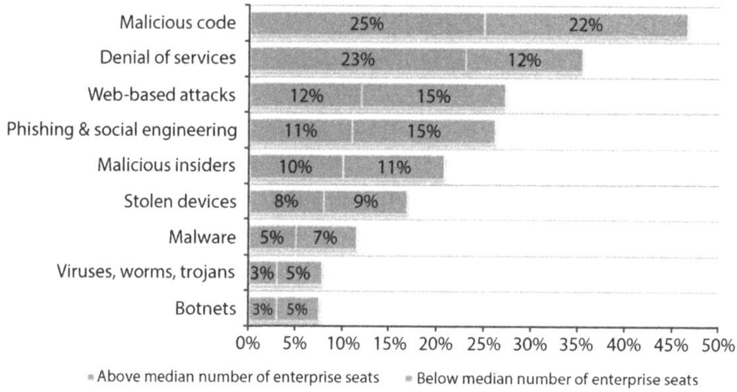

Figure 20.3 Types of cyberattacks on smart grids.

highlighting the areas most susceptible to cyberattacks and the potential impact on grid stability. In addition to these complexities, the threats are very broad and dynamic. New solutions may be discovered and exploited by malicious software. The final costs of cyberattacks in the energy sector can include many factors, including a direct impact on the system operator and customers. Estimating the costs of cyberattacks is much more complex than an algebraic equation because the components of the equation cannot be accurately measured, and its parameters are constantly changing. While the impact of cyberattacks on power grids is unclear, the Panman Institute reports that cybercrime in the electricity sector costs $80 million annually (Figure 20.3). Calculating the day-to-day costs of cyber intrusions, including disaster costs and human resources costs to respond to disasters and recovery, helps assess the cost of future intrusions and justifies the need for increased security and research and development measures. Another assessment of cyberattacks is their economic aspects, such as reduced production capacity and reduced sales (or operating costs of these attacks. In addition to events affecting information systems, there is evidence that threats to OT systems are rising. Cybersecurity experts agree that adopting new digital technologies, such as smart meters and improved control of renewables and other power sources, can challenge system reliability [27].

20.9.8 Physical Security

Over the past 32 years, the United States has suffered from the effects of several natural disasters that have destroyed basic equipment in power systems and caused significant economic losses. Between 1820 and 1980, the United States experienced 172 weather disasters, in which damages and

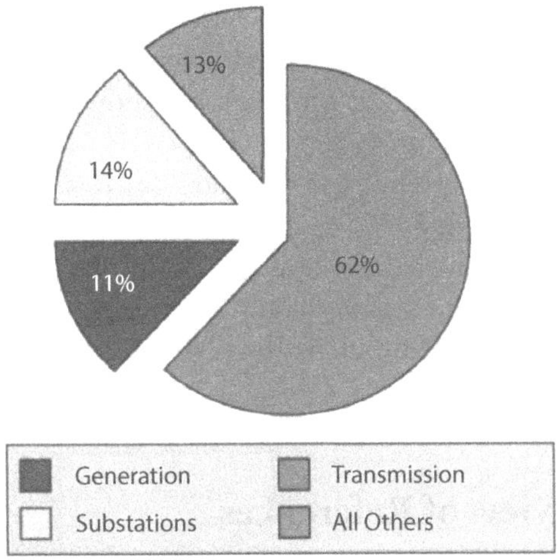

Figure 20.4 The rate of attacks on smart grids.

costs reached more than $1 billion. These include storms, droughts, heat, floods, and fires. In addition to these weather events, there are natural disasters like earthquakes. High-intensity earthquakes are very dangerous. These earthquakes can disrupt infrastructure and power supplies to subscribers. Physical security is a very important part of cybersecurity because

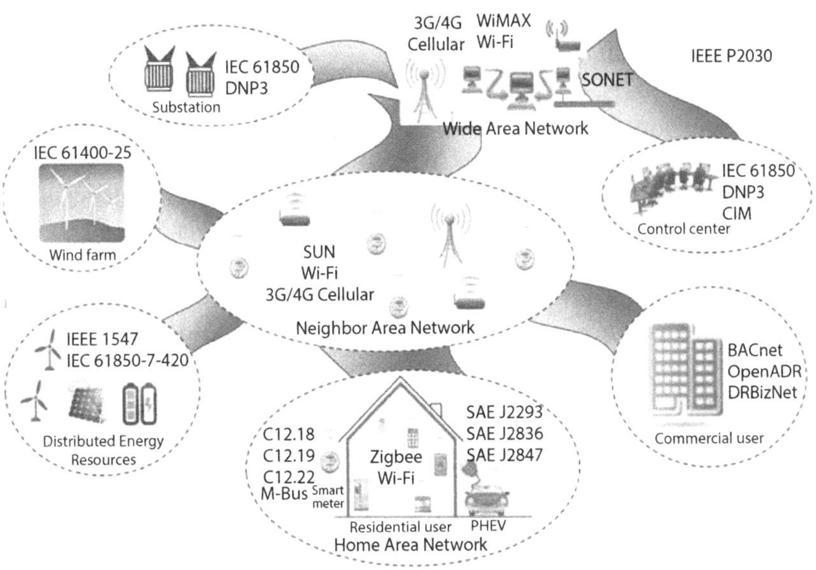

Figure 20.5 8030 standard and its scope in different parts of smart networks.

a cyberattack is much more likely if the attacker has physical access to the equipment. Attacks on physical equipment are not a new issue, and many countries have had to endure it for years. As shown in Figure 20.4, most attacks were on transmission systems.

There must be a comprehensive standard for dealing with these types of attacks. Standard IEEE 8030 is a complete reference for countering and enhancing security against these attacks. Each intelligent network infrastructure is defined in this standard, and each implementation must comply with this standard. Figure 20.5 shows the 8030 standards in different parts of SGs.

20.10 Overview of References

Many articles have been published on cyber and physical attacks. These articles offer many solutions for modeling cyber-physical attacks. Cyberattacks are often unrecognizable.

In reference [6], a new method for attacking the structure of smart meters is examined; this attack disrupts the measurements of smart meters. In this reference, the attacker can access the switches on the bus many times and simultaneously to the meters. For smart attack, reference [7] presents a robust method of defending against cyberattacks by injecting the wrong information. These types of attacks are carried out that are undetectable and undetectable. Reference [8] suggests a comprehensive approach to cyberattack resistance. This paper divides attacks on the power grid into three parts: cyberattacks, physical attacks, and coordinated cyber and physical attacks. The relationship between telecommunication infrastructure and optimal and secure operation of the network has also been investigated. Reference [9] introduces an integrated mixed integer linear programming method to minimize security risk and maximize access to electrical infrastructure and appliances. In reference [10], it is shown that attackers can damage the power system with various attacks, including attacks on network topology, attacks on system state estimation, attacks to increase operating costs or casualties, and attacks on the SCADA system.

In references [8] and [10], an attacker can compromise system security with an attack; the attacker can break the rules in three stages, infiltrate the network and access the operating systems, and perform a coordinated attack on critical components. The power system causes a total loss of the power system. In reference [11], a systematic approach to creating microgrids is presented. For this purpose, a sample system with optimal distribution of different types of DG units in a distribution system is first constructed.

DG units include solar cell sections, wind turbines, and biomass. Reference [12] aims to investigate the effects of LR[1] attacks. An LR attack is when the bass repeatedly attacks the flow of lines and measurements [13]. A comprehensive assessment of the security situation of the SG system is made with a focus on the sudden attack on switches. The attacker aims to turn off consecutive switches and island blackouts. Reference [14] presents the effects of a malicious LR attack on the performance and control of power systems. This attack increases the operating costs of the system. There is also good modeling for LR attacks. In reference [15], a possible cyberattack to increase the density of transmission lines is investigated, and a three-step saw problem-solving model is proposed. In reference [16], the attack on the network topology is analyzed practically by examining the amount of information required by the network. This paper proposes an attack model that requires reducing network information. Understanding the dynamics of power systems and reducing potential risks is essential to develop many critical infrastructures successfully. In reference [17], the system's strength against cyberattacks based on line flow has been investigated. In reference [12], an attack of incorrect information injection type in the distribution system has been investigated. It has been shown that an attacker can launch an effective attack on the system with little effort based on the information injected into the bass. In reference [18], a new attack strategy, FDI[2] has been suggested that its purpose is to distort the normal operation of a power system controlled by automatic voltage controls. Reference [19] has used nano grids to increase the security and reliability of networks, as well as the probabilistic behavior of uncertain distributed generation sources using scenario generation with cyber equipment. Reference [20] has analyzed a new type of coordinated cyber-physical attacks, which, if they occur, can have very severe adverse effects on the power grid. These attacks must be undetectable. It can be very important because these sources handle voltage and frequency control within the microgrid. Reference [21] investigates the effect of LR cyberattack on power grid reliability set. In the long run, the power grid's reliability will decrease despite cyberattacks. Reference [17] examines physical cyberattacks on power systems, and the nature of these attacks is undetectable and undetectable. These attacks remove power lines without detection. Power system attack equations are expressed as a complex linear optimization problem. Reference [13] states that an attacker can infiltrate smart measuring devices by injecting incorrect information to form a cyberattack. This attack can affect the estimation devices, and this

[1]Load redistribution attack

[2]False data injection attack

system cannot do its job properly during the attack. A cyberattack on all system equipment is impossible and difficult, so this article assumes that this attack is performed locally and affects part of the system. Reference [9] examines the effects of a physical cyberattack on SGs. The objective function of this paper is system operating costs. Analog metering systems are assumed to be attacked, and power transmission lines are overloaded. The technologies required to protect measuring devices installed in remote terminal units are discussed. Due to the very large and severe coordinated cyber and physical attack on Ukraine's power grid, this chapter focuses on the standards to combat such cyber-physical attacks in smart distribution networks. This article focuses on the standards to combat cyber-physical attacks in smart distribution networks. In reference [21], control systems have provided a suitable defense method for dealing with cyber-physical attacks. The system under study is micro-networks with DC links. In reference [22], using game theory, a suitable defense strategy to counter cyberattacks is presented. A two-step problem-solving model is used for LR attacks and a proper defense against this attack. In references [23–26], an undetectable attack model for a cyberattack on the network topology is seen. In this model, first, a cyberattack is performed, and then a cyberattack is performed to prevent the detection of a physical attack. In references [28, 29], the effect of cyberattack on power transfer between microgrids is investigated. By injecting incorrect information into the system, the attacker prevents the correct and optimal power exchange between the microgrids and increases the final cost for the consumer. In references [30–32], a creative inclination to estimate the state of microgrids is seen. Using this estimate increases the ability of the system to withstand cyberattacks. The resilience of intelligent power transmission systems against cyberattacks is discussed in reference [33]. Power transmission systems must have the necessary resilience against attacks. Some cyberattacks have a wide range, such as the Ukrainian cyberattack, so they must be tested for resistance and resilience to these threats. In references [34–40], cyberattack models are presented; this article examines cyberattacks and provides appropriate defense strategies to deal with this type of attack.

20.11 Summary and Conclusion

Many of the articles presented emphasize cyber and physical attacks on power networks. This dissertation discusses these attacks on smart nets and micro nets—types of cyberattacks and coordinated physical cyberattacks.

A cyberattack is described as an undetectable attack. For the system during 81 h, how to attack and the effect of the attack on the network are laid out.

References

1. M. Zand, M. A. Nasab, A. Hatami, M. Kargar and H. R. Chamorro, "Using Adaptive Fuzzy Logic for Intelligent Energy Management in Hybrid Vehicles," *2020 28th ICEE*, , pp. 1-7, doi: 10.1109/ICEE50131.2020.9260941. IEEE Index.
2. Ahmadi-Nezamabad, H., Zand, M., Alizadeh, A., Vosoogh, M., & Nojavan, S. (2019). Multi-objective optimization based robust scheduling of electric vehicles aggregator. *Sustain. Cities Soc.*, *47*, 101494.
3. Milani, O. H., Nguyen, T., Parekh, A., Cetin, A. E., & Prasad, B. (2023). 0537 Incident Hypertension Prediction in Obstructive Sleep Apnea using Machine Learning. *Sleep*, *46*(Supplement_1), A236-A237.
4. Zand, M., Nasab, M. A., Sanjeevikumar, P., Maroti, P. K., & Holm-Nielsen, J. B. (2020). Energy management strategy for solid-state transformer-based solar charging station for electric vehicles in smart grids. *IET Renew. Power Gener.*, *14*(18), 3843-3852.
5. Ghasemi, M., Akbari, E., Zand, M., Hadipour, M., Ghavidel, S., & Li, L. (2019). An efficient modified HPSO-TVAC-based dynamic economic dispatch of generating units. *Electr. Power Compon. Syst.*, *47*(19-20), 1826-1840.
6. Nasri, S., Nowdeh, S. A., Davoudkhani, I. F., Moghaddam, M. J. H., Kalam, A., Shahrokhi, S., & Zand, M. (2021). Maximum Power point tracking of Photovoltaic Renewable Energy System using a New method based on turbulent flow of water-based optimization (TFWO) under Partial shading conditions. In *Fundamentals and Innovations in Solar Energy* (pp. 285-310). Springer, Singapore.
7. Rohani, A., Joorabian, M., Abasi, M., & Zand, M. (2019). Three-phase amplitude adaptive notch filter control design of DSTATCOM under unbalanced/distorted utility voltage conditions. *J. Intell. Fuzzy Syst.*, *37*(1), 847-865.
8. Zand, M., Nasab, M. A., Neghabi, O., Khalili, M., & Goli, A. (2019, December). Fault locating transmission lines with thyristor-controlled series capacitors By fuzzy logic method. In *2020 14th International Conference on Protection and Automation of Power Systems (IPAPS)* (pp. 62-70). IEEE.
9. Z. Zand, M. Hayati and G. Karimi, "Short-Channel Effects Improvement of Carbon Nanotube Field Effect Transistors," *2020 28th Iranian Conference on Electrical Engineering (ICEE)*, Tabriz, Iran, 2020, pp. 1-6, doi: 10.1109/ICEE50131.2020.9260850.
10. Tightiz, L., Nasab, M. A., Yang, H., & Addeh, A. (2020). An intelligent system based on optimized ANFIS and association rules for power transformer fault diagnosis. *ISA Trans.*, *103*, 63-74.

11. Zand, M., Neghabi, O., Nasab, M. A., Eskandari, M., & Abedini, M. (2020, December). A hybrid scheme for fault locating in transmission lines compensated by the TCSC. In *2020 15th International Conference on Protection and Automation of Power Systems (IPAPS)* (pp. 130-135). IEEE.
12. Zand, M., Nasab, M. A., Khoobani, M., Jahangiri, A., Hosseinian, S. H., & Kimiai, A. H. (2021, February). Robust speed control for induction motor drives using STSM control. In *2021 12th Power Electronics, Drive Systems, and Technologies Conference (PEDSTC)* (pp. 1-6). IEEE.
13. Sanjeevikumar, P., Zand, M., Nasab, M. A., Hanif, M. A., & Bhaskar, M. S. (2021, May). Spider community optimization algorithm to determine UPFC optimal size and location for improve dynamic stability. In *2021 IEEE 12th Energy Conversion Congress & Exposition-Asia (ECCE-Asia)* (pp. 2318-2323). IEEE.
14. Azimi Nasab, M., Zand, M., Eskandari, M., Sanjeevikumar, P., & Siano, P. (2021). Optimal planning of electrical appliance of residential units in a smart home network using cloud services. *Smart Cities, 4*(3), 1173-1195.
15. Nasab, M. A., Zand, M., Padmanaban, S., Bhaskar, M. S., & Guerrero, J. M. (2022). An efficient, robust optimization model for the unit commitment considering renewable uncertainty and pumped-storage hydropower. *Comput. Electr. Eng., 100*, 107846.
16. Azimi Nasab, M., Zand, M., Padmanaban, S., & Khan, B. (2021). Simultaneous long-term planning of flexible electric vehicle photovoltaic charging stations in terms of load response and technical and economic indicators. *World Electr. Veh. J., 12*(4), 190.
17. Zand, M., Nasab, M. A., Padmanaban, S., & Khoobani, M. (2022). Big data for SMART Sensor and intelligent Electronic devices–Building application. In *Smart buildings digitalization* (pp. 11-28). CRC Press.
18. Padmanaban, S., Khalili, M., Nasab, M. A., Zand, M., Shamim, A. G., & Khan, B. (2022). Determination of Power Transformers Health Index Using Parameters Affecting the Transformer's Life. *IETE J. Res.*, 1-22.
19. Sanjeevikumar, P., Samavat, T., Nasab, M. A., Zand, M., & Khoobani, M. (2022). Machine learning-based hybrid demand-side controller for renewable energy management. In *Sustainable Developments by Artificial Intelligence and Machine Learning for Renewable Energies* (pp. 291-307). Elsevier.
20. Khalili, M., Ali Dashtaki, M., Nasab, M. A., Reza Hanif, H., Padmanaban, S., & Khan, B. (2022). Optimal instantaneous prediction of voltage instability due to transient faults in power networks taking into account the dynamic effect of generators. *Cogent Eng., 9*(1), 2072568.
21. Asadi, A. H. K., Jahangiri, A., Zand, M., Eskandari, M., Nasab, M. A., & Meyar-Naimi, H. (2022, January). Optimal Design of High Density HTS-SMES Step-Shaped Cross-Sectional Solenoid to mechanical stress reduction. In *2022 International Conference on Protection and Automation of Power Systems (IPAPS)* (Vol. 16, pp. 1-6). IEEE.

22. Nasab, M. A., Zand, M., Hatami, A., Nikoukar, F., Padmanaban, S., & Kimiai, A. H. (2022, January). A Hybrid Scheme for Fault Locating for Transmission Lines with TCSC. In *2022 International Conference on Protection and Automation of Power Systems (IPAPS)* (Vol. 16, pp. 1-10). IEEE.
23. Dashtaki, A. A., Khaki, M., Zand, M., Nasab, M. A., Sanjeevikumar, P., Samavat, T., ... & Khan, B. (2022). A Day Ahead Electrical Appliance Planning of Residential Units in a Smart Home Network Using ITS-BF Algorithm. *Int. Trans. Electr. Energy Syst., 2022.*
24. Rastgoo, Samaneh, Zahra Mahdavi, Morteza Azimi Nasab, Mohammad Zand, and Sanjeevikumar Padmanaban. 2022. "Using an Intelligent Control Method for Electric Vehicle Charging in Microgrids" *World Electr. Veh. J.*, 13, no. 12: 222. https://doi.org/10.3390/wevj13120222.
25. Seresht, R. M., Miri, M., Zand, M., Azimi Nasab, M., Sanjeevikumar, P., & Khan, B. (2023). Frequency control scheme of an AC Islanded microgrid based on modified new self-organizing hierarchical PSO with jumping time-varying acceleration coefficients. *Cogent Eng.*, *10*(1), 2157982.
26. Padmanaban, S., Samavat, T., Nasab, M. A., Nasab, M. A., Zand, M., & Nikokar, F. (2023). Electric Vehicles and IoT in Smart Cities. *Artificial Intelligence-based Smart Power Systems*, 273-290.
27. Padmanaban, S., Nasab, M. A., Shiri, M. E., Javadi, H. H. S., Nasab, M. A., Zand, M., & Samavat, T. (2023). The Role of Internet of Things in Smart Homes. *Artificial Intelligence-based Smart Power Systems*, 259-271.
28. Ahmadi, M., Rastgoo, S.,Mahdavi, Z., Nasab, M.A., Zand, M., Sanjeevikumar, P.,Khan, B.: Optimal allocation of EVs parking lots and DG in micro grid using two-stage GA-PSO. *J. Eng.* 2023, e12237 (2023). https://doi.org/10.1049/tje2.12237
29. Zand, M., Nasab, M. A., Padmanaban, S., Maroti, P. K., & Muyeen, S. M. (2023). Sensitivity analysis index to determine the optimal location of multi-objective UPFC for improvement of power quality parameters. *Energy Rep.*, *10*, 431-438.
30. Sagar, A., Kashyap, A., Nasab, M. A., Padmanaban, S., Bertoluzzo, M., Kumar, A., & Blaabjerg, F. (2023). A Comprehensive Review of the Recent Development of Wireless Power Transfer technologies for Electric Vehicle Charging Systems. *IEEE Access.*
31. Nasab, M. A., Zand, M., Dashtaki, A. A., Nasab, M. A., Padmanaban, S., & Blaabjerg, F. (2023). Uncertainty compensation with coordinated control of EVs and DER systems in smart grids. *Solar Energy*, *263*, 111920.
32. Nikoufard, M., & Hatami, M. (2016). Ultra-short novel transverse magnetic to transverse electric polarization rotator in hybrid integration of InGaAsP/silicon-on-insulator technologies. *Indian J. Phys.*, *90*, 211-217.
33. Padmanaban, S., Samavat, T., Nasab, M. A., Nasab, M. A., Zand, M., & Nikokar, F. (2023). Electric vehicles and IoT in smart cities. *Artificial Intelligence-based Smart Power Systems*, 273-290.

34. Samavat, T., Nazari, M., Ghalehnoie, M., Nasab, M. A., Zand, M., Sanjeevikumar, P., & Khan, B. (2023). A Comparative Analysis of the Mamdani and Sugeno Fuzzy Inference Systems for MPPT of an Islanded PV System. *Int. J. Energy Res., 2023.*
35. Padmanaban, S., Nasab, M. A., Shiri, M. E., Javadi, H. H. S., Nasab, M. A., Zand, M., & Samavat, T. (2023). The role of internet of things in smart homes. *Artificial Intelligence-based Smart Power Systems*, 259-271.
36. Zand, M., Nasab, M. A., Hanif, H. R., Ghahri, M. R., Padmanaban, S., & Khan, B. (2023). A hybrid scheme for optimal performance of photovoltaic system converters with multilayer structure. *The J. Eng., 2023*(1), e12218.
37. Padmanaban, S., Nasab, M. A., Samavat, T., Nasab, M. A., & Zand, M. Securing Smart Power Grids Against Cyber-Attacks. In *IoT and Analytics in Renewable Energy Systems (Volume 1)* (pp. 17-36). CRC Press.
38. Padmanaban, S., Nasab, M. A., Samavat, T., Zand, M., Nasab, M. A., & Hashemi, E. Cyber Security in Smart Energy Networks. In *IoT and Analytics in Renewable Energy Systems (Volume 2)* (pp. 309-325). CRC Press.
39. Padmanaban, S., Nasab, M. A., Samavat, T., Zand, M., & Nasab, M. A. Artificial Intelligence Techniques for Smart Power Systems. In *IoT and Analytics in Renewable Energy Systems (Volume 1)* (pp. 107-123). CRC Press.
40. Nikoufard, M., & Hatami, M. (2018). Analysis of ultra-compact TE to TM polarization rotator in InGaAsP and SOI technologies. *Optik, 153*, 9-15.

21

Reinforcement Learning Method Optimizes the Planning and Design of the Solar Energy Systems

R. Dhanasekaran[1]*, S. Sreenatha Reddy[2], K. Santhi[3], Ch. Nagaraja Kumari[4] and Saba Noureen[4]

[1]*Department of Mechanical Engineering, Adhiyamaan College of Engineering, Tamil Nadu, India*
[2]*Department of Mechanical Engineering, Guru Nanak Institute of Technology, Telangana, India*
[3]*Department of Electrical and Electronics Engineering, Adhiyamaan College of Engineering, Tamil Nadu, India*
[4]*Department of Electrical and Electronics Engineering, Guru Nanak Institute of Technology, Telangana, India*

Abstract
Now, the world is moving toward renewable energy sources; solar energy has an important role as a prominent and sustainable solution to meet the increasing energy demands of the current situation. Anyway, the maintenance, operation, and management of solar energy systems are crucial roles with several challenges, including intermittency, grid integration, maintenance, and optimization of energy production. Artificial intelligence (AI) technologies, such as machine learning, deep learning, and predictive analytics, have demonstrated a remarkable potential to revolutionize the solar industry. By analyzing tremendous measures of historical and real-time data, AI algorithms can predict solar irradiance, weather conditions, and energy consumption patterns with unprecedented accuracy. These predictive capabilities enable solar power plant operators to optimize their energy generation and grid integration strategies, thus reducing backup power sources and minimizing overall operational costs. By using anomaly

**Corresponding author*: dr.r.dhanasekaran@gmail.com

O.V. Gnana Swathika, K. Karthikeyan, Milind Shrinivas Dangate and Nicoletta Ravasio (eds.)
Biomass and Solar-Powered Sustainable Digital Cities, (363–370) © 2024 Scrivener Publishing LLC

detection algorithms, AI can identify faulty components or performance deviations, ensuring prompt intervention and minimizing the overall system time. In addition to operational optimization, AI plays a vital role in improving the design and planning of solar systems. AI algorithms can analyze geographic data, historical energy consumption patterns, and economic factors to recommend optimal locations for solar installations and suggest appropriate system configurations.

***Keywords*:** Artificial intelligence, reinforcement learning, optimization, planning, design, operation, solar energy

21.1 Introduction

Solar energy is the energy that comes from the sun's radiation, captured by photovoltaic (PV) cells (solar panels), and converted into usable electricity. The working principle of solar energy is converting the sunlight into electricity through PV effect. PV cells are made from semiconductor materials, like silicon. When sunlight (photons) falls on the PV cells, it excites electrons in the semiconductor, generating an electric current. This is known as PV effect. The excited electrons create a voltage difference between the front and back of the PV cell, resulting in electric field. Due to this electric field, the electrons can move in a specific direction, generating direct current (DC) electricity [1]. The DC electricity needs to convert into alternating current (AC). This is standard form of electricity; it is used in houses. An inverter is used to perform the conversion. The AC electricity generated by the PV panels can be used in power applications and devices within building. Excess electricity cans feedback into the grid [2]. Artificial intelligence (AI) techniques such as machine learning (ML) and reinforcement learning (RL) are now integrated into solar panels to improve performance and efficiency is shown in Figure 21.1.

Artificial intelligence can analyze historical and real-time data such as weather conditions, solar panel efficiency, and energy generation pattern. The energy generation pattern to predict how much energy is producing

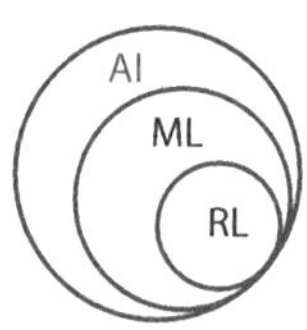

Figure 21.1 The relationship between AI, ML, and RL.

by a solar panel. This helps us in optimizing energy consumption and grid integration. AI algorithms can dynamically adjust the orientation and tilt of solar panel to maximize energy capturing and it depends upon the weather conditions and sun's position. AI can monitor the performance of individual panels and detect the faults. It is also predicting when maintenance is needed. In this article, we will study about the solar energy using RL [3].

21.2 Objectives and Methodology

Reinforcement learning can be applied to optimize the use of solar energy systems by learning how to make decisions that maximize energy capture and usage efficiency. This involves training algorithms to control variables such as the angle of solar panels or energy storage based on environmental conditions and energy demand (Figure 21.2). The goal is to find optimal strategies that balance energy production and consumption with maximum efficiency and cost saving.

21.3 Solar Power Plant Operations Using RL

As opposed to other types of ML, RL allows an agent to learn how to make decisions by interacting with the environment and maximizing the reward signal. In the context of a solar power plant, the environment includes factors like sunlight, intensity, temperature, cloud cover, and energy demand. The agent, in this case, could be an AI program or controller responsible for adjusting various parameters of solar power plants [4].

21.3.1 Solar Energy Production with RL

The production of RL depends on weather conditions, energy demand, and the state of the battery. The RL agent can adjust parameters like the tilt angle of the solar panel, the orientation, and the amount of energy sent

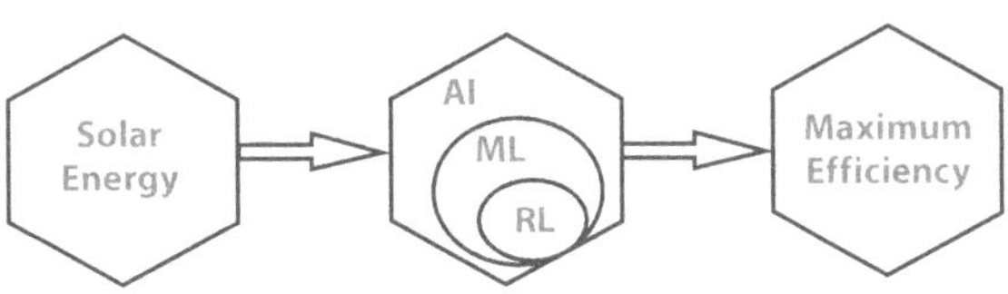

Figure 21.2 A pictorial view of solar energy with AI.

to the grid or stored in the battery. The rewards could be higher when energy production is close to demand or the battery is optimally charged. The excess energy goes into used energy, or the battery is overcharged [5]. The RL agent learns from different outcomes for example: it will learn to adjust the panel angle to capture more sunlight during certain hours and adjust it differently during other times. RL allows continuous changes when the weather suddenly changes when the RL agent can quickly adjust the panel angle to minimize the impact on energy production. The RL based solar energy systems can integrate into the smart grid where they communicate with other energy sources and consumers to optimize overall energy distribution and consumption. A SoftMax actor-critic agent optimizes the energy production by the dual axis of solar panels [6].

21.3.2 Improving Solar Panel Efficiency Using RL

Solar energy provides a sustainable and pollution-free way to harness energy directly from the sun. The efficiency of solar panels depends on their orientation and tracking algorithms. The efficiency of a solar panel is affected by various environmental factors such as sky irradiance, ground reflection, and cloud coverage [7]. To improve the efficiency of solar panels, the hardware must be of the best quality. The panels may include tilt angles and azimuth angles, and it is essential to have a sensor hardware tracking algorithm that calculates the sun's relative position in the sky throughout the day [8]. The reward design function provides feedback to the RL agent. The output of energy is based on the function of the panels. Over time the energy must be maximized, so rewards reflect the amount of energy. Choose the RL algorithm, like deep Q networks and proximal policy optimization to train agents [9]. By choosing this RL algorithm, the environment receives rewards and the energy is produced. This interacting algorithm or RL agent increases the energy rewards by adjusting panel parameters. The agent uses the past data and improves the experience. Solar panel efficiency can change over time due to factors like weather patterns. RL system is designed to continuously adapt and learn from the new data to increase performance [10].

21.3.3 Energy Control in Microgrid Using RL

In solar microgrids, the generation of energy from solar panels can vary due to factors like weather conditions and daytime. To make more efficient use of this energy, the system should include information such as current solar generation, energy consumption, battery charge level, and grid electricity prices.

The RL can achieve goals like electricity bills, maximizing self-consumption, and battery health. The energy stored in the battery, and the amount of energy supplied to the grid, determine how much can be used locally. This decision affects the battery state and grid interaction, and overall energy is balanced. The rewards are used for reducing energy costs or maintaining battery charges within a desired range. Negative rewards are given for inefficient or excessive battery depletion [11]. The RL algorithm learns from the trial and error method. It investigates various actions and studies the results.

Over time, it identifies which action is more favorable to rewards. The algorithm starts making better decisions based on the previous data. If they sell excess energy to the grid during peak hours, they will receive higher rewards. The RL learning adjusts strategy to minimize the rewards in the new conditions. The RL algorithm learns dynamically during the battery's charging and discharging cycles, based on energy generation and consumption, to prevent excessive charging and extend the battery's lifespan [12].

21.3.4 Solar Irradiance Prediction Using RL

Solar irradiance is the total amount of solar energy that reaches the earth's surface within a given period, and it is measured in watts/m^2. The prediction of solar irradiance is dependent on time, prevailing weather conditions, and past data. The terms direct normal irradiation, diffuse horizontal irradiation, and global horizontal irradiation are the most used types of solar radiation [13]. Choosing a suitable RL algorithm like DQN depends on factors such as the complexity of the problem and the availability of data. Historical solar irradiance data, weather forecasts, and energy generation records are needed for the training of the model that is time-consuming. Solar irradiance prediction requires real-time data because of the changing weather conditions. This RL model should have to make quick decisions to optimize energy generation [14].

21.4 Future Scopes of AI in Solar Power

The future scope of AI technology is enhancing solar energy performance. AI-driven analytics can identify and diagnose system problems that could lead to malfunction or failure. This helps solar energy producers reduce repair and maintenance costs by identifying problems early and preventing them from occurring. AI optimizes solar panel efficiency manages energy storage and grid management. AI of solar systems is expanded by making them more efficient and reliable [15].

21.5 Conclusion

The performance of solar panels can be improved with RL. AI techniques have been widely employed in various fields such as generation, prediction, solar irradiance, and forecasting. AI-based simulation techniques provide a powerful alternative approach, offering several advantages. RL mainly depends upon features like panel orientation, tracking algorithms, weather conditions, state of the battery, and historical data. The RL model is learning from previous outcomes and improving the efficiency of the data. At a high-level, the agent learns from the following:

- The agent interacts with the environment and receives some rewards.
- The agent learns from the rewards and updating of selecting action in the given state.
- The process repeats.

Solar forecasting contributes to efficient energy management, grid state, and economic saving. Grid integration in solar energy refers to the process of connecting solar power systems to the electrical grid in a way that ensures system stability, efficiency, and reliability by utilization of both solar energy and grid infrastructures.

References

1. Neha Sehrawat, Sahil Vashisht, Amritpal Singh, Solar irradiance forecasting models using machine learning techniques and digital twin: A case study with comparison. *Int. J. Intell. Networks*, 4, 90-102, 2023.
2. Yuekuan Zhou, Siqian Zheng, Guoqiang Zhang, Machine-learning based study on the on-site renewable electrical performance of an optimal hybrid PCMs integrated renewable system with high-level parameters' uncertainties. *Renew. Energy*, 151, 403-418, 2020.
3. Ahmed, R, Sreeram, V., Mishra, Y., Arif, M.D., A review and evaluation of the state-of-the-art in PV solar power forecasting: Techniques and optimization. *Renew. Sustain. Energy Rev.*, 124, 109792, 2020.
4. Anwaar, F., Iltaf, N., Afzal, H., and Nawaz, R., HRS-CE: A hybrid framework to integrate content embeddings in recommender systems for cold start items. *J. Comput. Sci.*, 29, 9–18, 2018.
5. Yongfang Li. Molecular design of photovoltaic materials for polymer solar cells: Toward suitable electronic energy levels and broad absorption. *Acc. Chem. Res.*, 45, 723–733, 2012.

6. Eduardo F Camacho and Manuel Berenguel, Control of Solar Energy Systems. *IFAC Proceedings Volumes,* 45, 848–855, 2012.
7. Rustu Eke and Ali Senturk, Performance comparison of a double-axis sun tracking versus fixed PV system. *Sol. Energy,* 86, 2665–2672, 2012.
8. Leo, R., Milton, R.S., Sibi, S. Reinforcement learning for optimal energy management of a solar microgrid. *IEEE Global Humanitarian Technology Conference - South Asia Satellite (GHTC-SAS).* 2014.
9. Hossein Mousazadeh, Alireza Keyhani, Arzhang Javadi, Hossein Mobli, Karen Abrinia, and Ahmad Sharifi, A review of principle and sun-tracking methods for maximizing solar systems output. *Renew. Sustain. Energy Rev.,* 13, 1800–1818, 2009.
10. Sadegh Etemund and Nasser Mozayani, Using reinforcement learning to make smart energy storage sources in microgrid. *30th International Power System Conference (PSC), IEEE.* 136-141, 2015.
11. Seyyed Shahabaddin Hosseini Dehshiri and Bahar Firoozabadi, Comparison, evaluation and prioritization of solar photovoltaic tracking systems using multi criteria decision making methods. *Sustain. Energy Technol. Assess.,* 55, 102989, 2023.
12. Masoumeh Abdollahpour, Mahmood Reza Golzarian, Abbas Rohani and Hossein Abootorabi Zarchi, Development of a machine vision dual-axis solar tracking system. *Sol. Energy,* 169, 136-143, 2018.
13. Rahul G. Makade and Basharat Jamil, Statistical analysis of sunshine based global solar radiation (GSR) models for tropical wet and dry climatic Region in Nagpur, India: A case study. *Renew. Sustain. Energy Rev.,* 87, 22-43, 2018.
14. Harendra Kumar Yadav, Yash Pal, and Madan Mohan Tripathi, Short-term PV power forecasting using empirical mode decomposition in integration with back-propagation neural network. *J. Inf. Optim. Sci.,* 41, 25-37, 2020.
15. Byung-Ki Jeon and Eui-Jong Kim, Solar Irradiance Prediction Using Reinforcement Learning Pre-Trained with Few Historical Data. *Soc. Sci. Res. Network,* 1-19, 2022.

22

Photovoltaic-Thermal Solar-Assisted Heat Pump Systems for Building Applications: Integration and Design Methods

Krishna Kumba[1]*, Venkateswarlu Gundu[2] and R. Arul[1]

[1]*School of Electrical Engineering, Vellore Institute of Technology, Chennai, Vandalur, Chennai, Tamil Nadu, India*
[2]*Department of Computer Science and Engineering, Koneru Lakshmaiah Education Foundation, Vaddeswaram, Guntur, AP, India*

Abstract
In recent years, the demand for sustainable and environmentally friendly energy sources has been on the rise. Photovoltaic-thermal (PVT) solar energy is one such source that combines the benefits of both solar thermal and solar photovoltaic systems. Employing PVT solar-assisted heat pumps for residential and commercial buildings is a profitable technology that harnesses PVT energy for electricity and hot water. Therefore, a solar-assisted PVT system is described in this chapter. It provides a comprehensive overview of PVT solar-assisted heat pump systems, including their components, working principles, advantages, and challenges. Additionally, it covers integration, design methods, and life cycle costs such as investment, payback, and return on investment. Moreover, this chapter highlights the challenges of PV thermal-assisted systems.

Keywords: Photovoltaic thermal systems, solar-assisted heat pump, building applications, integration methods, design techniques heat pump, energy efficiency, environmental impact, technological readiness

22.1 Introduction

Solar power is a great source of clean energy that is readily available and has minimal costs and environmental impacts. Utilization of solar energy

**Corresponding author*: krishna.kumba@vit.ac.in

O.V. Gnana Swathika, K. Karthikeyan, Milind Shrinivas Dangate and Nicoletta Ravasio (eds.) Biomass and Solar-Powered Sustainable Digital Cities, (371–382) © 2024 Scrivener Publishing LLC

and environmentally friendly refrigerants, coupled with a range of photovoltaic-thermal (PVT) collectors, should be employed to regulate heat pumps (HPs) [1–4]. Heat pumps using vapor-compression technology are already considered to be the most suitable clean technology for buildings. A single device that combines solar photovoltaics and vapor-compression HPs enables more applications such as the creation of environmentally benign water heating, reducing heat transfer losses, and controlling incident solar heat. This system combines PV modules with a HP to generate both electricity and heat, resulting in improved production and efficiency. In comparison to other hot water innovations that rely on traditional or non-traditional energy sources, it is additional energy improvement and accurate. Solar-assisted HP system (SAHP) is a better choice than the traditional solar water heater when it comes to water heating [5]. It offers superior performance and requires less initial investment, even when weather conditions are unpredictable [6, 7]. Water is the working fluid in conventional water heating systems, while refrigerant is used in hybrid systems to transfer heat. Several approaches are presented in this chapter for integrating and designing SAHP systems in buildings.

22.2 Functioning of Photovoltaic and Thermal Heat Pump Systems

The illustration of the integrated photovoltaic and thermal HP system is shown in Figure 22.1. In the optimal performance of the PVT-SAHP system, certain key components are essential. These include solar collectors, heat exchanger (HX), compressors, condensers, and evaporators. The sun's

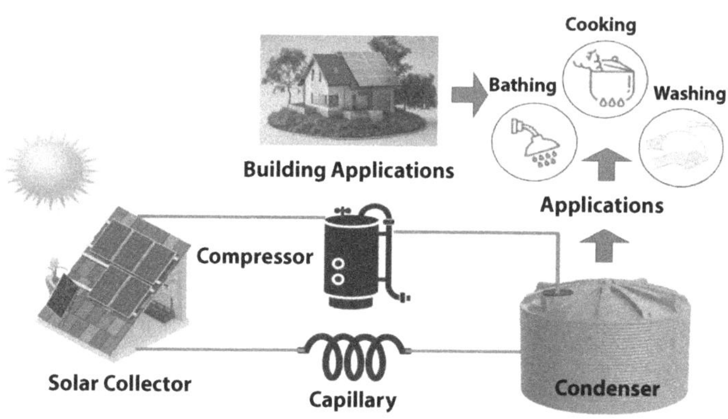

Figure 22.1 Integrated PV and heat pump schematic diagram.

energy is captured by the solar collectors, which subsequently transform it into heat that is transferred to the HX. The heat is transferred to the HP refrigerant by the HX. Refrigerant is compressed by the compressor, which raises its temperature and enables it to discharge the heat in the collector. To complete the cycle, the evaporator takes heat from the surrounding air or water and transfers it to the refrigerant. Combining these components in a PVT-SAHP can create a highly efficient and cost-effective way to heat our home or business [7, 8].

PVT air collectors utilize solar energy to produce both hot air and electricity simultaneously. Heat pump systems can be improved by integrating these collectors. It has been successfully used in subtropical climates where ambient temperatures do not exceed 35°C. When it comes to the functioning of photovoltaic and thermal HP systems, there are a few key things to keep in mind. First and foremost, both of these systems rely on the conversion of energy from one form to another to generate power. In the case of photovoltaic systems, this involves the use of solar panels to convert sunlight into electricity. Thermal HP systems, on the other hand, use refrigerant to extract heat from the air or ground and then convert it into usable energy [7].

One of the important factors to consider when it comes to these systems is efficiency. Both photovoltaic and thermal HP systems can be quite efficient, but it is important to choose the right system for your specific needs. For example, photovoltaic systems may be a better choice for homes or buildings that receive a lot of direct sunlight, while thermal HP systems may be more effective in areas with colder climates. Another important consideration is maintenance. Both photovoltaic and thermal HP systems require consistent maintenance in order to make sure they keep working properly. This may include cleaning solar panels, checking refrigerant levels, and replacing any worn or damaged components. Overall, the functioning of photovoltaic and thermal HP systems can be quite complex, but, with the right knowledge and support, these systems can provide a reliable and sustainable source of energy for your home or business.

22.3 Mathematical Formation of PV-Thermal-Assisted System

22.3.1 PV and Thermal Efficiency

The PV efficiency (η_{pv}), thermal efficiency (η_{the}), and complete efficiency (η_o) of PV plus thermal collector can be obtained by the following relation [7].

$$\eta_{pv} = \frac{P_{pv}}{A_{pv}G} \tag{22.1}$$

$$\eta_{the} = \frac{Q_{eva}}{A_{pv}G} \tag{22.2}$$

$$\eta_o = \eta_{pv} + \eta_{the} \tag{22.3}$$

Here,
P_{pv} = electric power generated from the module,
A_{pv} = area of the collector,
G = solar irradiance on the collector, and
Q_{eva} = fraction of solar radiation that the heat transfer fluid can recuperate.

The HP performance can be calculated using Equation (22.3). The coefficient of performance (COP) is by dividing the total heat absorbed by the refrigerant (Q_{hp}) from all ambient sources by the amount of solar energy (P_{el}) [8].

$$COP = \frac{Q_{hp}}{P_{el}} \tag{22.4}$$

Similarly, the energy efficiency ratio (EER) assesses chiller and HP accomplishment. The EER is the ratio between the amount of heat extracted by the chiller (Q_{ch}) and the amount of solar power absorbed (P_{el}). It is used when chillers and HPs operate in the opposite direction:

$$EER = \frac{Q_{ch}}{P_{el}} \tag{22.5}$$

Equations (22.1) to (22.5) can be integrated across the period under consideration in order to calculate the daily, monthly, and annual effectiveness of the PVT and HP. The life cycle cost (LCC) can be estimated using Equation 22.6.

$$LCC = C_{initial} + C_{operation} + C_{main} + C_{replacement} + C_{residual} \tag{22.6}$$

Installation, testing, and transport of the HP system are all part of the development process. We can calculate using the following equation:

$$C_{initial} = C_{equipment} + C_{installation} + C_{testing} + C_{tranportation} \tag{22.7}$$

The following equations can be used for calculating energy consumption, maintenance costs, and replacement costs. Here, r and n are the refrigerants and running time (h).

$$C_{operation} = C_{operation} * \left(\frac{1-(1+r)^{-n}}{r} \right) \tag{22.8}$$

$$C_{maintenance} = C_{maintenance} * \left(\frac{1-(1+r)^{-n}}{r} \right) \tag{22.9}$$

$$C_{replacement} = C_{replacement} * (1+r)^{-n} \tag{22.10}$$

In order to estimate the residual cost, payback, and return on investment of the HP system, the following relationship can be used [7]:

$$C_{residual} = \frac{Initial\ cost - Scrap\ cost}{Life\ span} \tag{22.11}$$

$$Payback = \frac{Investment\ of\ heat\ pump}{Cost\ of\ energy\ savings} \tag{22.12}$$

$$\text{Return on investment} = \frac{Cost\ of\ energy\ savings}{Investment\ of\ heat\ pump} * 100 \tag{22.13}$$

22.4 Heat Pump Systems with Solar PV and Thermal for Building Applications

22.4.1 Direct-Expansion Systems

Direct-expansion (DX) heating pump systems are the first type of SAHP. This configuration has been developed and studied by various authors over the years. It combines conventional PVT collectors with HP. The most straightforward configuration is the direct-expansion (DX) SAHP, which primarily makes use of domestic hot water (DHW) production. The PVT collectors in DX-PVT-SAHP systems serve as HP evaporators, supplying solar PV energy to the mechanism. During the phase-change operation of the PVT collector-evaporator, HP refrigerant streams through the heat absorber and extracts heat [1, 2].

22.4.2 Single-Source Systems

The DX-PVT-SAHP is used to heat water for a groundbreaking application that combines hybrid PVT with HPs. It is the first and simplest of its kind, with the hybrid collector serving as the PV solar energy source for the system, as illustrated in Figure 22.2. This system is capable of producing hot water at various temperature levels for space heating or DHW purposes, depending on the design. However, it is important to note that cooling is not available in this configuration.

22.4.3 Dual-Source Systems

The second evaporator can be used in parallel with solar PV and thermal collectors in the DX-PVT-SAHP system to provide a second heat source. It is illustrated in Figure 22.3. In circumstances where the PV energy recovery is insufficient to run the system, the additional heat source, typically air, can enhance the system's performance [11, 12]. PVT refrigerant division is usually linked together with an air source HX. Here, the temperature of the evaporator, the surrounding temperature, and heat gain all affect their ability to evaporate, and, as a result, the refrigerant flows through each circuit. When PV energy is accessible, more refrigerant evaporation occurs on the PVT side. The second source's capacity for evaporation, however, rises with greater ambient temperatures (air or ground sources) but at the cost of PVT [9]. Some researchers have experimented with operating the machine inversely bypassing the solar circuit and shrinking the air-source HX for the water refrigerating on the customer side.

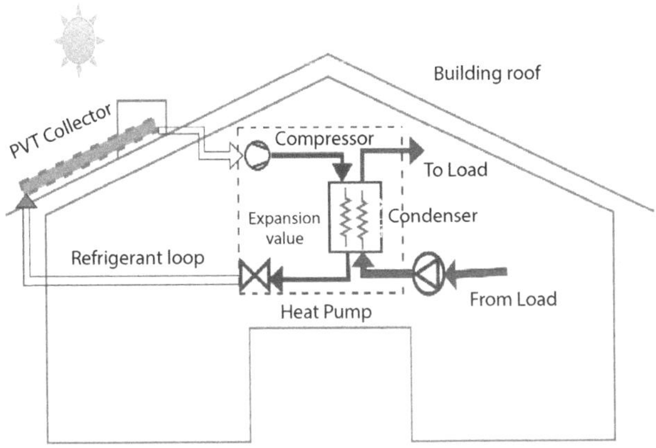

Figure 22.2 DX-PVT-SAHP of a single PV source.

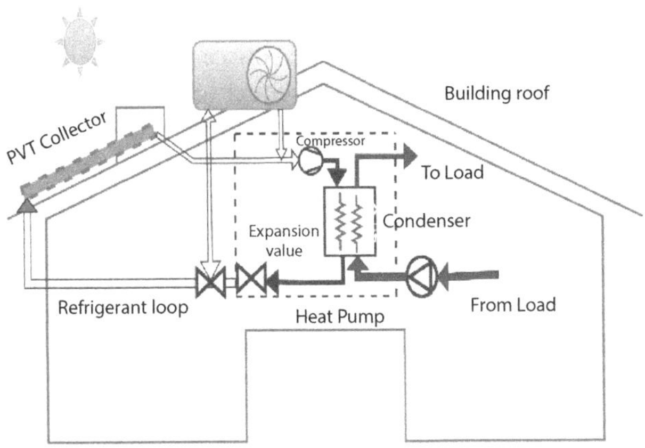

Figure 22.3 DX-PVT-SAHP of a dual PVT source.

22.4.4 Indirect-Expansion (IDX) Systems

The IDX-PVT-SAHP system uses a hybrid collector that is not directly connected to the HP evaporator. Instead, a middle intermediary layer is placed between the PV and refrigerant circuits. This allows for more flexibility in system operation. The HP and PVT collectors are separate, and a water storage system can be used to reduce the need on PV solar energy obtainability. On the other hand, the storage can be located on the user side after the HPs. Indirect-expansion systems' important characteristics include the mutual positioning of the collectors, HX, and extra thermal storage. Here, HP systems can be coupled with solar PV and thermal collectors as a single or additional heat source [11].

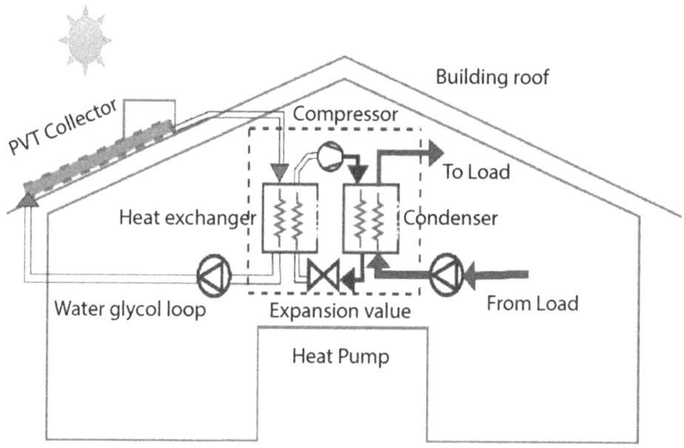

Figure 22.4 IDX-PVT-SAHP of a single source.

22.4.5 Single-Source Indirect-Expansion Systems

The IDX-PVT-SAHP system of a single source is depicted in Figure 22.4 [9]. The IDX-PVT-SAHP is a straight sequel to the DX-PVT-SAHP. It is sometimes referred to as a series or serial in literary works. After PVT collectors have heated a stream of clean water or water mixed with antifreeze, the equipment is heated using a water-to-gas evaporator. Between the solar PV collectors and the HP, the intermediate HX's presence reduces performance losses brought on by erratic weather. By using an HX between the system and solar PV collector loop, the system management can be improved, and a suitable working fluid can be chosen instead of using HP refrigerant in both primary and secondary circuits.

22.4.6 Dual-Source Indirect-Expansion Systems

Compared with single-source IDX-PVT-SAHP systems, the dual-source indirect expansion is novel and intriguing. This includes combining a PVT system with an additional heat source, such as ambient air or the earth, and using two different evaporators in the device. In air-source or water-source HP, PVT collectors are frequently utilized as a supplemental energy source. This can provide DHW, heating, and cooling under several circumstances. Here, two sources are often linked in parallel to form an HP refrigerant circuit when air-source HP is used with solar PV and thermal collectors, as depicted in Figure 22.5 [9, 10]. As seen in Figure 22.6, a combined dual-source water or air evaporator, generally known as a hybrid dual-heat-source evaporator, can be a heat source for external air and water source PVT. Dual-source evaporators, also known as composite dual heat source

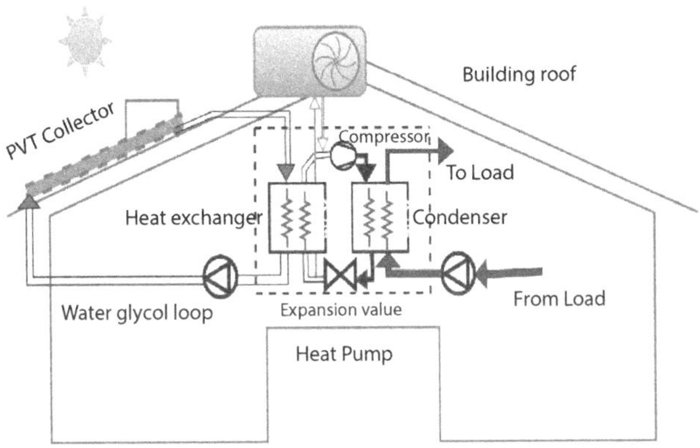

Figure 22.5 IDX-PVT-SAHP of a dual source.

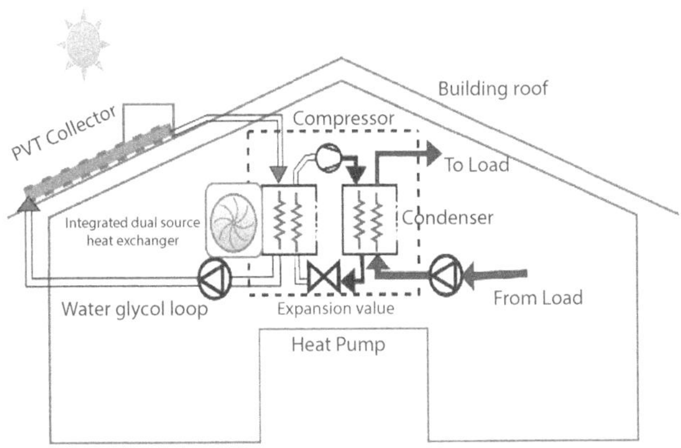

Figure 22.6 Integrated air/water evaporator of IDX-PVT-SAHP.

evaporators, maximize both water-source and external air sources. This type of evaporator can harness two different heat sources. As depicted in Figure 22.7, the evaporator portion of a conventional water-source HP can additionally have a geothermal loop coupled to the solar and thermal system. In certain situations, the PV solar may require an anti-freeze mixture and a middle HX to detach the solar and geothermal components. This was suggested by Bakker *et al.* [15]. A water thermal storage could also be used, as seen in Figure 22.8. Here, both series and parallel configurations have been extensively studied. The survey accomplished by the International Energy Agency revealed that the best configuration in terms of seasonal performance factor is a glazed PVT collector operating. This works parallel

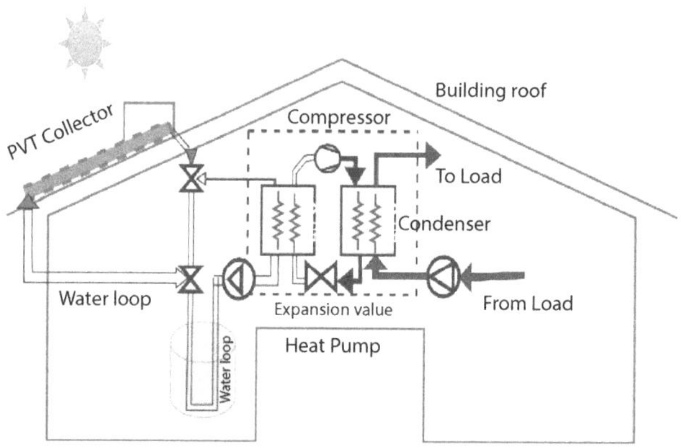

Figure 22.7 PVT and geothermal HX of dual-source IDX-PVT-SAHP.

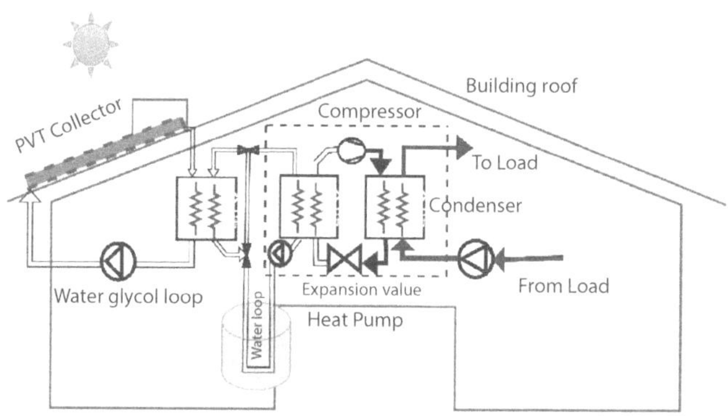

Figure 22.8 PVT and geothermal HX with intermediate HX.

with a ground source HP system. The series design, however, is advantageous when more straightforward borehole HX and rejuvenation are required [11–15].

22.5 Challenges of PV-Thermal-Assisted System

Several challenges arise when it comes to PVT-SAHP. One of the biggest challenges is ensuring that the PVT-SAHP system is designed properly so that it can effectively capture and convert solar energy into usable heat. This requires careful attention to several aspects, including the extent and placement of the PV panels, the kind of HP being used, and the overall design of the system. Another challenge is ensuring that the system can operate efficiently and reliably over time. This requires ongoing maintenance and monitoring, as well as careful attention to the various components of the system, including the solar panels, the HP, and the various pipes and ducts that are used to transfer heat. Finally, there is the challenge of ensuring that the system is safe and that it complies with all relevant regulations and standards. To ensure that the system is running safely and effectively, this calls for knowledge of the system's design and installation as well as continual monitoring and testing.

22.6 Conclusion

Photovoltaic-thermal SAHP systems offer significant benefits, including reduced energy costs, increased efficiency, and reduced environmental impact. Through careful attention to these challenges, these systems can deliver these benefits over the long term with confidence. A dual-source indirect-expansion system, a hybrid of single-source and DX systems, represents the most profitable solution to provide all building thermal requirements such as hot water, heating, and cooling without the main drawbacks that make them inapplicable. In addition, there is a need to improve the technological readiness of dual-source HPs and solar PV and thermal collectors with unaffected lamination to maximize their implementation and minimize their costs.

References

1. James, A., Mohanraj, M., Srinivas, M., Thermal analysis of heat pump systems using photovoltaic-thermal collectors: a review. *J. Therm. Anal. Calorim.*, 144, 1–39, 2021. https://doi.org/10.1007/s10973-020-09431-2.
2. Das D, Kalita P, Roy O., Flat plate hybrid photovoltaic-thermal (PV/T) system: a review on design and development. *Renew. Sustain. Energy Rev.*, 84, 111–30, 2018.
3. K. Kumba, S. P. Simon, K. Sundareswaran, P. S. R. Nayak, K. A. Kumar and N. P. Padhy, Performance Evaluation of a Second-Order Lever Single Axis Solar Tracking System, *in IEEE J. Photovoltaics.*, 12, 5, 1219-1229, 2022. doi: 10.1109/JPHOTOV.2022.3187647.
4. Kumba, K, Simon, S. P, Sundareswaran, K, and Nayak, P. S. R., Comparative Evaluation of Second-Class Lever Principle Based Single-Axis Solar Tracking System and Conventional System. *Distrib. Gener. Altern. Energy J.*, 38(05), 1559–1584, 2023. https://doi.org/10.13052/dgaej2156-3306.3859.

5. M.S. Buker, S.B. Riffat., Solar assisted heat pump systems for low temperature water heating applications: a systematic review., *Renew. Sustain. Energy Rev.*, 55, 399-413, 2016. doi: 10.1016/j.rser.2015.10.157.
6. Z. Wang, P. Guo, H. Zhang, W. Yang, S. Mei, Comprehensive review on the de- velopment of sahp for domestic hot water. *Renew. Sustain. Energy Rev.*, 72, 871–881, 2017.
7. Dhillon, B.S., *Life Cycle Costing for Engineers* (1st ed.). CRC Press, 2009. https://doi.org/10.1201/9781439816899.
8. Xu, Guoying, Zhang, Xiaosong, Deng, Shiming., A simulation study on the operating performance of a solar-air source heat pump water heater. *Appl. Therm. Eng.*, 26, 1257–65, 2006.
9. J. Cai, J. Ji, Y. Wang, F. Zhou, B. Yu, A novel pv/t-air dual source heat pump water heater system: dynamic simulation and performance characterization. *Energy Convers. Manage.*, 148 635–645, 2017.
10. Alessandro Miglioli, Niccolò Aste, Claudio Del Pero, Fabrizio Leonforte, Photovoltaic-thermal solar-assisted heat pump systems for building applications: Integration and design methods. *Energy Built Environ.*, 4, 1, 39-56, 2023. https://doi.org/10.1016/j.enbenv.2021.07.002.
11. K. Martin-Escudero, E. Salazar-Herran, A., Campos-Celador, G. Diarce-Belloso, I. Gomez-Arriaran, Solar energy system for heating and domestic hot water supply by means of a heat pump coupled to a photovoltaic ventilated façade, *Solar Energy*, 183, 453-462, 2019. https://doi.org/10.1016/j.solener.2019.03.058.
12. Al-Waeli AHA, Sopian K, Kazem HA, Chaichan MT., Photovoltaic/ thermal (PV/T) systems: status and future prospects. *Renew. Sustain. Energy Rev.*, 77, 109–30, 2017.
13. Meena, C.S, Raj, B.P, Saini, L., Agarwal, N., Ghosh, A. Performance Optimization of Solar-Assisted Heat Pump System for Water Heating Applications. *Energies,* 14, 3534, 2021. https://doi.org/10.3390/en14123534.
14. Chenni RA, Makhlouf M, Kerbache T, Bouzid A., A detailed modeling method for photovoltaic cells. *Energy*, 32, 1724-1730, 2007.
15. M. Bakker, H. Zondag, M. Elswijk, K. Strootman, M. Jong., Performance and costs of a roof-sized pv/thermal array combined with a ground coupled heat pump, *Sol. Energy*, 78 (2), 331–339, 2005.

23

Critical Review: IoT-Based Data Logger for SAPV System

Kshitij Tiwari[1], Anil Kumar Jha[1], Manish Kumar[1], Ayush Shankar[1], Aditi Verma[1] and O.V. Gnana Swathika[2*]

[1]*School of Electrical Engineering, Vellore Institute of Technology, Chennai, India*
[2]*Centre for Smart Grid Technologies, Vellore Institute of Technology, Chennai, India*

Abstract
This paper focuses on previous works that are done in the field of stand-alone photovoltaic (SAPV) systems and IoT-based data loggers. Different types of data loggers are discussed and are further analyzed as to how they can be developed. The challenges of monitoring SAPV systems and the potential benefits of using IoT data loggers are elaborated as well. Furthermore, a study of a cloud-based observation system for SAPV systems was presented. The outcomes indicate that the cloud-based monitoring system is a great technology for improving the reliability and efficiency of SAPV systems. In the next part of the review, the use of IoT-based data loggers for monitoring SAPV systems is found to be very advantageous. The use of IoT-based data loggers for monitoring SAPV systems is evolving in nature. Future research in this area should focus on developing more affordable and reliable data loggers and on developing more sophisticated data analysis techniques.

Keywords: Data logger, stand-alone photovoltaic, IoT, cloud

23.1 Introduction

Solar photovoltaic (PV) systems are becoming increasingly popular as a way to generate clean and renewable energy. Stand-alone PV (SAPV) systems are those that are not connected to the grid, and they are often used in remote or off-grid locations. SAPV systems can be used to provide

**Corresponding author*: gnanaswathika.ov@vit.ac.in

O.V. Gnana Swathika, K. Karthikeyan, Milind Shrinivas Dangate and Nicoletta Ravasio (eds.) Biomass and Solar-Powered Sustainable Digital Cities, (383–426) © 2024 Scrivener Publishing LLC

power for a variety of applications, such as lighting, water pumping, and telecommunications.

The performance of SAPV systems relies on a number of variables, including the sun's irradiance, the surrounding temperature, and the presence or absence of shade. To maintain a consistency in operation of SAPV systems, it is important to monitor their performance.

Internet of Things (IoT)–based data loggers are a promising technology for monitoring SAPV systems. IoT data loggers can collect data on a variety of parameters, such as the ambient temperature, battery voltage, load current, and sun irradiation. These data can be used to track the performance of the system and to identify any potential problems.

In this paper, we review the previous works in the field of SAPV systems and IoT-based data loggers. We discuss the different types of data loggers that have been developed; the challenges of monitoring SAPV systems include the following.

The integration of Simulink and Simscape models with real-world interactions through Real Time simulation lab software platform (RT-LAB), fully integrated with MATLAB/Simulink®, is highlighted in this work. The system architecture, as demonstrated in the OP4510 setup, allows engineers to swiftly design and validate real-time applications regardless of complexity.

Further experiments are planned using the RT-LAB for experiments using rapid control prototyping and hardware-in-the-loop (HIL). RT-LAB serves as a flexible and scalable platform, catering to various industries including power systems, power electronics, aerospace, and automotive sectors, offering solutions across multiple domains [1].

Prototypes consist of basic elements such as a 20*4 LCD display, ATmega 328, ACS756 50-A current sensor, voltage divider circuit, push button switch, 12-V cooling fan, and SD card module. This logger prototype takes voltage, current, power, and energy as reading. Trigger switch is used to run through steps in the logger as initiating SD card, and starting and stopping logging data. Having linear output voltage that is proportional to the current makes it easy to interface with a microcontroller or other data acquisition system [2].

A system has been analyzed using Raspberry Pi where it has been used to monitor the required hardware readings. There are two types of panels in this system, i.e., high-voltage and low-voltage panels. The system is capable of delivering a total power of 22 KW by operating a total of 8 h in a day. The discussed analysis system on IoT uses an industry's conventional meters along with Raspberry Pi, infrastructure, and visualization system, i.e., display such as laptop, mobile, and desktop. Raspberry Pi and Node.js programming are used to interface with Raspberry Pi. It helps to collect data from the company's existing energy meters and store them locally using Grafana, giving access via any locally connected device. The system has been found useful in monitoring the daily basis energy model of the implemented device and has also been able to help the idea of lowering energy usage [3].

There are Projects all about using a wireless system to keep a close watch on how the solar panels are working. Many companies rely on solar energy these days, and having accurate, real-time data about their solar panels is crucial. In the past, there their issues with technology and getting precise real-time data, making it tough to monitor solar panels effectively. So, they came up with a wireless system that can monitor solar panels anytime and anywhere. Special sensors that connect to Wi-Fi using a device called Arduino is used [4].

The emergence of various computing technologies has led to the development of advanced simulation tools that offer improved performance and cost-effectiveness. One such tool is the MATLAB/Simulink environment compatible with the open real-time simulation program RT-LAB. Using this technology, we brought about a revolutionary shift in model-based design by enhancing performance. Its flexibility and scalability have made it a preferred choice for implementing in numerous systems and control applications, effectively boosting simulation capabilities when required. The core components of a real-time simulator include the real-time distributed simulation package and algorithmic toolboxes. RT-LAB executes Simulink block diagrams on a PC cluster, whereas algorithmic toolboxes aid in designing intricate electrical circuits and some controllers. In numerous engineering models, real-time simulation and HIL applications are essential, especially in the context of power electronics and electrical systems. This article offers a thorough analysis of Open Platform for Advanced Linear Real-Time applications (OPAL-RT) technology advancements, stressing its groundbreaking qualities that have the potential to greatly advance the creation of ideal digital systems for industrial usage. The platform uses cutting-edge design techniques and technology, which find application across various technical HIL simulations, citing representative projects as examples [5].

Hybrid renewable energy systems (RESs) are becoming increasingly popular as a way to meet the world's growing energy needs. However, it can be difficult to visualize the results of impact analyses for these systems, which is why visualization techniques have become increasingly important. Over the years, many visualization techniques have been developed to address this challenge. These techniques have been the focus of research by many scientists and engineers around the globe. This review paper gives a thorough rundown of the most recent developments in the application of immersive visualization (IV) techniques for RES development. IV tools allow users to interact with and explore data in a more intuitive way than traditional 2D or 3D visualization techniques. This makes them ideal for understanding the complex dynamics of RESs.

The review article discusses the advantages of RES creation using current or potential tools, as well as the challenges that still need to be addressed. It also provides recommendations for future research in this area [6].

To create an innovative, cost effective monitoring and logger system for observing pico solar PV systems, IoT–based VIEW™ protocol is used here. The prototype consist of ESP32 MC module, DHT11 sensor (humidity and temperature), light dependent resistor (LDR) (ambient light), MAX471 (current sensor), and cloud storage server to save the data obtained from server to make it accessible remotely [7].

23.2 Grid Connected PV System Opal RT

This comprehensive overview of the RTDS (Real-Time Digital Simulator), known as the simulator called OPAL-RT, highlights their features, advantages, and usages, particularly in the electrical system domain for designing, analyzing, and also for testing of the system. The paper covers various aspects, including software, hardware, input output system, modeling, and HIL capabilities. This simulator capitalizes on the remarkable progress in computer system storing capacity and computing speed over the past decades. The result is unique capabilities such as real-time model-in-the-loop and HIL capabilities. Cost-effectiveness of the OPAL-RT Simulator further boosts its popularity among researchers. The paper underscores the OPAL-RT simulator in experimental research for power systems analysis. Using the RT-Lab software, it also presents a semi-simulation experiment involving a PV system. The key tools of OPAL-RT, namely, RT-LAB and ARTEMIS, are summarized, showcasing their applications in power electronics. Ongoing advanced research aims to enhance the speed, flexibility, and scalability of real-time simulators for predicting current and future power demands more effectively. The interrelation between MATLAB/ Simulink and eMEGAsim (RT-LAB) is highlighted as a prominent feature of the OPAL-RT simulator. The paper demonstrates the integration by showcasing a PV system connected to the grid using eMEGAsim. It serves as a valuable reference for the emerging new energy sector and microgrids. Various uses of eMEGAsim properties such as RT-LAB, ARTEMIS, and RTELAB are considered effective tools for analyzing, testing, and designing power systems and power electronics. All the features also find applications in diverse areas like automotive defense and aerospace as well as academic and research. This paper concludes by asserting that, in future, real-time simulators are pivotal for the various technical fields [8].

The growing prominence of generation of solar energy in the past 10 years has underscored the importance of understanding and modeling the behavior of PV cells. Effective and precise mathematical modeling and simulation of solar cells have gained significance for various applications, including their integration into microgrids and the broader grid as a distributed energy resource (DER). The paper introduces a custom, unified, cross platform, real-time simulation model designed for solar cells, modules, and arrays. The focus is on characterizing their performance in comparison to existing models. The paper begins by detailing the development of a simpler PV model, an enhancement over previous single-diode models. This simplification not only streamlined the model but also reduced the hardware resources necessary for real-time simulation on the OPAL-RT platform. The study investigates whether this simplification affects the accuracy of performance predictions, with a benchmark provided by the PV model from MathWorks. Once the newly proposed model was successfully verified against the benchmark, its performance was analyzed on two distinct real-time hardware platforms: the Opal RT OP5030 Simulator with OP4520 expansion unit and the Typhoon Harware-in-loop 603 Simulator. The paper highlights that the results obtained from the proposed model by real-time simulation were consistently reproducible on both hardware architectures. This consistency was attributed to the meticulous efforts made to ensure nearly identical implementations on both platforms. The study showcases that the model's accuracy and performance are maintained even with the simplification, and it demonstrates the model's cross-platform applicability and reliability [9].

This paper focuses on the enhancement of the electrical system's efficiency and reliability through renewable energy integration (RE) into microgrids. The effective controlling strategies is highlighted by using the power electronic (PE) converters as crucial for establishing a successful connection between renewable energy sources (RESs), loads, and the grid available in both modes, i.e., islanded (IS) and grid-tied (GT). Study employs an microgrid (AC) setup with PV modules and system of energy storage and conducts analyses in MATLAB or Simulink and with HIL implementations. The paper's objectives encompass a complete examination of power distribution and matching within the AC microgrid, particularly during IS and GT modes. The research investigates the roles and impacts of various control levels or controllers within the control strategies. The findings highlight that proper power sharing in the AC microgrid hinges on the correct selection of design parameters and effective controller performance. This study is centered around achieving precise power balance in microgrids. It investigates various aspects of AC microgrid architecture, including renewable energy sources, energy storage systems (ESSs), power demand, and transmission networks. The research underscores the importance of efficient energy management, system stability, and the control framework for PE converters in AC microgrids. The primary objective is to comprehensively examine the control structure across different operational scenarios and clarify the distinct roles of each controller within the system. To conduct this research, an experimental setup was developed using MATLAB or Simulink and implemented through HIL techniques with RT Lab and OP5600, allowing for realistic emulation and assessment of microgrid performance. The study reveals that every controller within the microgrid converter control system is crucial for maintaining power equilibrium while meeting energy demands and ensuring system stability. In "IS mode 3," precise controller operation is vital for proper energy distribution, whereas "GT mode" primarily relies on controllers input control current limit (ICCL) and power management control (PMC). The ICCL regulator manages fluctuations in PV output due to changing solar irradiance, and the voltage-to-frequency (V/f) regulator is essential for voltage and frequency control in Isochronous mode (IS). The PMC controller selects the correct input to provide the current reference. This investigation emphasizes the interdependence of these controllers; any inaccuracy in one controller can significantly impact the performance of the entire control system [10].

An innovative live modeling and mockup methodology takes center stage, specifically tailored for the meticulous regulation of both active and reactive power within the intricate landscape of a NPC (neutral point clamped) multilevel VSI (voltage source inverter). The study introduces a dissociated control model, founded on precise terminal current reading from the VSI. This avant-garde research conducts its simulation endeavors with precision and finesse, harnessed through the utilization of cutting-edge OPAL RT and digital simulators. Furthermore, it embarks on an exhaustive exploration of fully closed-loop models, placing substantial emphasis on the intricacies of the feedback loop governing the dynamic adjustments of active (p) and reactive (q) power. The research does not merely hint at the potential; it boldly and resoundingly demonstrates that real-time modeling and prototyping represent an invaluable avenue for the development and rigorous validation of practical PE devices, poised for seamless integration within the intricate tapestry of power networks. The paper's central narrative revolves around the real-time modeling and simulation of active and reactive power control within the NPC multilevel voltage source inverter. It ingeniously adopts a forward-thinking feed-forward control approach grounded in terminal current measurements. To further optimize simulation efficiency, a methodical and controlled switching function strategy is elegantly implemented, thereby accommodating complex power system models with larger time steps. The dynamic realm of real-time simulations is expertly navigated through the utilization of the OPAL RT and digital simulators, both of which exhibit notable and comparable performance metrics. These groundbreaking findings not only underscore the immense potential of real-time simulation but also stand as a resounding testament to its pivotal role in propelling the development and rigorous validation of PE devices that are destined to shape the landscape of real-world power network applications [11].

This paper introduces an approach to designing and implementing a reliable controller model for grid connected PV systems, i.e., fractional-order proportional-integral (FO-PI) model. The paper's focus is on ensuring the controller's robustness in the face of varying constraints, achieved through an analysis of sensitivity functions, gain, and phase margins within both the loops. Because a physical FO-PI controller is unavailable, and owing to the intricate nature of grid-tied PV systems, the study introduces an unconventional hardware implementation method. It leverages a factor-based approach, utilizing an RC ladder–type network for the inner loop of FO-PI controller, resulting in the development of an emulator circuit. This emulator circuit comprises the FO-PI controller, including factors and proportional gain, the inverter's transfer function, and a filter. Validation of this emulator circuit's performance is carried out through real-time simulation using OPAL-RT. Subsequently, the FO-PI component of the emulator circuit is integrated into a desktop target, establishing a HIL setup via OPAL RT. The paper's primary contributions go beyond merely developing a robust FO-PI controller; it explores the implementation of this controller through a HIL setup. This innovative FO-PI controller is designed to manage both the inner current loop and the outer voltage loop within a grid-connected PV system. The design methodology meticulously considers key factors such as gain crossover frequency, phase margin, and resilience against system uncertainties. For a practical assessment of the control system's performance, the paper introduces an emulator circuit for the inner current loop. This circuit undergoes rigorous testing within a customized HIL platform tailored for grid connected PV systems. The results of these experiments confirm the enhanced robustness of the proposed controller and its effectiveness in handling variations in parameters and uncertainties within the dynamic environment of grid connected PV systems [12].

The paper addresses the challenge posed by integrating solar PV systems into distribution networks, which leads to bidirectional power flow. This necessitates modifications to DOCR (directional overcurrent relay) settings for maintaining perfect relay co-ordination. Adjusting relay settings for varying PV plant capacities can be complex and failing to do so can lead to relay malfunctions. To tackle this, the paper presents an innovative adaptive protection scheme for DOCRs in distribution systems with PV integration. The proposed method involves receiving real-time data about the PV plant's capacity through a communication channel. These data are then utilized to update the trip characteristics of the relays accordingly. The effectiveness of the method is verified using a 34-bus distribution system as a test case and has shown accurate performance. To validate the proposed approach, HIL testing is conducted using an Arduino Uno microcontroller board as a relay in conjunction with the OPAL-RT platform. The primary concern addressed is the bidirectional current flow caused by PV injections of varying capacities, which disrupts relay coordination during faults in the distribution system. The adaptive protection setting, as suggested in the paper, mitigates the risk of relay malfunction under such conditions. By dynamically adjusting relay characteristics based on the PV plant's capacity, the method ensures that the DOCRs operate effectively. This adaptation is significant in both PV-connected and disconnected states. The paper's findings demonstrate the method's capability to select suitable relay characteristics for different PV output levels and fault locations, leading to improved DOCR performance. The proposed approach is also shown to be robust against potential influencing control strategies. The results of the study highlight the method's competence in enhancing relay coordination in distribution systems with integrated PV. The comprehensive validation, including comparative results with HIL testing, confirms the viability of the proposed approach for potential applications in distribution systems. In summary, the paper introduces an adaptive protection scheme that effectively addresses DOCR setting challenges caused by bidirectional power flow from PV integration in distribution networks [13].

The paper focuses on addressing power outage challenges faced by utility companies, especially in delivery circuits with high reusable energy penetration. To comprehend impacts of error conditions on the power grid, the study compares different simulation approaches while considering fault currents from DERs. The objective is to accurately analyze the contributions of DER systems to fault currents and their influence on coordinated protection schemes. The study employs MATLAB/Simulink/RT-Labs for modeling and simulation of a reduced-order distribution system. It introduces fault conditions at various bus locations and assesses the response under steady-state and dynamic scenarios. Real-time power HIL simulations are also used to enhance model fidelity. Additionally, physical hardware PV inverters are incorporated to validate the results. Findings indicate all ways strictly correlate under steady-state conditions. However, capturing transient responses, such as fault current rise and settling times, proves challenging with PV models. The physical behavior of the PV inverter is best represented through PHIL simulations. The paper highlights that dynamic simulation tools provide valuable insights into transient behaviors, aiding in selecting appropriate protection devices like relays, fuses, and reclosers. The study underscores the benefits of each simulation tool. Incorporating a physical PV inverter in PHIL simulations allows for a comprehensive study of circuit behavior during fault conditions, offering detailed insights. While PHIL simulations provide the most accurate results, they require additional setup effort. For cases where approximate fault dynamics suffice, the Simulink option is attractive due to its efficiency. Moreover, the study demonstrates the feasibility of programming realistic response dynamics, such as voltage ride-through and brief termination, in simulated PV inverters. In conclusion, the article contains a comprehensive comparison of distribution system modeling tools with a focus on high PV penetration and fault conditions. It underlines the significance of dynamic simulations for capturing transient responses and highlights the trade-offs between simulation accuracy, complexity, and computational efficiency across different approaches [14].

The employment of next-generation systems grounded on unorthodox energy sources has brought significant changes to the electrical infrastructure. Distributed generation and network automation have been introduced, with information systems playing a crucial role in monitoring and controlling these power systems. The continuous improvement of SCADA systems has enabled remote network operation. A noteworthy addition to SCADA systems is the concept of state estimation, which aids in automation and diagnostic tasks. Real-time simulation is essential for engineers to implement unique way effectively in automation, monitoring, and control systems. This work points the importance of real-time simulation in validating newer automation designs. A SCADA system utilizing ETAP Real-Time facilitates state approximation implementation and is verified using OPAL-RT technologies for real-time emulation of the network. The integration of these platforms in real time enables simultaneous signal processing and acquisition. ETAP Real-Time developers have created a platform that employs calculation engines for applying the state estimator, a mathematical model reflecting the current network conditions. The system's validation involved modeling the power system and performing real-time emulation. The use case involves a SCADA system capable of detecting measurement inconsistencies and conducting predictive study on the network. Online approximations play a role in implementation of automation and control systems, allowing mathematical models to be compared with real systems. Considering the ongoing progress in automatic monitoring and control systems dependent on information systems and new technologies, RT simulation is pivotal for validation and configuration testing. It offers the ability to execute these tests on a network emulation using OPAL RT. With tools like the state estimator and network simulation, early detection and mitigation of exceeding loads and voltages, and other irregular system phase become feasible. This approach accelerates system implementation, reduces risks, and identifies issues in SCADA configuration through a network replica [15].

This paper introduces a methodology and algorithm designed to validate the performance of grid connected PV system. The study focuses on observing how changes in irradiance affect PV arrays and subsequently impact the load side of the converter, all facilitated through real-time simulation. This approach enables the testing of system behavior in the presence of uncertainties in PV power production while also evaluating the controller's ability to maintain a consistent DC bus voltage. The methodology employs the OPAL-RT platform as a HIL setup. The system is modeled using OPAL-RT's MATLAB-based library, with PE gears directly applied in the FPGA segment of platform. Concurrently, the controller and mathematical PV array model are implemented in the CPU section. By utilizing this real-time platform, the paper showcases how irradiance changes influence various system variables. Moreover, the presented HIL simulation platform serves to verify the effectiveness of the system control algorithm's design. As microgrids and renewable resources, particularly PV arrays, gain prominence, the need for reliable and efficient test beds becomes vital for researchers in this domain. This paper addresses this need by proposing an algorithm that validates the performance of a grid-connected PV system while also examining the effects of irradiance changes on the system's converter and load side. The integration of OPAL-RT as a HIL platform, along with its dual implementation of power units in FPGA and controller/mathematical models in the CPU, enables effective real-time simulation. The results from these simulations validate the controller's efficacy in achieving the system's desired objectives [16].

This research introduces an enhanced mathematical approach to establish the active and reactive power (P-Q) capability curve for grid-integrated solar PV inverters. The aim is to optimize the injection of reactive power into the distribution grid while maintaining the PV system's operation at the maximum power point, considering various environmental conditions. The proposed methodology involves evaluating all feasible operating points within the stable operational range of the system. The study employs a software-in-the-loop condition using the OPAL-RT OP4510 RT digital simulator. This simulator integrates the coordination of several control loops, including the dc-link voltage regulator, MPPT controller, and real and imaginary power controller. The objective is to authenticate the enhanced mathematical methodology's effectiveness in managing reactive power compensation and active power generation for the grid-integrated PV system. The research demonstrates the methodology's validity through a software-in-the-loop simulation environment and by comparing measured active power injections with the derived P-Q values [17].

The article discusses the advancements in electric vehicle (EV) charging techniques, particularly focusing on wireless power transfer methods. The motivation behind this development is the pursuit of fully autonomous vehicles, which has led researchers to explore wireless charging options to address the limitations of grid-dependent charging systems. Traditional EVs relied solely on grid power, which increased energy consumption from the grid and faced challenges related to vehicle weight and battery size.

Here, the articles shows enhanced systems that employ PE converters to interface PV power with the EV within a residential grid setup. This bidirectional system not only allows excess PV power to be fed back to the grid but also enables EV charging using grid power when solar power is unavailable. A bidirectional cycloconverter is employed to control power flow by adjusting the frequency of operation. To validate the proposed system, the authors created an experimental prototype integrated with a real-time simulator. They compared simulation results with experimental data, demonstrating the system's smooth operation across various conditions, both with and without solar power. The system achieved an efficiency of approximately 85%, highlighting its effectiveness. An intelligent controller was utilized to manage different operating modes seamlessly, ensuring uninterrupted EV battery charging. The alignment between observed results and expected outcomes underscores the viability of the grid-connected wireless EV battery charging system [18].

Here, we see RT simulation design of a LT microgrid configuration, which encompasses reusable energy sources, a battery ESS, and a modular multilevel converter (MMC)–distribution static compensator (DSTATCOM). The main goal is to facilitate reactive power reimbursement to provide both the dispersal network and utility loads. The modeling and simulations are executed using MATLAB/Simulink, and the real-time testing framework is built around an OPAL-RT simulator. The study explores various operational aspects of the microgrid system, including grid-feeding, grid-supporting, grid-following, and grid-forming behaviors of the microgrid inverters, all through offline and real-time online simulations. The operation of the MMC-DSTATCOM is analyzed under different scenarios involving reactive power control, encompassing both grid-following and grid-forming modes. The focal challenges addressed is achieving accurate real-time simulations when working with switching PE converter designs and their controls while minimizing computation overruns within the RT-Lab platform. The findings of the research highlight the successful implementation of microgrid and switch converter operations via live simulations, even with a notably low step size of 25 μs. The MMC-DSTATCOM effectively tracks its reactive power reference and fulfills the reactive power demands of the microgrid system across various scenarios during real-time simulations. Future work is anticipated to involve the adoption of this real-time simulation testbed for power HIL testing of a 100-kVA–rated three-phase MMC prototype hardware operating in a DSTATCOM mode [19].

The paper addresses the issue of potential fluctuations in minimal power supply grids that integrate DERs, particularly during peak DER generation and minimum load requirements. These situations often result in power quality (PQ) difficulties at the PCC (point of common coupling), including voltages sags, voltages swells, and harmonic. These problems can lead to discontinuous and unregulated power supply to loads. To mitigate these concerns, voltage regulation at the PCC becomes crucial. The paper proposes a way to control PCC potential by linking reusable energy to filter to the grid (RC filter). Filters reduce harmonics to an extent that is in line with the IEEE standards. The study includes a relative study between energy from sun with the help and without MPPT to assess their compatibility for grid integration. Through MATLAB simulations, the effectiveness of the proposed structure is demonstrated. Studies evaluate the different systems for integration of grid, analyzing the Total Harmonic Distortion (T.H.D) values for both. Integration of solar PV modules into the low voltage grid reduces harmonics through a simple low-pass RC filter. Stabilizing resistor in series with capacitor contributes to harmonic reduction and voltage stabilization. The bi-directional converters, together with a battery cells storage system, enable autonomous power management and help regulate voltages at the PCC. Charging modes are presented graphically, showcasing the benefits of the battery storage system. The system is also capable of compensating for voltage sags using renewable energy sources and battery converters, as illustrated in fault scenarios. Overall, the study provides a way to control PCC potential using renewable energy sources and an RC filter, reducing harmonics and ensuring compliance with PQ standards. The combination of bidirectional converters, battery storage, and autonomous power management contributes to energy savings and improved PQ, particularly beneficial for distributed generators like rooftop solar panels [20].

This paper outlines the process of conducting automatic HIL examinations on protecting relay using the simulators, which is based on real-time power system. These tests serve two main purposes: validating the presentation of new relay beforehand they are deployed in electrical substations and enhancing power system reliability. The paper provides an overview of the RTDS simulator's software, hardware, and applications. It also presents digital imitation design of power grid that acts as a standard for HIL tests involving the SEL-421 distance relay. The main advantage of RT digital simulation tools like RTDS is their ability to make accurate model of various power network configurations, facilitating comprehensive studies and analyses within power systems. One distinctive capability of these simulators is their capacity to perform HIL tests on physical protection and control equipment under realistic power system operating conditions. This attribute has led to their rapid global adoption. Moreover, the continuous interaction with real-time simulations allows electrical utilities and universities to use these simulators for education, training engineer and technician to enhance their understanding of power systems operations. Developing rigorous HIL tests for protective relays using script files is crucial for assessing their performance across diverse operational situations and circumstances. These tests help ensure the reliability of power supply while safeguarding the electrical organization. RTDS, being a broadly accessed device, finds applications in electrical places, manufacturing factories based on electrical system, research centers, and campuses. Notably, in Mexico, the Testing Lab of Equipment's and Measurements employs the RTDS simulator to validate and certify the performance of innovative protective relays. In summary, this paper highlights the significance of HIL tests for protective relays using RTDS simulations in ensuring power system reliability and enhancing engineer and technician training [21].

This paper presents two advanced and easy simulation tools designed for analyzing power systems: PE and drive systems. One is PSB (power system block) set that is an offline mockup tools. The other tool is the Real-Time Simulator Hyper Sim. The paper demonstrates their capabilities through an application example involving an induction machine AC drive. The MathWorks PSB is tailored for power electronics and electric drive examination and modeling. It features controlling inconstant step integration approaches, making it well-suited for accurately simulating complex systems commonly found in the electric drive industry. However, this accuracy comes at the cost of increased computation time, which escalates with the system's complication and dimensions. Alternatively, the real-time simulator Hyper Sim employs static integration time steps and parallel computing. This enables it to provide both accuracy and high computational efficiency for heavy industrial system containing numerous electric drive and intricate device. Hyper Sim offers real-time and accelerated-time simulation modes, making it adaptable to various scenarios. In summary, the paper presents two robust simulation tools: the power system block set (PSB) for precise offline simulations, particularly for complex electric drive systems, and the real-time simulator Hyper Sim, which balances accuracy and computational speed for extensive industrial setups containing multiple electric drives and intricate components. The tools' capabilities are showcased by implementation example involving an induction machine AC drive [22].

23.3 IoT Advancements and Implementation in SAPV Monitoring

The simple prototype like normal data logger: Consisting of an Arduino board as its main processor it also employs a voltage sensor, a temperature sensor (LM35DZ), a real-time clock (DS3231), a memory unit (Micro SD card module), a controlling unit, and web applications. Here, the PV panel relates to voltage and temperature sensor that are again connected to Arduino. Arduino processes the data and displays it on the LCD screen in real time. It also saves the readings to SD card on module. Arduino has its own DC supply. Also, the data on memory are sent to cloud to for remote access and storage. A separate web application was developed and HTTP was used to proceed [23].

A data logger is based on IoT which specifically used Lora WAN communication to reduce the power that was consumed in monitoring system on the use of other protocols. Prototype uses a HelTec ESP32 LoRa, ACS 712 Current Sensors, 12-V light for loads. The operation of this data logger is as simple as previous data loggers discussed. The data are received from the sensors and are sent to LoRa module for processing. The esp. module displays the same data on the OLED screen and, afterward, sends it to ThinkCloud using Wi-Fi. The sensors here and the module are provided with separate power supply. Here, for load, we have taken several 12-V lights for observing different ranges. ThingSpeak is a cloud-based IoT analytics software for gathering, picturing, and analyzing live information streams [24].

A monitoring mechanism and data logger for a PV plant: Here, other than the normal sensors and cell, a new way is introduced to process the data and its storage. This proposed design includes Arduino Mega as well as Raspberry Pi for the data procession and further execution. This design offers two platforms for access of the data: one is the My Sequel cloud storage that is equipped with numerous features of PHP, CSS, HTML, and JavaScript enabling it to display the data in phone, tabs, or PCs along with their curves or graphs over time. The other platform is the touch screen display fit in the box. Also, this prototype uses a virtual button on platforms to control the relays enabling a mobile access over the system remotely. For the offline use, the data are stored in USB that is connected to Raspberry Pi ports and stored as .csv files [25].

Also, the proposed design incorporates an esp32, at Mega, Arduino board, ZMPT101B module, ACS712 current sensor, clamp meter. Here, we have two readings for the same data of voltage and current: one is through the sensors and other one is through the clamped meter. Here, the data after being read is sent to esp32, which, like any data logger, forwards it to cloud for further processing. The result of the sensed and measured readings was compared and found to be approximately the same. A mobile application is designed to display real-time data to the user [26].

Creating and trying a smart IoT-based data logger that was mainly created from and was using Open-Source Tools for its normal functionality: For this purpose, an Arduino™ UNO board along with a real-time clock for sync purpose and two external ADCs of 18-bits resolution for complying with the IEC61724 precision criteria and sensors signal conditioning for adapting and filtering unwanted signal, and a module to minimize energy usage on an extra PCB board was added to terminate the limitation of simple Arduino board was used. Also, low-cost sensors were used. ThingSpeak™, an open IoT data platform was used. Mobile data communication was selected due to its wider area of users. An Ethernet Shield stacked to board helped Arduino to connect over internet. Mobile app was used to access the cloud data and for observations. Data logger was reliable and was able to reconnect in case of power shortage or failures. On comparison of the received data and that of 22-bit resolution high precision data logger, the readings were reliable and complied with IEC standard requirements of accuracy [27].

A low-cost DAQ device for monitoring the parameters of SHS prototype PV system: The prototype used an Arduino board with ATmega 2560 chip microcontrollers and normal parameters sensors, real-time clock

(DS1307), SD card, and its plug-and-play module along with an LCD to display the data. This prototype has used three voltage (A1, A3, and A7) and thee current (A0, A2, and A6) sensors to measure the current and voltage at PV panel output, battery charge, and DC load. Two temperature sensors (one for front temperature of panel and other for ambient temperature and humidity) of DS18B20 are used on front and the back of the file with time stamp. Separate 9-V supply has been provided for Arduino board to run. A reset button has been provided for restart. The results were observed and device was found reliable [28].

An original and better-quality form of this solar Arduino™ economical data logger based on IoT: Here, the Arduino™ Ethernet (Rev 3) board was used, which was compatible to connect to internet and at low cost than other options available. TL-MR3020 router of TP-LINK was nominated due to its cost effectiveness. Here, the clock is coordinated in the initial of routine or when reset button is pressed and at 00 AM once per day. Internet connectivity provides the capability to synchronize the clock from NTP server. After the connection is set up transfer of data takes place. MySQL database in main server was used as storage, and an open cloud service ThingSpeak was used as secondary to visualize and analyze live data. Data visualization in real time and remotely was one of the many perks of this prototype [29].

A remote monitoring system for PVs with the following: 1) corrosion and noise-resistant temperature sensor; 2) long-term continuous data logging; 3) high-temperature endurance; and 4) ease of retrofitting into existing systems. Sensors are attached to module to transfer the digital temperature data. A modern, cheap, and innovative microcontroller board armed by the Wi-Fi module that is used in the advanced solution script is used in transmission and entered in a cloud spreadsheet with remote view access. An alert is sent to handler in case temperature reaches outside a specific value using email or in case of an error. No batteries have been used in this design instead a small solar panel provides with the power that is needed for the system to sustain [30].

A design prototype is based on a microcontroller STM32. Here, the solar panels are connected to chargers/controllers that are again in connection with accumulators. The inverter is connected to the accumulator and house supply. First, the microcontroller senses the current from the panels, and, again, the voltages and current are measured from the accumulators' values that are processed, and Linux-based single board computer is used to send the data to main server. HSLR 10-P SP33 sensors were used to measure current. The inverter was turned off by force in case of accumulators' voltage dropped low, i.e., to 10 V/8 V [31].

The paper intending to design an IoT platform for application based on mainly renewable energy. Here, a case of agro food application is taken. Different things were proposed in this paper including a solar dryer, a compact meteorological station, a software for interaction with the data, etc. The solar dryer here used a hybrid source of power and a backup for temperature stability. Also, these systems are equipped with remote relays for controls. Weather station was designed to measure ambient temperature, moisture, compression, route, and wind speed at 2 to 10 m. Protocol was used for exchange of data. MySQL server was used for data storage. A grid-associated PV system was used for a case, whereas LA [16] and LV [17] sensing devices were used for measuring the AC and DC current and voltages, ESP as remote terminal unit for low cost, and ADSL internet Wi-Fi router was used for higher bandwidth connection speed. The graphical user interface (GUI) was designed using Node-red and JS [32].

A chapter talking about the incorporation of IoT in monitoring system for PV system: Also, how integration of AI can help in automation and adaptation. It talks about different layers of architecture for the incorporating IoT in PV systems. The four levels are IoT hardware, communication layer, cloud infrastructure, and data analytics. It also explains the fields in PV system, which can be benefitted from IoT as detecting the fault early, giving early maintenance warning, thus reducing the overall downtime for a system. Also enabling it can optimize the system to a great extent by automatically configuring the system based on weather condition and system's health [33].

This paper intends to build a monitoring system for PV system on least expenses, which can monitor different parameters of panel continuously. The prototype is based in Arduino nano. For the data transfer to internet transmitter, NodeMCU V3.0 was used. An ACS712 current sensor was used and basic voltage divider theory was used for the measurement of voltage. ThingSpeak, an open-source cloud storage, was used. Overall conclusion comes to a point that this prototype was almost like that discussed in most of the papers above [34].

This work intends to design a low-cost prototype for monitoring of PV systems based on IoT. The design is mainly based on Arduino mega 2560. An ESP-01 Wi-Fi unit has been used for transfer of the information to host computer. A current (ACS712), voltage, and temperature (LM335) sensor is equipped in the system. A LCS 4*16 is present to display the live values of the sensing devices at site. In the process of data logging for this prototype, first ESP01 and LCD are initiated and the module is connected to router, and, then, the values are read and transferred and displayed on loop. This prototype brings in the benefits of reduced human intervention, low cost, easy implementation, and real-time monitoring [35].

A design based on Arduino and ESP32 MC: A special software "Blynk" has been used, and the code is written Arduino IDE. The PV panel is connected to voltage and current sensors that read the data and send the values to Arduino. NodeMCU ESP32 that is connected to internet reads the values from Arduino. The next stage in this prototype is Blynk software to display all the recorded values in curves and graphs in cloud. The Blynk server is reliable for the transfer and communication [36].

A wireless cost-efficient PV monitoring has been designed. A set of microcontrollers, photoresistors, current sensor, ADCs, solid state relays, and ZigBee (XBee) advanced wireless microcontroller was used for the design prototype to make it wireless, and LabVIEW-grounded data logging software was used for registering the sensor values. Different parameters were measured including OC voltage, SC current, ambient temperature, and cell temperature. Also, real-time analysis is in the form of graph and curves, which was done for amperage–potential and also P-V characteristics [37].

Also, a prototype of real-time IoT-based solar monitoring system: For the hardware parts, ESP8266-based NodeMCU controller, a digital humidity sensor, temperature sensor (DHT22), current sensor (ACS712), impedance divider sensor, and a photo diode sensor. A real-time clock, ADC, and SD card module have also been used. An open-source cloud software ThingSpeak has been used. The system initiates and checks for Wi-Fi access availability, and, then, all the sensors will check for RTC modules time. All the sensor data are written on SD card. The microcontroller unit with combined Wi-Fi opening is employed to transfer the same information to the cloud. This process keeps on in loop till power off the device. ThingSpeak platform can be used to analyze the data graphically and make sense out of it [38].

23.4 Photovoltaic System and IoT Innovations

This paper has been held up for a design to reduce the energy consumption in the PV system in such a way that the system to be designed and used on the demands of consumer requirements; the main system used is consists of voltage sensor, current sensor, and a battery charging capacity with ATmega328 MCU. The entire data acquired are saved in an SD card, which was extracted for monitoring within every 2 min of interval. All the information is changed from analog to digital using ESP32-S2, which is used as a minimal powered data logger. Here, the data logger offers dual mode, i.e., deep sleep and web server mode, accessed using a toggle button. The deep sleep mode will log the data after every 60 s, and the web server mode gives access to accessed device connected to the local network to monitor or log the data [39].

Here, a PV array that is having a capacity of 10 kWp, battery, and a connection to a grid network. The system named PVGCS system was installed on an area of 150 m^2 near to a laboratory. The PV system is consists of 56 Weibel Solar W1750 PV modules. Each power module gives a SC current of 5.35A, an OC potential of 44.5 V, a maximum power point (IMPP) current of 4.96 A, and a maximum power point (VMPP) voltage of 36.3 V. In this environment, a 10-kW peak o/p is possible in the PV array. The energy obtained is forwarded to the close by loads, and the remaining power is fed to the grid, i.e., approximately 5 kW. The power of the storage batteries included with the plant is a 48-V 24-cell series connected to 800-Ah capacity. The unit is equipped with a nominal Hoppecke OPzS 800 lead-acid battery with a capacity of 800 Ah. For 10 h of discharge, the discharge capacity is estimated to be maximum 915 Ah with respective 2-V voltage. There are two arrays each providing 5-kW power and, if arrangement results surpass the local load, then additional power would be supplied to charge battery. The second array always wants to charge the battery when the battery is fully charged [40].

With an observable growing interest in the field of solar SAPV systems, it can also be noticed that, as there occurs a drip in potential that is quite dependent to temperature instead of increasing current, hence the main objective here is supposed to be the size of the model compared to the whole PV system. This voltage at this decrease will affect the strength of the DC string as well as the AC component voltage. The findings of the system are compared to the methods employed, mean square error, the root mean square error, the error of the mean bias (MBE), and the standard deviation error (SDE) between simulated test intensities that are reduced independently of addition in radiation. Hence, the system size can be reduced from 2.8% to 6%, and the difference in load interruption due to sudden disturbances such as clouds passing over the PV system is more than 0.14 s (seven cycles); if it should be specified module, then the temperature should be measured accurately [41].

In this case, the PV systems have been used with the help of monitoring by a MPPT (maximum power point tracking) system, where the base model is based on a FLC (fuzzy logic controller) that is comparatively easy to set, works well for nonlinear systems, and does not require exact parameters for the system. PLECS software has been used to help in the designing of DC-to-DC converter for the system, which is further used as an impedance between the source and the load element of system. The system was then linked to PV module and tried for a number of environmental conditions and compared with other PV systems to obtain the maximum power that is built on FPGA (field programmable gate array), and the name of this method is P&O (perturbation and observation) method. The results proved that the system has less number of oscillations, and a quality of maximum power and high response was achieved [42].

Power loss analysis works in an efficient way for keeping an analysis for large power plants and optimizing hygiene routine, but it is not suitable for remote data logging of isolated PV systems that are used in poor rural-areas as it is expensive and prone to radiation to irradiance sensor. The first method of determining pollution levels is the power loss calculation, where the quantified electrical current is related to the anticipated current calculated based on output of radiation. For direct comparison, measure the pollution level but, for these two cases, the cost of pollution levels, sensors for measuring radiation level, pollution levels, or even atmospheric constraints is too innegligible to be afforded by compact SAPVs, and the radiation sensing device is vulnerable to pollution. It affects the soiling estimation at remote locations. Artificial neural network (ANN) is a prototype that can accurately estimate the pollution loss of the PV power output using radiation signals or the pollutant characteristics; ANN is applied to environmental factors for monitoring the derating factor. They also are being used to predict the PV damping in the form of a function [43].

Today, PV systems can translate up to 0.2 part of solar power into electrical energy. The residual energy is used in the PV module (~0.1 of total) or transformed to unused heat (0.7 of unit), which can increase the present temperature of PV cells and reduces the solar-to-electricity translation competence. In proportion of greater than 3:1 (thermal electricity), here, the unused heat offers upon an important chance to reprocess the waste power. About 10% in the US, primary energy usage includes space and water heating in urban buildings. Majority of this energy comes from upper end emitted fossil fuels. The SunPower solar cell exhibited an efficiency of 57.4% at ambient temperature and an electrical efficiency of 12.3% with maximum temperature at 80°C. Given this performance, it shows a cumulative PVT of 226 kWh/year for solar power at 5.5 kWh/sq m/day (kWh) electrical energy and 603 kWh of thermal, thus avoiding natural gas consumption of kWh 1280 and 183.8 kg CO_2 releases. PVT solar gatherer cascaded along parabolic concentrator with phase changing constituents achieving 40% to 50% solar to heating efficiency [44].

In this system, a PV system has been designed in such a way that it develops an algorithm for MPPT. The system is in command of IoT control system, where the DC voltage and current parameters are measured by a real-time Arduino board. The converter used is a modified ZETA that is kept between solar module and variable. The Pulse Width Modulation (PWM) is generated using the same IoT module for the maximum power required MPPT and further used to calculate the duty ratio of the power switch under several isolation levels. All the data fed to the Arduino are stored in a cloud after being recorded through the system. In the discussed system, IoT-based PSO (particle swarm optimization) and MPPT technique are being implemented, which are suitable for the consumption of wireless networks, energy management, etc., of the given PV system. The key function of PSO-IoT–based procedure is transmission of the power to load from the PV module. For continuous and max power extraction from each PV modules for a maximum power tracking, PV current is altered. The PSO-IoT–based MPPT algorithm can be used to set the desired inductor current, which will then be maintained by the system. This can be useful for ensuring that PV system operates at its maximum power point [45].

A research study in Malaysia found that the accumulation of dust and other particles on PV systems can result in significant power losses. A new mathematical model was developed to correlate level of soiling with area of the PV module. The model revealed that soiling can reduce the output power of PV systems by an average of 26.22% in Setiu, Malaysia, over a period of 12 months. Soiling occurs when dust and other particles collect on the surface of a PV module. These particles can obstruct sunlight from reaching the solar cells, which reduces the amount of electricity that the PV system can generate. The severity of the soiling issue can vary accordingly with topography of PV system. For example, PV systems in areas with dusty or sandy conditions are more likely to be affected by soiling than those in areas with clean air. In addition to soiling, there are other factors that can affect the resultant of a solar system. These factors include the efficiency of the PV modules, the efficiency of the inverter, and the losses due to cable resistance. When estimating the output power or yield of a PV system, it is important to consider all of these factors [46].

The building is a system of combined semi-transparent PV (BiSPVT) system that integrates translucent PV modules along with houses and enables heat energy harvesting and daylighting; the study discusses about the development of the thermal model for a 7.2-kWp BiSPVT system and the evaluation of the annual energy consumption of the system. The efficient performance of the SPV system was analyzed on the clock live, i.e., by balancing the system [with the system connected power conditioning unit

(PCU), non-DC loads, and battery storage). Mean installed capacity of the SPV system at peak capacity by the PCU was said to be 6.90%. The output produced by the BiSPVT system in a time of 1 year is reported to be 8,927.1 kWh. The use of translucent PV (SPV) modules in a structure has grown important in past period of time because they contribute to solar power absorption and have better electrical productivity compared to other PV units for the identical packing factor and with a comparatively higher isolation on the roof system [47].

The PVWPS+LIB facility for a set of rooftop systems consists of eight polycrystalline PV modules with a capacity of 305 Wpk. The modules are located on the west side of the roof, facing diagonally south at a tilt of 30°. This orientation and tilt provide the best conditions for solar power absorption and yearly production of energy. The modules are cascaded. A radiation sensor is attached on above side of opening beside the starting two solar modules. The sensing device also is attached to an aluminium plate that supports the modules, and it has the exact special arrangement as the PV modules. The ambient temperature recorded in space amid the first PV modules and electrical area situated in the back them is recorded. This is because the sensors have a limited length of cable, so they must be located near the maximum power point. A marketable general networking protocol, the UWP 3.0, acts as a junction among all devices that communicate using Modbus-RTU over RS485 and data attainment system (based on Modbus TCP/IP). The devices that do not have RS485 result are linked to ADAM 4017+ part, which is a 16-bit worldwide analog input unit along with Modbus output.

Ethernet port and Raspberry Pi were linked after discussion. The Raspberry Pi uses TCP/IP rules as client of UWP3 and transmits the information to the database server at a specified frequency. The data collected through Raspberry Pi and exchanged through its Wi-Fi port are established on the server via local Ethernet system [48].

The use of IoT solutions for maintaining and evaluating PV solar systems: IoT technology employs specialized sensors and connectivity for real-time monitoring and data analysis of these solar energy setups. By continually gathering data on parameters like performance, temperature, and efficiency, IoT enhances maintenance by spotting issues early and enabling quick action. This proactive approach ensures PV systems run at peak efficiency, minimizing downtime and maximizing energy production. Additionally, insights from IoT data analysis aid in optimizing system design and operation for long-term sustainability [49].

23.5 IoT in Hybrid Systems

Far-off safety and healthcare IoT application implementation using edge-based hybrid systems: Information is directly shifted from the sensor grid to the cloud through gateway to many traditional IoT applications for additional processing. A low-range IoT wireless protocol like Bluetooth Low Energy is an example of where this strategy may not always be appropriate. In this paper, a novel hybrid edge network system architecture is presented. An IoT gateway and hybrid routers are included. In order to increase their longevity, these routers have solar energy harvesters and support both BLE and LoRa wireless protocols. Using LoRa wireless technology, they may increase the range of short-range BLE networks and manage crucial edge computing functions, namely, early data handling. The IoT entry is adaptable and supports a number of IoT protocols. Advanced edge computing duties including data sorting out, storing, processing, user interface work, and cloud connectivity can all be carried out by it. The article offers three case studies on smart hospitals, wearable safety monitoring, and healthcare monitoring. These show the system's capacity for a range of IoT applications. Experimental analyses demonstrate that it achieves minimal delays of just 11.5 ms by processing data at the edge. Additionally, BLE coverage can go up to 2.4 km with the hybrid LoRa network [50].

Intelligent campus hybrid RES: Effective real-time HRES monitoring is crucial in this paper. It offers precise information that helps system administrators to evaluate performance and identify any unusual circumstances. A proposed IoT-based architecture for HRES is presented in this paper. Wind turbine, solar system, battery storing device, and one diesel generator are included in this architecture. Energy, data attainment, communication system, and application levels make up its four tiers of organization. The IEC 61850 standard is used in this work to develop communication models for HRES because HRES frequently uses a variety of communication technologies and lacks a consistent communication model. Information about electrical, status, and environmental factors makes up the categories of monitoring parameters [51].

Monitoring system for PV and thermoelectric generators utilizing an Arduino-based data attainment system method: The growth of a data attainment system for a hybrid PV-TEG is discussed in this study. The DAQ continually records data while monitoring PV-TEG system parameters like voltage, current, and power utilizing an Arduino Mega 328P board and ATmega 328P processor. The 200-W panels and 192 Thermoelectric generators (TEGs) are combined in the hybrid system through series and parallel connections to boost power output. These data are converted by the DAQ and kept on an SD card. The system's performance was evaluated, revealing key findings. The DAQ stored data at 1-s intervals and recorded maximum parameters for the hybrid system, PV subsystem, and TEG subsystem. Notably, the hybrid system produced a peak power of 191.1 W, representing a significant 15.38% improvement over the PV subsystem alone [52].

DC potential and current measuring instruments and data loggers are devices that quantify the current and voltage produced by wind turbines and save the measured results. These devices are used to trigger the mock load on the wind turbine control system to reduce the voltage that exceeds the safe limit, thereby saving electrical energy.

The research aimed to design, manufacture, and analyze a DC electrical potential-current measuring tool by means of an Arduino grounded data logger. The instrument was designed to be accomplished with quantifying the DC and potential produced by hybrid power plants (PLTH) and to use the metrology technique to analyze the extent of ambiguity of the measuring gadget. The study was conducted in three phases. In the first phase, the hardware of the instrument was designed and assembled. In the second phase, the software of the instrument was developed. In the third phase, the instrument was tested to determine its accuracy and reliability. The results of the research showed that the instrument was able to measure the DC and potential produced by PLTH with high accuracy and reliability. The degree of uncertainty of the instrument was found to be within the acceptable limits. This research has played an important role in the progress of DC electrical voltage-current measuring instruments. The instrument developed in this research can be used to improve the efficiency of wind turbine control systems and save electrical energy. The hardware design of the device began with testing each component to be used, designing the device, and creating the working system of the device. The sensor used was WCS1800 for the current detection and CR5310-300 as the voltage instrument, as inputs that would be processed by the Arduino Uno microcontroller. The results of the readings from sensors were analog signals so that the output of the current sensor was on pin A1, and the output of the input voltage sensor was on pin A0. Pins A0 to A5 were the ADC pins or input pins to convert analog signals into digital signals in the form of binary numbers. The binary numbers were then being processed by the Arduino to be converted to volts for the voltage sensor output and amperes for the current sensor output. The results of the conversion were displayed on the 16×2 LCD and stored in the SD card [53].

The system considered in this paper consists of three PV modules with equivalent power 480W, a wind turbine power of 400 W, a lead-acid battery with 30 Ah, an inverter, and controller. This paper explains the system approaching to a stand-alone HRES with a low level of complexity in MPPT and lower number of components comparatively to minimize the price and enhance the output power efficiency. A simple HRES normally has a PV system, a wind turbine, and a storage device. Here, an inverter is used to for DC-to-AC conversion after when the DC voltages have been controlled through a DC-to-DC converter and to alter the resultant voltage at the essential amount under several circuit situations. The battery's state of charge (SOC) talks about how much of battery's volume is required to safeguard the battery's health, span, and efficient working. The depth of discharge (DOD) of battery here shows part of volume used in a charge cycle. There are two modes of charging, i.e., float charging and bulk charging [54].

In this system, hybrid solar trackers are used for optimal and maximum solar energy efficiency. This system has not been maintained at much greater level and is not implemented in an adaptable manner. A simple and economical DAQ has been used for keeping it eco-friendly and to positively design for data attainment and automatic turning on a hybrid PV monitor. The results show the strength of sunlight at 3.5 V, and verge voltage is acting as the switching constraint of a hybrid PV tracker and is affecting LDR voltage. Data transmission systems with telemetry 3DR module for the maximum distance of the blocked area is 120 m^2, whereas the blocked area is 350 m^2 and can reach an efficiency of up to 51.56% as compared with power aberration of 183.88 Wh. The power output of the tracker PV is 540.47 Wh, and the energy usage of the chief unit is 71.68 Wh. Hence, the potential power storage is found out to be 468.79 Wh [55].

A comprehensive guide to PV/thermal (PV/T) systems, which are hybrid devices that convert solar radiation into electrical and also thermal energy: PV/T systems are becoming increasingly popular as a way to generate clean and sustainable energy, and this book provides the latest information on their design, operation, and applications.

The book begins with an overview of the PV/T technology, discussing its history, basic principles, and advantages. It then goes on to cover the different types of PV/T systems, as well as the materials and components that are used in their construction. The book also discusses the factors that affect the efficiency of PV/T systems, and how to optimize their performance.

Keeping in mind the technical aspects of PV/T systems, this book also discusses their applications in different sectors, such as residential, commercial, and industrial. It also provides a detailed overview of the current state of research in the field and the challenges that need to be addressed in order to make PV/T systems more widespread.

This book is an essential resource for anyone who is interested in learning more about PV/T systems. It is also a valuable reference for engineers, architects, and other professionals who are involved in designing and implementing RESs [56].

23.6 IoT in SAPV and Its Data Logging (An Analysis)

To reduce the energy consumption, the solar PV system is usually connected in modular form to make it versatile and flexible with deviation in load uses accordingly to the local utility grid. Constant data logging and behavior analysis of SAPV play a key function in getting a grip on its optimality. The main discussed system in the paper uses a voltage sensor, current sensor, and a battery charge ability by using ATmega328 MC. Data is stored in SD card at interval of a minute, which can be extracted by using a computer for monitoring and analysis purpose. All the data are converted from analog to digital using ESP32-S2, which is used as an ultra-low powered data logger. Here, deep-sleep mode and web-server mode are the two modes of the data logger that can be activated by the toggle switch. The deep sleep mode logs the data after every 60 s and the web server mode gives access to the authorized device connected to the local network to monitor/log the data [57].

Many houses have been installing PV modules to meet the energy requirements. The battery consists of a battery regulator to alter the SOC and also its maximum current; the battery controller operates as both in charge and in discharge operations. The battery controller will command the power to go between the solar panel arrangement and then to the battery. The installation of the PV panel with the batteries is only supposed to supply it to a set of household load. From the results, we can say that the values of loss of load probability (LOLP) or loss of energy expectation (LOEE) increased as the yearly mean decreased and as the mean time of repair increased. Yet, there is an increase though very minor for medium boosting in the failure rate in the panels and battery regulator and in the average repair time the battery controller between 6 h and 24 h. This observation provides us a way to make out that the operation of PV array with much more stability within a limited response of the operation time [58].

The stand-alone self-sustainable DC nano grid basically has PV panels as means of power supplies and will have battery bank for storage of power generated through the PV module. Other outlook through the given module is that a range of DC voltages with separate plugs to match DC loads and equipment consumes and matches up to the power generated. The system has a range of sensor points to provide temperature, potential, and amperage readings to have a record on the calculated operations of the system. The prototype is powered by IoT features, giving the user the option to remotely connect to a local computer to control and monitor the system and view live data on a cloud platform. This is a logistical test because it confines dimensions of gear that can be taken to the fish farms. The area where PV panels will be installed is constrained to certain parts of zinc roof where fish farms are approximately 1.5 km from the mainland and the operation. Zinc roofing is extremely sturdy and weathering resistant, making it perfect in fishing areas. But, it is also an excellent heat conductor, keeping the ceiling temperature up to 40° during the day. This heat be shifted to the PV panels, and, if the module heat exceeds a specific temperature, then the performance of crystalline PV panel will deteriorate [59].

In this work, a representative solar power system is designed and is used to supply power to a milk chiller with 200-W DC compressors using PVsyst software under different climatic conditions. Determining least number of solar panels and designing of batteries and the yearly solar fraction for five Indian climates using PVsyst software can also be used to obtain values in India. At first, it is considered that a milk cooler with a built-in ice bank tank for cooling heat energy storage is used to maintain the temperature. Icicles formed during sunlight ensure that the milk is frozen throughout. The second case was considered in a condition such as a milk chiller with battery backup and, for charging during sunlight, it is supposed to run compressor in all the first case that needed PV panels. It was found useful in battery backup and better solar fraction. Using a fixed power supply, the output power was simulated from the PV panel. In doing so, the power supply was changed in intervals of an hourly basis as projected by the PVsyst application considering peak voltage. From the testing, it was found that, considering the maximum capacity of a typical summer day, the amount of PV panels required decreased from four to two. The customized milk condenser was cooling milk from 37°C to 4°C within 2 h every day using the cooling power of the ice in the ice tank [60].

In this case, the PV system has been used to be operating in the form of conventional generating units as diesel generators where the outputs are in the form of AC, which is used because of lack of DC consumption equipment. In some of the noted locations, there exists a connection in the public electrical grid although supply is unstable. Here, the power conditioning unit is supposed to synchronize with the grid and switch to autonomous operation mode when the load is noted to be unstable. The work is expected to combine to a MPPT, an advance charge controller using a charge balance, an energy flow monitor combined with a data logger, and a single phase DC-AC, which will condition the system to an SAPV system. For switching purposes, a 300-Hz PWM switching is carried out by a single IGBT device. The voltage of the DC bus at the output of the MPPT is used to control the flow of energy from the battery, and, to do the same, FLC has been applied to compare the conventional supplies [61].

Consumers are pushing to seek for an alternative energy saucer such as reusable PV systems. Thus, solar PV systems come in the form of small and economical modules. Although designing PV systems proved to be a difficult task, not because of its complexity but because of the options available once, a stand-alone system is installed that does not rely on grid connection for its operation such as when the stored energy is exhausted for which they stop working for the solar panel array where it consists of four 320-W polycrystalline panels linked one after another. Based on the average area above the residential roofs, four frames are considered. The entire module can hold an area of approximately 8 m^2, considering buildings where the residential areas are supposed to hold the panels within range from 10 m^2 to 12 m^2, which is accessible among the housing buildings. The system will contain of an inverter, a charge controller, and a battery. A 1.2-kW AC inverter of a working potential of 90 V to 580 V is used. The inverter is selected based on resultant potential that can be found from panels when they are connected cascaded. A 26-V lithium-ion battery with 180 Ah has been chosen as the backup to provide power when there is no sunlight. The battery can store up to 3.6 kWh of energy. The battery can store up to 3.6 kWh of energy, 2.7-kW DC as charging energy and 0.3-K WAC as maximum output power. MPPT is provided for the battery charge/discharging of the monitoring system. Also, they consider the user load of lights, fans, and laptop with all the everyday capacities included [62].

In the discussed PV system, it has been integrated with a heat pump that seems to be another option for dropping the power usage, which is connected to cooling operations; it is also used to increase the use of renewable power among the electrical systems. The system represents an outlook on SAPV-HP system without the operation of the batteries to control two unlike algorithms for the given compressor in the system in the MPPT and the inverter control. Both algorithms show solar power fluctuations due to the clouds passing above the PV cells, and, due to this, 75% of clouds were to be considered during the recording of the data. The values of the core key performance indicators also called KPIs are employed for modeling of PV and HP systems; the parameters that were considered were performance ratio (PR), energy efficiency ratio (EER), and seasonal performance factor (SPF), which are determined under several weather conditions; some new KPIs were brought up for PV-HP systems. In some cases, it consisted of two thermally isolated rooms that were of the dimension 2.5 m × 2.5 m × 2.5 m, whereas the inner and outer parts of the commercial air-to-air heat pump would have been placed and powered up by a PV system without any batteries. By this, one room turns as the place to be chilled, whereas the further as the outer ambient [63].

The paper introduces the proposed solution, an IoT-based data attainment system. This system uses open-access software and cloud services, making it more accessible and cost-effective. It also provides clock based data monitoring. The discussed system is appropriate for harsh outdoor conditions and allows for remote monitoring and control using Wi-Fi-enabled switches. It claims to save energy and extend the life of sensors. The paper mentions an observation that, without any cleaning process, the generated power of PV cells declined by 13% in 28 days, emphasizing the need for efficient data acquisition and monitoring. The paper hints at a comparison between the proposed IoT-based system and existing data acquisition methodologies, suggesting that the proposed system offers advantages such as cost-effectiveness, reliability, and ease of use [64].

A work pointing to the importance of detecting faults in PV grids to prevent serious power output reductions and potential damage to PV modules: It suggests that implementing an automatic system for fault detection is necessary. The proposed system is based on IoT technology and LabVIEW, a graphical programming environment often used for data acquisition and control systems. A key hardware component is mentioned, which is centered on the Atmega328 system. This component is responsible for automatically reconfiguring a series-parallel (SP) solar PV array to a fully cross-linked array when partial shading conditions (PSC) are detected. This reconfiguration aims to maximize power generation. An automated observation system is established in the LabVIEW environment. This system evaluates the PV generator's DC side voltage drop losses and calculates a weighted value based on the faulty solar panels. The calculated fault information is transmitted to a remote station using an RF modem. Additionally, the built-in interface of LabVIEW provides a signal for the defective solar panel. The GUI indicator in LabVIEW helps manage and visualize the fault status of the PV panels. When a fault is detected, the PV system's damaged panels, the system generates a panel alert. The receiver section refers to NodeMCU as a component that permits the transmission of PV array fault status data via Internet connectivity. This suggests that the fault status can be accessed remotely over the internet. The 3 × 3 PV array's failure status is visualized using the Blynk app. It provides a dashboard that displays the condition of every array, allowing for easy monitoring and visualization of any faults [65].

This research paper mentions the development of the original data acquisition system using ATMEL MEGA 2560 and ATMEL 328 microcontrollers. This system is designed to collect data on various parameters related to PV panels and PVT systems. The data gathering system measures the air intake and exit temperatures, solar panel surface temperature, PV current, and PV voltage. The information gathered is contrasted with information from the reference power system. The purpose of this comparison is to show the effect of cooling (probably by a PVT system) on the electrical output of the solar panel. The experiments were conducted in Tiruchirappalli, Tamil Nadu, India, at fixed coordinates (latitude 11°N, longitude 79°E). This information is necessary to understand local environmental conditions that may affect solar panel performance. The data collection system was tested and monitored for 3 months from February to April 2017. This extended period allows the performance of the system to be evaluated under different seasonal conditions. The section concludes that the outcomes show satisfactory performance of the original PV/PVT data attainment system. This suggests that the system has successfully collected the required data and probably completed its task of estimating the heat and electrical efficiency of the PV panel [66].

This paper about a system designed to monitor stand-alone solar energy systems. These methods are frequently employed in distant locations or other scarce resource situations. The monitoring system uses modern technology called the IoT to gather and transmit data over the internet. The main idea is to make this tracking system cost-effective so that it can be used even in places with limited budgets. They have created something called a "power diagnostic node" that acts as a smart sensor. It measures important factors such as the electrical result of the PV panels, their temperature, and the condition of the batteries. To make everything work, they used a small computer known as an ESP 8266. This computer is responsible for collecting data and sending it to a central location where the users can access it. Data can be viewed remotely via the internet. One of the great features of this system is that it allows users to view and analyze data from anywhere there is an internet connection. In addition, it stores all the collected data, which is useful for identifying and solving problems with the solar energy system before they become bigger problems. In summary, this paper presents an inexpensive and practical solution for monitoring stand-alone solar energy systems. It uses IoT technology to deliver real-time data, making it especially valuable in areas where reliable solar power is critical [67].

This project's goal is to make the monitoring systems (MS) better for two solar power stations (GCPVS) connected to the grid at Renewable Energy Research Unit in Saharan region Adrar (URERMS). They want to fix the problems the existing systems have. To do this, they are creating a new and improved MS that can handle both solar stations, even though they work differently. This new MS comes with special LabVIEW-based software that lets you see real-time data and check how the solar stations are doing. It also shows data in 2D and 3D graphs, which helps us understand how the different parts of the solar system work together in real-time. The new software also makes sure that all the data collection units in the solar stations are working together at the same time. To be sure about how these solar stations are doing, they are calibrating the sensors to make sure that they give accurate measurements. They are also adding a PV-UPS (uninterruptible power supply) to keep the monitoring system running smoothly, and they have made a user-friendly screen that lets you see how the PV-UPS is performing [68].

This work focuses on the need to monitor and gather data from renewable energy sources, like solar panels. Monitoring helps us understand how these systems are working, find issues, and make them more efficient. Most commercial monitoring systems for solar panels are expensive and cannot be easily customized. This project suggests a solution: a cost effective, RT IoT system designed for small-scale solar power systems, like those used in homes or small businesses. This system can continuously measure things like voltage, current, and electrical power from the solar panels. It also keeps an eye on seven weather-related variables, such as temperature and sunlight. This technology is special because it checks all these parameters via the solar panels themselves, not just the inverter. It was created using open-source software, has a wireless internet connection, saves information remotely and in the cloud, and has a unique protocol to ensure that all devices are on the same time. They also conducted experiments and found that their system works and is reliable. The cool part is that this system can help with something called "fog and cloud computing" in solar power. It collects a lot of data over time, which can be used for machine learning. This means they can make smarter decisions and create more efficient solar power systems in the future [69].

This work introduces a cost-effective model for monitoring the highest power generated by a home-based solar power system using IoT technology. They first used MATLAB/Simulink to simulate the very popular tracking algorithms such as P&O, InCond, HC, VSS InCond, and FL, and, then, they applied them to an accessible microcontroller (Arduino). The system keeps an eye on various parameters like current, voltage, load current, load voltage, power at the highest power point, duty cycle, module temperature, and solar irradiance on the solar panel. Thanks to IoT, users can conveniently check how their solar installation is performing in real-time from anywhere and at any time, without extra effort or cost. This prototype is designed for home-based solar power systems, especially in remote areas. It also enables users to spot any irregularities in their system by monitoring changes in the maximum power produced through online checks. Experimental results demonstrate that the system works effectively. Additionally, this prototype is user-friendly, cost-effective, and time-saving and reduces manual effort. In the future, this monitoring system could be expanded by integrating algorithms for detecting and diagnosing faults [70].

To make solar panels work their best, it is important to keep an eye on how they are doing. We need to collect data from them to use in research and development at universities and research centers. In this project, they have created a way to watch these data from far away. We use special sensors to get the important information from the solar panels. Then, they use a tiny computer and a wireless connection to send these data to a bigger computer at our office. Once the data gets to our office computer, they process it and show it on a computer screen. We use a software called LabVIEW to make it easy to understand. This system lets us check how the solar panels are working in real-time, even if they are not near them. This helps us learn more about how the solar panels are performing [71].

A study compares data collected through IoT monitoring with data that are directly measured from PV panels. In this experiment, they set up two panels that use different methods to measure their output and efficiency. The findings reveal that there is a variance between direct measurement and IoT-based monitoring of 2.9708 watts in output and 3.085% in efficiency. But here is the interesting part: even though the numbers do not match exactly, the graphs look the same. This suggests that, with the right adjustments and light sensors, IoT monitoring can show us the same data as direct measurement. In simpler terms, this study tells us that IoT technology can give us accurate information about how our solar panels are doing, and that is important because it helps us make the most of solar energy and do our part to reduce carbon emissions [72].

The main focus here in article is the proposal and application of an IoT-based tracking system for solar power generation systems of an IoT-based tracking system for photo energy generation systems. PV solar power plants, which use the sun and its radiation to generate electricity, stand to benefit greatly from the adoption of IoT technology. Traditional monitoring and maintenance methods often require on-site inspections and manual data collection, making them time-consuming and prone to errors. However, with IoT, real-time data collection and remote control become feasible, improving overall efficiency. The proposed IoT system enables the continuous monitoring of critical parameters such as current as well as voltage within the solar power production system. This information is vital for assessing the system's performance, identifying potential issues, and optimizing energy production. The system employs hardware components like Arduino Uno and Raspberry Pi to facilitate data collection and processing [73].

A cost effective data logger was developed for SAPV systems in progressing nations. The data logger uses open source software along with hardware and is capable of measuring electrical and meteorological parameters. The data logger meets the precision necessities of the Universal Electrotechnical Commission (IEC) standard for PV observation systems.

The data logger can be installed in areas without telecommunications networks and stores data in an SD card. It requires minimal maintenance. A real SAPV system was installed at the IMDEA Water Institute facilities in Madrid, Spain, to test and validate the data logger under real conditions. The results showed that the data logger meets the IEC standard and can monitor all of the requested parameters with high accuracy and low power consumption [74].

A modern power management system is discussed for the SAPV system, involving different priorities of loads such as emergency, critical, essential, and convenient. An IoT-based EMS was developed to give easy and all-round function as load shedding, power management, and monitoring and controlling the SAPV block. An energy management system is computer-based system that collects, analyzes, and stores data related to energy consumption and generation. This information can help in energy optimization use, minimize costs, and improve efficiency. LabVIEW was used for predicting the load in system. Using all these, the overall system triggering was abled that isolated the loads and sent a warning mail to user in case of faults. A mobile application can be used to alter the loads, look into networks, and restore their working by the user. The proposed model is confirmed and tried on two-bus and three-bus SAPV system [75].

A study was done to develop a low-cost SCADA system for monitoring and controlling SPV systems that were used for the development of the system, i.e., web services were added with the old SCADA for strong supervisory control and monitoring. Basic analog current and voltage sensors were used to get data from panel. Also, an Arduino board was used to get data, and Raspberry Pi with a Node-red to help further processing the data and for storing and analyzing purpose Emoncms was used [76].

Solar PV systems have become progressively popular as a basis of clean, renewable energy. However, the performance of PV systems can vary because of many factors, such as the amount of sunlight, the temperature, and the shading. To ensure that PV systems are operating at their best, it is important to monitor them remotely and in real time.

The IoT can be employed to implement a remote and real-time analysis system for PV systems. The IoT is a network of physical objects that are connected to the internet and can gather and transmits and receive data. In situation of PV systems, the IoT is employed to collect data related to the efficiency of the system, namely voltage, current, and power output. This information is then sent to a remote server for monitoring and analysis.

The study proposed a distant and on the clock monitoring system for PV systems using IoT. The system consists of three main components:

- a data acquisition device that collects data from the PV system;
- a data gateway that transfers the data to a remote server; and
- a smartphone application that displays the data from the remote server.

The study showed that the intended system was able to collect data from the PV system with 98.49% accuracy. The data were also sent to the remote server with an average transmission time of 52.34 s. The outcome of the study suggests that the discussed design can be an answer to advanced remote and on the clock observation of PV systems [77].

The rising demands of sun's energy, growth in energy costs, and general expenses are the reasons why energy monitoring is important. In this work, an IoT-based remote real-time monitoring of energy has been developed to monitor solar power generation.

Many current and potential instruments are combined along a flexible microcontroller to aggregate information. The aggregated data are then visualized, and the power produced for a provided dispersed production system is evaluated using an IoT analytics platform. This is beneficial for constant remote analysis and improving the overall outcome of power plants. With the help of remote observation, we can identify variances that deviate the working of solar panels. The results show that the developed monitoring system is more efficient and has lower error rates than conventional energy monitoring systems [78].

23.7 Other Works in the Field

The IoT is rapidly gaining traction in the medical sector, thanks to the advent of ultra-fast fifth- or sixth-generation mobile networks, AI, and Big Data analytics. Healthcare industry can make far reaching changes in coming years when these powerful technologies combine with IoT. The development of IoT in healthcare will help in transforming the way patients are taken care of and treated from a distance far of. This will expand the efficiency of healthcare employees and free up their time to focus on other tasks.

The study discusses the state-of-the-art research on IoT and healthcare while focusing on hardware requirements, complexity, and challenges.

Here are some of the hardware requirements for IoT-based healthcare applications:

- Sensors to collect data from patients and devices
- Connectivity to transmit data to the cloud
- Computing power to store and analyze data
- Software to visualize and interpret data

The complexity of IoT-based healthcare applications can be daunting. Some of the challenges include the following:

- Integrating different types of devices and sensors
- Safeguarding the security and discretion of patient information.
- Managing huge amount of data generated by IoT devices
- Developing user-friendly interfaces for patients and health-care workers

Despite the encounters, the advantages of IoT cannot be ignored. IoT can help to advance patient attention, reduce cost, and improve the overall quality of healthcare systems [79].

To develop future PV solar energy projects, it is crucial to have accurate knowledge of the available solar resources. There are numerous methods for estimating the amount of solar radiation incident in a specific location. However, most of them are prohibitively costly or lack the supreme qualities for effective observation of a specific PV system.

To give a talk about this issue, this paper presents an IoT-based electronic device that can estimate global energy in PV applications. Due to its several characteristics, it is ideal for measuring PV systems, including being low-cost and reliable. The device was tested for determining global horizontal irradiance, and it achieved a correlation coefficient of R2 = 0.994.

It can be used to monitor the performance of the system and identify any potential problems. The data collected by the device can also be used to enhance the design and efficiency of future PV systems [80].

This paper focuses on the usage of modern computing techniques for making prognosis about the amount of power produced by PV units. The authors are benefiting from concept of neural networks for coming to the outcome. To do this, they will build an advanced meter which would be attached to a minimal power PV panel. This advanced meter will be collecting data and sending it independently to distant server employing low-range IoT connection. These data will then be handled for making prognosis regarding quantity of power generated.

The authors will use non-linear autoregressive neural networks called as NARX alongside MATLAB and ThingSpeak IoT information detention. The outcomes are good for open loop arrangement, but, for closed loop, it can be enhanced. Suggestions for further developments are provided [81].

23.8 Conclusion

In this paper, we have reviewed the previous works in the field of SAPV systems and IoT-based data loggers. We have discussed the different types of data loggers that have been developed, the challenges of monitoring SAPV systems, and the potential benefits of using IoT data loggers.

We have also presented a study of a cloud-based observation system for SAPV systems. The outcomes indicate that cloud-based monitoring system is a great technology for improving the reliability and efficiency of SAPV systems.

The use of IoT-based data loggers for monitoring SAPV systems has a number of advantages. First, it allows for real-time monitoring of the system's performance. This can help to identify any potential problems early on, before they cause major damage. Second, it allows for remote monitoring of the system. This can be useful for systems that are located in remote or difficult-to-access locations. Third, it can be used to collect data on the system's performance over time. These data can be used to improve the design and operation of the system.

The use of IoT-based data loggers for monitoring SAPV systems is just started, but it can change the way that these systems are operated and maintained. Future research in this area should focus on developing more affordable and reliable data loggers and on developing more sophisticated data analysis techniques.

References

1. Buraimoh, E. and Davidson, I.E., 2022, June. Laboratory procedure for real-time simulation experiment of renewable energy systems on OPAL-RT digital simulator. In *2022 10th International Conference on Smart Grid (icSmartGrid)* (pp. 221-226). IEEE.
2. Ariffin, M.R., Shafie, S., Hasan, W.Z.W., Azis, N., Yaacob, M.E. and Supeni, E.E., 2022. Development of Stand-Alone DC Energy Datalogger for Off-Grid PV System Application based on Microcontroller. *Pertanika J. Sci. Technol.*, 30(1).
3. Mudaliar, M.D. and Sivakumar, N., 2020. IoT based real time energy monitoring system using Raspberry Pi. *Internet Things*, 12, p.100292.
4. Shaukhi, A.S.A. and Bakar, M.I.A., 2021. Development of Wireless Photovoltaic (PV) Monitoring System. *J. Eng. Technol.*, 9(1), pp.29-35.
5. Mohanty, A., Viswavandya, M., Ray, P.K. and Mohanty, S., 2018, June. Literature survey on OPAL-RT Technologies with Advance features and Industrial applications. In *2018 1st International Conference on Advanced Research in Engineering Sciences (ARES)* (pp. 1-5). IEEE.
6. Rozmi, M.D.A.B., Thirunavukkarasu, G.S., Jamei, E., Seyedmahmoudian, M., Mekhilef, S., Stojcevski, A. and Horan, B., 2019. Role of immersive visualization tools in renewable energy system development. *Renew. Sustain. Energy Rev.*, 115, p.109363.
7. Hasan, S.T., Shompa, M.S., Rahman, M.A., Rasel, M.A., Apu, M.R.H. and Rahman, M.A., 2022, December. IoT Based Solar Power Monitoring & Data Logger System. In *2022 IEEE International Women in Engineering (WIE) Conference on Electrical and Computer Engineering (WIECON-ECE)* (pp. 182-187). IEEE.
8. Noureen, S.S., Roy, V. and Bayne, S.B., 2017, November. An overall study of a real-time simulator and application of RT-LAB using MATLAB simpowersystems. In *2017 IEEE green energy and smart systems conference (IGESSC)* (pp. 1-5). IEEE.
9. Jia, X., Adhikari, P.M. and Vanfretti, L., 2020, August. Real-Time simulation models for photovoltaic cells and arrays in Opal-RT and Typhoon-HIL. In *2020 IEEE Power & Energy Society General Meeting (PESGM)* (pp. 1-5). IEEE.
10. Tasnim, M.N., Ahmed, T., Ahmad, S. and Mekhilef, S., 2023, March. Hardware in The Loop Implementation of The Control Strategies for The AC-Microgrid in OPAL-RT Simulator. In *2023 IEEE IAS Global Conference on Renewable Energy and Hydrogen Technologies (GlobConHT)* (pp. 1-6). IEEE.
11. Delavari, A., Brunelle, P. and Kamwa, I., 2018, October. Real-time closed-loop PQ control of NPC multi-level converter using OPAL-RT and speedgoat simulators. In *2018 IEEE Electrical Power and Energy Conference (EPEC)* (pp. 1-5). IEEE.

12. Panda, S., Biswal, G.R., Pradhan, J.K., Subudhi, B. and Mallick, A., 2021. Development of Prototype Fractional Order Controller for a Grid-Tied Photovoltaic System. *IEEE Trans. Circuits Syst. II Express Briefs*, 69(4), pp.2171-2175.
13. Mishra, P., Pradhan, A.K. and Bajpai, P., 2018, December. Adaptive relay setting for protection of distribution system with solar PV. In *2018 20th National Power Systems Conference (NPSC)* (pp. 1-5). IEEE.
14. Darbali-Zamora, R., Hernandez-Alvidrez, J., Summers, A., Gurule, N.S., Reno, M.J. and Johnson, J., 2019, June. Distribution feeder fault comparison utilizing a real-time power hardware-in-the-loop approach for photovoltaic system applications. In *2019 IEEE 46th Photovoltaic Specialists Conference (PVSC)* (pp. 2916-2922). IEEE.
15. Luna, E.G., Manrique, R.F. and Bocanegra, E.L.P., 2018, August. Monitoring and Control System Using ETAP Real-Time on Generation Plant Emulation Using OPAL-RT. In *2018 IEEE ANDESCON* (pp. 1-6). IEEE.
16. Ghanbari, N. and Bhattacharya, S., 2019, September. Hardware-in-the-loop implementation of a grid connected pv system. In *2019 IEEE Industry Applications Society Annual Meeting* (pp. 1-6). IEEE.
17. Dinkhah, S., 2019. *Vehicle to grid (V2G) for reliable microgrid operation: Voltage/frequency regulation and real-time OPAL-RT digital simulation* (Doctoral dissertation).
18. Subudhi, P.S., Padmanaban, S., Blaabjerg, F. and Kothari, D.P., 2022. Design and implementation of a PV-fed grid-integrated wireless electric vehicle battery charger present in a residential environment. *IEEE J. Emerging Sel. Top. Ind. Electron.*, 4(1), pp.78-86.
19. Ertay, M.M., Chowdhury, D., Biswas, M.M. and Ginn, H.L., 2022, September. Real-Time Simulations for Testing of a Low-Voltage Microgrid with MMC-DSTATCOM. In *2022 IEEE 20th International Power Electronics and Motion Control Conference (PEMC)* (pp. 522-529). IEEE.
20. P. K. Madaria, M. Bajaj, S. Aggarwal, and A. K. Singh, "A gridconnected solar PV module with autonomous power management," in *Proc. IEEE 9th Power India Int. Conf. (PIICON)*, Feb. 2020, pp. 1–6, doi: 10.1109/PIICON49524.2020.9113065.
21. R. Iracheta-Cortez and N. Flores-Guzman, ""Developing automated Hardware-In-the-Loop tests with RTDS for verifying the protective relay performance,"" *2016 IEEE 36th Central American and Panama Convention(CONCAPAN XXXVI)*, San Jose, 2016, pp. 1-9."
22. Kaddouri, A., Khodabakhchian, B., Dessaint, L.A., Champagne, R.,Snider, L., ""A new generation of simulation tools for electric drives and power electronics,"" Power Electronics and Drive Systems, 1999. PEDS'99. *Proceedings of the IEEE 1999 International Conference on*, voU,no., ppJ48, 354 vol. I, 1999.
23. Haque, M.N.M. and Hossain, T., 2022. Smart Solar Data-Logger System. *AIJR Proceedings*, pp.9-14.

24. Talib, A.N., Hafeez, K., Mnati, M.J. and Khan, S.A., 2022, June. Design and Implementation of New Battery Monitoring System for Photovoltaic Application. In *2022 4th Global Power, Energy and Communication Conference (GPECOM)* (pp. 1-7). IEEE.
25. Rerhrhaye, F., Lahlouh, I., Ennaciri, Y., Benzazah, C., El Akkary, A., Sefiani, N. and Bouftane, S., 2022. IoT-Based Data Logger for solar systems applications. In *ITM Web of Conferences* (Vol. 46, p. 01003). EDP Sciences.
26. Omar, M., Qaiser, M., Usman, M., Tajamal, K., Larkin, S., Raw, B. and Khan, S., 2022. Solar remote monitoring unit. *Int. J. Eng. Technol.*, 11(2), pp.78-83.
27. López-Vargas, A., Fuentes, M. and Vivar, M., 2018, June. On the application of IoT for real-time monitoring of small stand-alone PV systems: Results from a new smart datalogger. In *2018 IEEE 7th World Conference on Photovoltaic Energy Conversion (WCPEC) (A Joint Conference of 45th IEEE PVSC, 28th PVSEC & 34th EU PVSEC)* (pp. 605-607). IEEE.
28. Sugiartha, N., Sugina, I.M., Putra, I.D.G.A.T., Indraswara, M.A. and Suryani, L.I.D., 2018, December. Development of an arduino-based data acquisition device for monitoring solar PV system parameters. In *International Conference on Science and Technology (ICST 2018)* (pp. 995-999). Atlantis Press. *2018.*
29. López-Vargas, A., Fuentes, M. and Vivar, M., 2018. IoT application for real-time monitoring of solar home systems based on Arduino™ with 3G connectivity. *IEEE Sens. J.*, 19(2), pp.679-691.[2018]
30. Priyadarshi, S., Bhaduri, S. and Shiradkar, N., 2018, June. IoT based, inexpensive system for large scale, wireless, remote temperature monitoring of photovoltaic modules. In *2018 IEEE 7th World Conference on Photovoltaic Energy Conversion (WCPEC)(A Joint Conference of 45th IEEE PVSC, 28th PVSEC & 34th EU PVSEC)* (pp. 0749-0752). IEEE.
31. Nehovski, N.V., Tomchev, N.N., Djamiykov, T.S. and Asparuhova, K.K., 2018, May. Datalogger for Small Solar Systems. In *2018 IX National Conference with International Participation (ELECTRONICA)* (pp. 1-3). IEEE.
32. Ziane, A., Dabou, R., Necaibia, A., Rouabhia, A., Bouchouicha, K., Sahouane, N., Lachtar, S., Bouraiou, A. and Larbi, A.A., 2022, November. IoT Platform For Online Monitoring Of Renewable Energy Systems. In *2022 3rd International Conference on Embedded & Distributed Systems (EDiS)* (pp. 55-60). IEEE.
33. Malik, A., Haque, A. and Kurukuru, V.B., 2022. IoT-Based Monitoring and Management for Photovoltaic System. *Fault Analysis and its Impact on Grid-connected Photovoltaic Systems Performance*, pp.291-318.
34. Sachan, P., Singh, A. and Bahar, T., 2022, December. IOT Based PV Monitoring System. In *2022 4th International Conference on Advances in Computing, Communication Control and Networking (ICAC3N)* (pp. 2382-2384). IEEE.

35. Hamied, A., Mellit, A., Zoulid, M.A. and Birouk, R., 2018, November. IoT-based experimental prototype for monitoring of photovoltaic arrays. In *2018 International Conference on Applied Smart Systems (ICASS)* (pp. 1-5). IEEE.
36. Gimin, N.A., Zambri, N.A., Yi, S.S. and Mustafa, F., 2022. IoT Solar Panel Tracking For IV And PV Characteristic. *Progress in Engineering Application and Technology (PEAT)*, 3(1), pp.411-418.
37. Khan, M., Iqbal, J., Ali, M., Muhmmad, A., Zahir, A. and Ali, N., 2022. Designing and implementation of energy-efficient wireless photovoltaic monitoring system. *Trans. Emerging Telecommun. Technol.*, 33(2), p.e3685.
38. Abdullah, B.U.D. and Lata, S., 2022, November. IoT-Based Solar Monitor System. In *2022 International Conference on Computing, Communication, and Intelligent Systems (ICCCIS)* (pp. 997-1002). IEEE.
39. Almonacid-Olleros, G., Almonacid, G., Fernandez-Carrasco, J.I., Espinilla-Estevez, M. and Medina-Quero, J., 2020. A new architecture based on IoT and machine learning paradigms in photovoltaic systems to nowcast output energy. *Sensors*, 20(15), p.4224.
40. Kebede, A.A., Berecibar, M., Coosemans, T., Messagie, M., Jemal, T., Behabtu, H.A. and Van Mierlo, J., 2020. A techno-economic optimization and performance assessment of a 10 kWP photovoltaic grid-connected system. *Sustainability*, 12(18), p.7648.
41. Chandra, S., Agrawal, S. and Chauhan, D.S., 2020. Impact of Photovoltaic Module Temperature on Size and Voltage Stability, a Case Study in Indian Climate. *Appl. Sol. Energy*, 56, pp.324-333.
42. Mahmood, M.H., Ali, I.I. and Ahmed, O.A., 2020, July. Design and implementation of intelligent MPPT based on FPGA for PV system. In *IOP Conf. Ser.: Mater. Sci. Eng.* (Vol. 881, No. 1, p. 012138). IOP Publishing.
43. Heinrich, M., Meunier, S., Samé, A., Quéval, L., Darga, A., Oukhellou, L. and Multon, B., 2020. Detection of cleaning interventions on photovoltaic modules with machine learning. *Appl. Energy*, 263, p.114642.
44. Widyolar, B., Jiang, L., Brinkley, J., Hota, S.K., Ferry, J., Diaz, G. and Winston, R., 2020. Experimental performance of an ultra-low-cost solar photovoltaic-thermal (PVT) collector using aluminum minichannels and nonimaging optics. *Appl. Energy*, 268, p.114894.
45. Priyadarshi, N., Padmanaban, S., Holm-Nielsen, J.B., Bhaskar, M.S. and Azam, F., 2020. Internet of things augmented a novel PSO-employed modified zeta converter-based photovoltaic maximum power tracking system: hardware realisation. *IET Power Electron.*, 13(13), pp.2775-2781.
46. Jamil, W.J., Rahman, H.A., Shaari, S. and Desa, M.K.M., 2020. Modeling of soiling derating factor in determining photovoltaic outputs. *IEEE J. Photovoltaics*, 10(5), pp.1417-1423.
47. Mishra, G.K. and Tiwari, G.N., 2020. Performance evaluation of 7.2 kWp standalone building integrated semi-transparent photovoltaic thermal system. *Renew. Energy*, 146, pp.205-222.

48. Gimeno-Sales, F.J., Orts-Grau, S., Escribá-Aparisi, A., González-Altozano, P., Balbastre-Peralta, I., Martínez-Márquez, C.I., Gasque, M. and Seguí-Chilet, S., 2020. PV monitoring system for a water pumping scheme with a lithium-ion battery using free open-source software and IoT technologies. *Sustainability*, 12(24), p.10651.
49. Kusznier, J. and Wojtkowski, W., 2021. IoT solutions for maintenance and evaluation of photovoltaic systems. *Energies*, 14(24), p.8567.
50. Wu, F., Qiu, C., Wu, T. and Yuce, M.R., 2021. Edge-based hybrid system implementation for long-range safety and healthcare IoT applications. *IEEE Internet Things J.*, 8(12), pp.9970-9980.
51. Eltamaly, A.M., Alotaibi, M.A., Alolah, A.I. and Ahmed, M.A., 2021. IoT-based hybrid renewable energy system for smart campus. *Sustainability*, 13(15), p.8555.
52. Saleh, U.A., Jumaat, S.A., Johar, M.A. and Jamaludin, W.A.W., 2021, September. Photovoltaic-thermoelectric generator monitoring system using arduino based data acquisition system technique. In *2021 IEEE International Conference on Artificial Intelligence in Engineering and Technology (IICAIET)* (pp. 1-6). IEEE.
53. Murti, S., Megantoro, P., Silva, G.D.B. and Maseleno, A., 2020. The design and analysis of DC electrical voltage-current datalogger device implemented on wind turbine control system. *J. Robot. Control (JRC)*, 1(3), pp.75-80.
54. Shaqour, A., Farzaneh, H., Yoshida, Y. and Hinokuma, T., 2020. Power control and simulation of a building integrated stand-alone hybrid PV-wind-battery system in Kasuga City, Japan. *Energy Rep.*, 6, pp.1528-1544.
55. Abadi, I., Setyawan, E.H. and Ardiansyah, M., 2020, June. Design Of Automatic Switching System And Data Acquisition On Mobile Hybrid Solar Tracking System For Stand Alone Small PV With Reflector. In *IOP Conf. Ser.: Earth Environ. Sci.* (Vol. 520, No. 1, p. 012001). IOP Publishing.
56. Hussein, A.H.A.W.Ě., Kazem, A., Chaichan, M.T. and Sopian, K., *Photovoltaic/Thermal (PV/T) Systems.*
57. Rehman, A.U. and Iqbal, M.T., 2020. Design of an Ultra-Low Powered Data-Logger for Stand-Alone PV Energy Systems. *Eur. J. Electr. Eng. Comput. Sci.*, 4(6).
58. Quiles, E., Roldán-Blay, C., Escrivá-Escrivá, G. and Roldán-Porta, C., 2020. Accurate sizing of residential stand-alone photovoltaic systems considering system reliability. *Sustainability*, 12(3), p.1274.
59. Quek, Y.T., 2020, November. Implementation of an IoT-based DC nanogrid in an offshore fish farm. In *2020 IEEE REGION 10 CONFERENCE (TENCON)* (pp. 1147-1152). IEEE.
60. Sidney, S., Thomas, J. and Dhasan, M.L., 2020. A standalone PV operated DC milk chiller for Indian climate zones. *Sādhanā*, 45(1), p.110.
61. Krauter, S. and Araújo, R.G., 2020, November. New power conditioning unit incorporating charge controller, energy flow monitor, data Logger, DC/AC

converter for standalone and combined PV-Diesel operation. In *Sixteenth European Photovoltaic Solar Energy Conference* (pp. 2575-2577). Routledge.
62. Akshai, K.N.B. and Senthil, R., 2020, August. Economic evaluation of grid connected and standalone photovoltaic systems using PVSyst. In *IOP Conf. Ser.: Mater. Sci. Eng.* (Vol. 912, No. 4, p. 042074). IOP Publishing.
63. Lorenzo, C., Narvarte, L., Almeida, R.H. and Cristóbal, A.B., 2020. Technical evaluation of a stand-alone photovoltaic heat pump system without batteries for cooling applications. *Sol. Energy*, 206, pp.92-105.
64. Gupta, V., Sharma, M., Pachauri, R.K. and Babu, K.D., 2021. A low-cost real-time IOT enabled data acquisition system for monitoring of PV system. *Energy Sources Part A*, 43(20), pp.2529-2543.
65. Samkria, R., Abd-Elnaby, M., Singh, R., Gehlot, A., Rashid, M., Aly, M.H. and El-Shafai, W., 2021. Automatic PV grid fault detection system with IoT and LabVIEW as data logger. *Comput. Mater. Contin.*, 69, pp.1709-1723.
66. Franklin, J.C., Chandrasekar, M. and Mattius, D.A., 2021. Development of cost effective data acquisition system to evaluate the performance of solar photovoltaic thermal systems. *J. Sol. Energy Eng.*, 143(1), p.011003.
67. Mahamudul Hasan, M., Hossain, A., Akhi, N.A. and Abdur Razzak, M., 2021. Design and implementation of a low-cost power diagnosis node for monitoring the stand-alone PV system in the mockery of IoT. In *Advances in Automation, Signal Processing, Instrumentation, and Control: Select Proceedings of i-CASIC 2020* (pp. 1107-1116). Springer Singapore.
68. Dabou, R., Bouraiou, A., Ziane, A., Necaibia, A., Sahouane, N., Blal, M., Khelifi, S., Rouabhia, A. and Slimani, A., 2021. Development of autonomous monitoring and performance evaluation system of grid-tied photovoltaic station. *Int. J. Hydrogen Energy*, 46(59), pp.30267-30287.
69. Melo, G.C.G.D., Torres, I.C., Araújo, Í.B.Q.D., Brito, D.B. and Barboza, E.D.A., 2021. A low-cost IoT system for real-time monitoring of climatic variables and photovoltaic generation for smart grid application. *Sensors*, 21(9), p.3293.
70. Rouibah, N., Barazane, L., Benghanem, M. and Mellit, A., 2021. IoT-based low-cost prototype for online monitoring of maximum output power of domestic photovoltaic systems. *ETRI J.*, 43(3), pp.459-470.
71. Al Aish, A.A., Shehab, W.F.A., Alshrari, M.I. and Nabhan, S., *Real-Time Performance Monitoring of PV Solar Module based on Wireless Data Acquisition System.*
72. Bow, Y., Taqwa, A., Dewi, T. and Sasmanto, A.A., 2021. Feasibility Study and Design of IoT-based Monitoring for Remote PV System. *Technology Reports of Kansai University (TRKU)*, 63(1).
73. Samosir, A.S., Rozie, A.F., Purwiyanti, S., Gusmedi, H. and Susanto, M., 2021, October. Development of an IoT Based Monitoring System for Solar PV power Plant Application. In *2021 International Conference on Converging Technology in Electrical and Information Engineering (ICCTEIE)* (pp. 82-86). IEEE.

74. López-Vargas, A., Fuentes, M., García, M.V. and Muñoz-Rodríguez, F.J., 2019. Low-Cost datalogger intended for remote monitoring of solar photovoltaic standalone systems based on Arduino™. *IEEE Sens. J.*, 19(11), pp.4308-4320.
75. Swathika, O.G. and Hemapala, K.T.U., 2019. IOT based energy management system for standalone PV systems. *J. Electr. Eng. Technol.*, 14, pp.1811-1821.
76. Aghenta, L.O. and Iqbal, M.T., 2019, May. Development of an IoT based open source SCADA system for PV system monitoring. In *2019 IEEE Canadian Conference of Electrical and Computer Engineering (CCECE)* (pp. 1-4). IEEE.
77. Priharti, W., Rosmavati, A.F.K. and Wibawa, I.P.D., 2019, November. IoT based photovoltaic monitoring system application. In *J. Phys. Conf. Ser.* (Vol. 1367, No. 1, p. 012069). IOP Publishing.
78. Khan, M.S., Sharma, H. and Haque, A., 2019, February. IoT enabled real-time energy monitoring for photovoltaic systems. In *2019 International Conference on Machine Learning, Big Data, Cloud and Parallel Computing (COMITCon)* (pp. 323-327). IEEE.
79. Quasim, M.T., Khan, M.A., Abdullah, M., Meraj, M., Singh, S.P. and Johri, P., 2019, December. Internet of things for smart healthcare: a hardware perspective. In *2019 First International Conference Of Intelligent Computing And Engineering (ICOICE)* (pp. 1-5). IEEE.
80. JI, F.C., 2019. Development of a utility model for the measurement of global radiation in photovoltaic applications in the internet of things (IoT). *Electronics*, 8(3), p.304.
81. Rogier, J.K. and Mohamudally, N., 2019. Forecasting photovoltaic power generation via an IoT network using nonlinear autoregressive neural network. *Procedia Comput. Sci.*, 151, pp.643-650.

Index

Abbreviations, 58–59
Abstract, 57–58, 81, 147, 163, 255, 339–340
Acknowledgments, 186
Acquisition, 322
Additives, 1–2, 8
Advantages of using renewable energy, 69–70
Analytic hierarchy process, 31, 35
Applications and functions of supercapacitor, 167
  energy storage technology, 167–168
  power electronic device, 170
  renewable integration, 170
  stabilization and traction, 170–171
  uninterruptible power supplies (UPSs), 168
Applications of bio-ML, 62–69
  biomass renewable energy systems, 68
  geothermal renewable energy systems, 67–68
  green-hydrogen renewable energy systems, 69
  hydro renewable energy systems, 66
  solar renewable energy systems, 65
  tidal renewable energy systems, 66–67
  wind renewable energy systems, 63–65
Artificial fuel, 33–34, 39–40
Artificial intelligence, 363, 364, 365, 366, 367, 368
Autonomous, 326–327, 332
Battery management system (BMS), 286–287
Benefits and implications for smart city energy management, 311–314
Benefits of IoT in smart parking systems, 106
  IoT in smart grids, 110
  IoT sensors for smart city monitoring, 106
  processes in smart grids, 110–111
  smart energy management systems, 111–112
  smart grids, 110
  smart offices, 106, 107, 108
  smart transportation, 108–109
  smart waste, 112–113
  some of the solutions for smart city traffic control, 109–110
  technologies in smart grids, 111
  useful reports from IoT sensors, 106
Bio-crude yield, 13–14, 19
Biodiesel, 243–248, 251
Bioenergy, 31
Biofuel, 231–232, 234, 241
Bioinspired-based smart energy management system, 70–72
Biomass, 21–28, 43–45, 55, 123–128, 130–134, 136–145
Biomolecule, 244
Bio-oil, 1–8, 10–11, 231–232, 234–239, 241
Bio-ultra carbo-fluids, 44–45, 54

Bio-ultracarbofluids, 221–222, 226–228
Biowaste, 21–23
Blockchain, 323
Building applications, 371
Buildings, 372

Capital optimization, 343
Cellulosic fermentations to ethanol (CE), 124
Challenges and future directions, 313–314
Cloud, 383, 386, 396–401, 403, 405, 409, 411, 413, 417–418, 425
Coefficient of performance (COP), 374
Collectors, 373
Combined grid-connected microgrid with demand response model, 267
Comparative analysis with traditional forecasting methods, 310–311
Comparative study of topologies, 175
Compressors, 372
Concepts of security in telecommunication systems, 343–344
Conclusion, 74–75, 117–118, 157, 281
Conclusion and future scope, 184–186
Condensers, 372
Conflict of interest, 186
Converter technologies in the power electronic field, 174
  converters for AC source AC applications, 183
  converters for AC source DC applications, 175–176
  converters for DC source AC applications, 177
    inverter topologies with multiple levels, 178–179
    neutral point clamped topologies, 178
  converters for DC source DC applications, 179
    isolated converters, 181–183
    non-isolated converters, 179–181
Cost effective, 373

Data centers and server, 97
  algorithm for smart parking system, 105
  applications and user interfaces, 100–101
  communication in the system, 103
  data cleansing, 98
  data compression, 98–99
  energy efficiency in IoT applications, 99
  energy-efficient servers, 99
  light control, 104
  reduction of data volume by removing redundant information, 99
  remote monitoring and management, 104
  sensor parking and management tools for city parking, 105
  smart homes and home automation, 101, 102, 103
  smart parking ecosystems, 105
  smart roads, 104
  smart street lighting system, 103
  system operation, 104
  ventilation and cooling of servers and data centers, 100
  virtualization of servers, 97–98
Data collection and preprocessing, 309
Data logger, 383–384, 396–398, 401, 406, 408–409, 415, 418–424
Dexterity, 324
Diesel fuel, 43–44, 46, 48, 54
Diesel fuel replacement, 222
Diffuse horizontal irradiation, 367
Direct current, 364
Direct expansion (DX), 376
Domestic hot water (DHW), 376

Economy, 231
Efficient combination, 22
Electrocatalytic hydrogenation (ECH), 123–124, 134, 144
Emulsification, 2, 10
Energy, 123–129, 131–134, 138–145
Energy consumption, 363, 364, 365, 366
Energy demand, 363, 365
Energy efficiency ratio (EER), 374
Energy management, 287, 290, 301, 303
Engine, 197–207
Enhanced pelican optimization algorithm, 268
Environment, 197–199, 200, 204, 206
Environmental, 371
Environmental pollution, 222
Evaluation and testing, 309–310

Fuel additive, 231
Fuzzy, 321–322, 325, 327, 332
Fuzzy logic-based assortment of finest compromise solution, 267

Gas-to-liquid, 198, 204
Geographic data, 364
Global horizontal irradiation, 367
Global warming, 209–211
Glycerol, 210, 212–217, 219
Granular, 322
Grid connected microgrid, 261

Heat exchanger (HX), 372
Heat pumps, 372
Historical backgrounds, 61–62
Human health, 198, 201, 205–206
HVAC, 239
Hybrid systems, 372
Hydrocarbon, 43–46, 54, 222

Ignition engines, 244
Incorporating multiscale decomposition, 309
Indirect expansion (IDX), 377
Information security, 344
Information security management system 1 (ISMS), 345
- attack on the architecture of the intelligent measurement system, 348, 349, 350
- comparison of security of telecommunication systems, 348
- cyber security, 351, 352, 354
- information security management standard, 345–346
- internet of things, 350–351
- physical security, 354, 355, 356
- security in smart grid telecommunication systems, 346–347
- threats in the smart power grid, 347–348

Information technology security, 344–345
Interpretation of comparative results, 311
Introduction, 59–61, 82, 147–148, 164–165, 258
- bulk data at the heart of the internet of things architecture, 149–150
- comprehensive framework based on the internet of things for smart homes, 150
- daylight sensor, 86–87
- infrastructure of information technology for energy data collection, 82, 83, 84, 85
- meter reading information collection service, 85
- mobile device sensors, 87
- network technologies and communications, 87–88
- smart water sensor, 85–86
- temperature and humidity sensor, 86
- the role of objects in the internet of things, 148–149

IoT, 383–385, 386, 396–400, 403–405, 408–409, 411–414, 416–419, 421–425
Irradiation, 367

Keywords, 58, 164

Layer of communication equipment and sensors, 113, 114, 115
application layer, 116–117
information management layer, 115–116
Life cycle cost (LCC), 374
Lipid, 243–244, 246–247
List of symbols and abbreviations, 256
Long short-term memory (LSTM) networks, 308
Low-power Wi-Fi (Halow) protocol, 88
6LoWPAN protocol, 91
802.15.4 IEEE protocol, 90–91
cloud computing, 95, 96, 97
management and storage systems for data, 93, 94, 95
MQTT protocol, 91, 92, 93
RPL protocol, 90
zigbee protocol, 88–89

Machine learning, 363, 364, 365
Macroalgae, 13–15, 17, 19
Mathematical modeling of demand response, 263
Maximum power point tracking (MPPT), 287, 327–328
Methanol, 209–210, 215–219
Methodology: multiscale LSTM-based deep learning approach, 307–310
Methyl ester, 198
MS, 13–14, 18

NMR, 14, 18–19

Objectives and structure of the chapter, 307
Oil, 13–14, 17, 20
Opportunities and challenges of future research, 72–74
energy management systems (EMSs), 73
hybridizing bioinspired approaches, 72–73
internet of energy (IoE), 73–74
Overview of references, 356, 357, 358

Palm oil, 1–3
Payback, 375
Photovoltaic, 318, 322
Photovoltaic-thermal (PVT), 371
Plant oil, 209–210
Pollution, 197–199, 204
Power converter topologies for energy storage, 172
active power converter topology, 173–174
passive power converter topology, 172–173
semi-active power converter topology, 173
Predictive, 329–330, 332
Previous research, 156–157
Problem formulation, 260
PV efficiency, 373
Pyrolysis, 1–5, 7–8, 10–11, 34–36, 38–40, 123–132, 134, 144–145, 231–232, 234–239, 241, 243–244, 246–247, 249, 251, 252

Quantitative metrics: mean absolute error and root mean squared error, 310–311

Rechargeable battery, 292
Reduce subscriber shutdown time, 341
References, 76–80, 118, 119, 120, 121, 122, 157, 158, 159, 160, 161, 186–195, 282, 359, 360, 361, 362

Refrigerant, 373
Regression, 329
Regulatory applications and their use in health services, 154
  architecture of software-oriented networks, 155–156
  challenges, 154
  concerns about healthcare, 154–155
  SDN for cloud service providers, 156
Reinforcement learning, 363, 364, 365, 366, 367, 368
Renewable, 44–45, 318, 319, 325, 327, 329–330, 332
Renewable biomass, 209–210, 219
Renewable energy sources, 363
Results and discussion, 274
Return on investment, 375
Running time, 375
Rural electrifications, 22

Smart grid, 340
Smart home, 150–151
  beneficial program, 151
  cloud, 151
  operational structure for personal medical services based on the Internet of Things, 154
  security in the Internet of Things, 152, 153, 154
  third party, 151
  user interfaces, 152
Solar energy, 363, 374
Solar forecasting, 368
Solar irradiance, 363, 367, 368, 374
Solar panel, 364, 365, 366, 367, 368
Solar panel efficiency, 364, 366, 367
Solar power plants, 365
Solar-assisted heat pump, 371
Solar-assisted HP system (SAHP), 372
Stabilising additive, 222
Stand-alone photovoltaic, 383, 407–408, 410, 412, 419, 421, 423–424
Summary and conclusion, 358
Sun's radiation, 364
Super capacitor, 291
Supercapacitor fundamentals, 166

Technique for order of preference by similarity to ideal solution, 35
Temperature, 373
The benefits of smart grids, 340–341
The benefits of the smart distribution network for subscribers, 341, 342, 343
The role of deep learning in smart city energy management, 306–307
Thermal efficiency, 373
Training and validation strategies, 309

Water heating, 372
Weather conditions, 363, 364, 365, 366

Also of Interest

Other Books by the Same Editors

Smart Grids as Cyber Physical Systems Volume 1: Artificial Intelligence, Cybersecurity, and Clean Energy for Next Generation Smart Grids, Edited by O.V. Gnana Swathika, K. Karthikeyan, P. Sanjeevikumar, ISBN: 9781394261697. With this first volume focusing on artificial intelligence, cybersecurity, and clean energy for next generation smart grids, this groundbreaking two-volume set provides a comprehensive overview of the fundamental security of supervisory control and data acquisition (SCADA) systems, making it an essential reference for practitioners in industries focusing on adaptive configuration and optimization in smart grid systems.

Smart Grids as Cyber Physical Systems Volume 2: Smart Grids Paving the Way to Smart Cities, Edited by O.V. Gnana Swathika, K. Karthikeyan, P. Sanjeevikumar, ISBN: 9781394261697. With this second volume focusing on smart cities, smart vehicles, and next-generation infrastructure, this groundbreaking two-volume set provides a comprehensive overview of the fundamental security of supervisory control and data acquisition (SCADA) systems, making it an essential reference for practitioners in industries focusing on adaptive configuration and optimization in smart grid systems.

SMART GRIDS FOR SMART CITIES VOLUME 1, Edited by O.V. Gnana Swathika, K. Karthikeyan, and Sanjeevikumar Padmanaban, ISBN: 9781119872078. Written and edited by a team of experts in the field, this first volume in a two-volume set focuses on an interdisciplinary perspective on the financial, environmental, and other benefits of smart grid technologies and solutions for smart cities.

SMART GRIDS FOR SMART CITIES VOLUME 2: Real-Time Applications in Smart Cities, Edited by O.V. Gnana Swathika, K. Karthikeyan, and Sanjeevikumar Padmanaban, ISBN: 9781394215874. Written and edited by a team of experts in the field, this second volume in a two-volume set focuses on an interdisciplinary perspective on the financial, environmental, and other benefits of smart grid technologies and solutions for smart cities.

INTEGRATED GREEN ENERGY SOLUTIONS VOLUME 1, Edited by Milind Shrinivas Dangate, W. S. Sampath, O. V.Gnana Swathika, and Sanjeevikumar Padmanaban, ISBN: 9781119847434. This first volume in a two-volume set presents the state of the art for the concepts, practical applications, and future of renewable energy and how to move closer to true sustainability.

INTEGRATED GREEN ENERGY SOLUTIONS VOLUME 2, Edited by Milind Shrinivas Dangate, W. S. Sampath, O. V. Gnana Swathika, and Sanjeevikumar Padmanaban, ISBN: 9781394193660. This second volume in a two-volume set continues to present the state of the art for the concepts, practical applications, and future of renewable energy and how to move closer to true sustainability.

Check out these other related titles from Scrivener Publishing

SMART GRIDS AND INTERNET OF THINGS, Edited by Sanjeevikumar Padmanaban, Jens Bo Holm-Nielsen, Rajesh Kumar Dhanaraj, Malathy Sathyamoorthy, and Balamurugan Balusamy, ISBN: 9781119812449. Written and edited by a team of international professionals, this groundbreaking new volume covers the latest technologies in automation, tracking, energy distribution and consumption of Internet of Things (IoT) devices with smart grids.

SMART GRIDS AND GREEN ENERGY SYSTEMS, Edited by A. Chitra, V. Indragandhi and W. Razia Sultana, ISBN: 9781119872030. Presenting the concepts and advances of smart grids within the context of "green" energy systems, this volume, written and edited by a global team of experts, goes into the practical applications that can be utilized across multiple disciplines and industries, for both the engineer and the student.

SMART GRIDS AND MICROGRIDS: Concepts and Applications, Edited by P. Prajof, S. Mohan Krishna, J. L. Febin Daya, Umashankar Subramaniam, and P. V. Brijesh, ISBN: 9781119760559. Written and edited by a team of experts in the field, this is the most comprehensive and up to date study of smart grids and microgrids for engineers, scientists, students, and other professionals.

MICROGRID TECHNOLOGIES, Edited by C. Sharmeela, P. Sivaraman, P. Sanjeevikumar, and Jens Bo Holm-Nielsen, ISBN: 9781119710790. Covering the concepts and fundamentals of microgrid technologies, this volume, written and edited by a global team of experts, also goes into the practical applications that can be utilized across multiple industries, for both the engineer and the student.

INTEGRATION OF RENEWABLE ENERGY SOURCES WITH SMART GRIDS, Edited by A. Mahaboob Subahani, M. Kathiresh and G. R. Kanagachidambaresan, ISBN: 9781119750420. Provides comprehensive coverage of renewable energy and its integration with smart grid technologies.

Encyclopedia of Renewable Energy, by James G. Speight, ISBN: 9781119363675. Written by a highly respected engineer and prolific author in the energy sector, this is the single most comprehensive, thorough, and up to date reference work on renewable energy.

Green Energy: Solar Energy, Photovoltaics, and Smart Cities, edited by Suman Lata Tripathi and Sanjeevikumar Padmanaban, ISBN: 9781119760764. Covering the concepts and fundamentals of green energy, this volume, written and edited by a global team of experts, also goes into the practical applications that can be utilized across multiple industries, for both the engineer and the student.

DESIGN AND DEVELOPMENT OF EFFICIENT ENERGY SYSTEMS, Edited by Suman Lata Tripathi, Dushyant Kumar Singh, Sanjeevikumar Padmanaban, and P. Raja, ISBN: 9781119761631. Covering the concepts and fundamentals of efficient energy systems, this volume, written and edited by a global team of experts, also goes into the practical applications that can be utilized across multiple industries, for both the engineer and the student.

INTELLIGENT RENEWABLE ENERGY SYSTEMS: Integrating Artificial Intelligence Techniques and Optimization Algorithms, Edited by Neeraj Priyadarshi, Akash Kumar Bhoi, Sanjeevikumar Padmanaban, S. Balamurugan, and Jens Bo Holm-Nielsen, ISBN: 9781119786276. This collection of papers on artificial intelligence and other methods for improving renewable energy systems, written by industry experts, is a reflection of the state of the art, a must-have for engineers, maintenance personnel, students, and anyone else wanting to stay abreast with current energy systems concepts and technology.

SMART CHARGING SOLUTIONS FOR HYBRID AND ELECTRIC VEHICLES, Edited by Sulabh Sachan, Sanjeevikumar Padmanaban, and Sanchari Deb, ISBN: 9781119768951. Written and edited by a team of experts in the field, this is the most comprehensive and up to date study of smart charging solutions for hybrid and electric vehicles for engineers, scientists, students, and other professionals.

Printed in the USA/Agawam, MA
September 16, 2026

910661.146